Reconnecting the City

Reconnecting the City
The Historic Urban Landscape Approach and the Future of Urban Heritage

Editors

Francesco Bandarin

UNESCO Assistant Director-General for Culture and
Professor of Urban Planning at the University Institute of
Architecture of Venice, Italy

and

Ron van Oers

Vice Director, World Heritage Institute of Training and Research for
Asia and the Pacific, China (WHITRAP)

WILEY Blackwell

Contents

Acknowledgements   xi

Preface   xiii

Contributors   xix

About the Companion Website   xxix

Introduction: Urban Conservation and the End of Planning   1
  Francesco Bandarin
  Post-War Attempts to Reconnect the City   3
  Contemporary Views on Urbanism and Landscape   7
  Repositioning Urban Conservation, Reconnecting the City   11

SECTION 1   The Layered Dimension of Urban Conservation   17

1. **Archaeology: Reading the City through Time**   19
    Tim Williams
    Introduction   19
    Problems and Issues   21
    Challenges to Presenting Archaeological Sites in Modern Urban Landscapes   25
    Preservation *in situ* and Mitigation Strategies   30
    Approaches and Potential   35
    Archaeological Knowledge and Its Potential Impact on Urban Communities   37
    Conclusion   44

2. **How Geology Shapes Human Settlements**   47
    Claudio Margottini and Daniele Spizzichino
    Introduction   47
    Clay-Based Human Settlements   49
    Soft Rock-Based Human Settlements   59
    Hard Rock-Based Human Settlements   67
    Time Variability and Complex Urban Environments   79
    Conclusions   84

3. **Morphology as the Study of City Form and Layering**   85
    Stefano Bianca
    Introduction   85
    Origins and Implications of the Term Morphology   86
    The Scope of Urban Morphology   87
    Methodology and Procedures   88
    Advantages and Problems of the Urban Morphology Approach   94
    Relevance within the Historic Urban Landscape Concept   98

Interview – *Searching for a Chinese Approach to Urban Conservation*   103
Wang Shu

Case Study – *Bologna: From Urban Restoration to Urban Rehabilitation*   107
Patrizia Gabellini

4. **Historic Cities and Climate Change**   113
    Anthony Gad Bigio
    The Emerging Challenges   113
    Exposure of World Heritage Cities to Multiple Hazards   115
    Historic Cities and Urban Resilience   119
    Historic Cities and Climate Change Mitigation   121
    Historic Cities and Climate Action Plans: The Case of Edinburgh, Scotland   122
    Risks   123
    Actions   123

    Interview – *Looking at the Challenges of the Urban Century*   126
    Filipe Duarte Santos

5. **The Intangible Dimension of Urban Heritage**   129
    Rohit Jigyasu
    Introduction   129
    Defining Intangible Values in Historic Urban Landscapes   130
    Urbanisation Processes and Impacts on Intangible Values   135
    Recognition of Intangible Values in Existing Urban Management
        Systems   136
    Documentation and Impact Assessment of Intangible Heritage Values   138
    'Heritage' – Elitist or Inclusive?   139
    Role of Intangible Heritage in Building Disaster Resilience of Cities   142
    Integrating Intangible Heritage Values in Urban Planning and Management   142
    Mainstreaming Intangible Heritage Through Sustainable Livelihoods and
        Cultural Tourism   143
    Redefining the Role of Professionals   144

    Interview – *Interpreting Cultural Landscapes as Expressions of Local Identity*   145
    Lisa Prosper

    Case Study – *The Traditional Chinese View of Nature and Challenges
        of Urban Development*   148
    Feng Han

6. **Planning and Managing Historic Urban Landscapes**   161
    Francesco Siravo
    Integrated Planning   161
    Key Aspects of Analysing and Planning Historic Urban Landscapes   163
    Governance: The Case for Public Management in Historic Urban Areas   168
    What Kind of Public Institution?   169
    Organisational Framework of the Conservation Agency   170
    Participatory Planning and Implementation Strategies   171
    Conclusion   172

Interview – *The Challenge of Urban Transformation*   176
Mohsen Mostafavi

7. **Cities as Cultural Landscapes**   179
    Ken Taylor

    Reflections   179
    A Paradigm Shift   180
    The Cultural Landscape Model: Landscape as History and Expression of
       Human Values and Identity   183
    Bangkok and the Chao Phraya River   186
    Canberra   187
    Cultural Landscape Characteristics   187
    Urban Identity, Plurality, Sustainable Development Tools for Urban
       Landscape Planning and Conservation Practice   190
    Tools   192
    Conclusion   202

SECTION 2   Building the Toolkit   203

8. **Evolution of the Normative Framework**   205
    Jukka Jokilehto

    Introduction   205
    Early Appreciation of Historic Townscape   205
    The Development and Impact of Modern City Planning   206
    Development of Instruments for Urban Conservation   209
    International Recognition of Historic Urban Areas   211
    How Normative Frameworks Respond to the Challenges of Change
       Caused by Urban Development   213
    New Tools for the Management of the Historic Urban Landscape   216

9. **Civic Engagement Tools for Urban Conservation**   221
    Julian Smith

    Introduction   221
    Ways of Seeing   222
    Cultural Mapping   224
    The Concepts of Equilibrium and Resilience   226
    Sustainable Diversity   229
    Influences of Civic Engagement: Towards Community-Based Design
       and Development   231
    Conclusion   235

    Interview – *Listening to the People, Promoting Quality of Life*   240
    His Highness the Aga Khan

    Case Study – *Valuing Cultural Diversity*   245
    Richard A. Engelhardt

10. Knowledge and Planning Tools   **249**
Jyoti Hosagrahar

Introduction   249
Mapping, Measuring, and Visualising the Urban Landscape   250
Reading and Interpreting the Urban Landscape   251
Protecting, Enhancing, and Improving the Urban Landscape   257
Traditional and Customary Systems of Management   260
Contextualising the Historic Urban Landscape Approach   260

Case Study – *Reading the City of Tokyo*   *261*
Hidenobu Jinnai

11. The Role of Regulatory Systems   **269**
Patricia O'Donnell

Defining Regulatory Systems   269
Legal Regulations Directly Addressing Public and Private Lands   270
Legal Regulations with Indirect Influence on Urban Heritage   275
Conclusion   278

Interview – *Constructing Cultural Significance*   *279*
Rahul Mehrotra

12. Devising Financial Tools for Urban Conservation   **283**
Donovan Rypkema

Introduction   283
Why are Financial Tools Required?   284
What Do Financial Tools Do?   286
What are the Characteristics of the Most Effective Financial Tools?   287
What are Some Examples of Financial Tools and How Do They Work?   288
Conclusion   290

Case Study – *A User's Guide for Heritage Economics*   *291*
Christian Ost

Case study – *The World Bank's Tools for Urban Conservation*   *297*
MV Serra

13. Researching and Mapping the Historic Urban Landscape   **301**
Michael Turner and Rachel Singer

Introduction   301
The Diverse City   303
Methodologies and Tools   305
The Role of University Research   309
The Role of UNESCO Chairs   310
The Role of Category 2 Centres (C2C)   310
Conclusion   311

Interview – *Heritage and the Metropolis*   *313*
Rem Koolhaas

Conclusion: The Way Forward: An Agenda for Reconnecting the City   317
Ron van Oers

Managing the City as a Living Heritage   317
Identity and Sense of Place   318
Local Heritage and Corporate Image   319
The City as Repository of Urban Experiences   321
Integrating Disciplines and Professional Practices   322
Future Challenges of Urban Conservation   324
The Critical Path: Historic Urban Landscape Action Plan   326
Historic Urban Landscape: A Stepped Approach   326
Interdisciplinary Context and Operational Coordination   328
A 20-Point Research Agenda for Planners and Designers   329

Index   333

Acknowledgements

This edited book on the on-going process of elaboration and implementation of the 2011 *Recommendation on the Historic Urban Landscape*, as developed and promoted by UNESCO since 2005, is part of an international effort to adapt urban conservation to the operational realities of the twenty-first century in which cities have assumed a critical role in human development.

A broad and growing coalition of professionals, decision-makers and community representatives in all parts of the world is participating in this process. It would be impractical to name all of them here.

We would like however to thank all our colleagues at the UNESCO Headquarters and in the Field Offices for their continued support and encouragement.

Various professionals in different parts of the world have provided critical reflections and observations on the Historic Urban Landscape as a process or product. We would like to acknowledge, in particular, the contribution of: Joseph King and Gamini Wijesuriya at ICCROM in Rome; Gustavo Araoz, Kristal Buckley and Sheridan Burke at ICOMOS; Stefania Abakerli and Guido Licciardi at the World Bank in Washington; Jeffrey Soule at the American Planning Association; Ana Pereira-Roders in Eindhoven; Marie-Theres Albert in Cottbus; Sarah Semple, Andreas Pantazatos, David Petts and Seif al-Rashidi in Durham; Karel Bakker in Pretoria; Alfredo Conti in La Plata; Muhammad Juma in Zanzibar; Susan Fayad in Ballarat; Louise Cox in Sydney; Jian Zhou in Shanghai; Lynne DiStefano in Hong Kong; Ayesha Pamela Rogers and Nadeem Tarar in Rawalpindi; Nobuko Inaba in Tokyo; Augusto Villalon in Manila; Christopher Young in London; Birgitta Ringbeck in Berlin; Jad Tabet in Beirut; Marc Breitman in Paris; Daniele Pini in Ferrara; Paolo Ceccarelli in Milan; Heleni Porfyriou in Rome; Pietro Laureano in Florence; Sophia Labadi in Canterbury; Lynn Meskell in Stanford; Paola Falini in Rome; Alessandro Balducci in Milan.

We would like to thank them all sincerely for their involvement and their dedication to the cause of urban conservation and we look forward to continued collaboration and expansion of the Historic Urban Landscape network.

Our final thank you, goes to all the contributors of this book: in total, 30 people (including the editors) offered a contribution to the reflection: 17 for the essays, 6 for the case studies and 7 for the interviews. While their texts have been in some cases revised before being integrated into the book, we have done our best to respect the meaning of the original. The editors have inserted most of the quotes at the beginning of the chapters. The interviews have been conducted and drafted by the editors and revised by the interviewee. Errors or imprecisions remain, of course, our full responsibility.

Francesco Bandarin, *Paris*
Ron van Oers, *Shanghai*
May 2014

Preface

Francesco Bandarin and Ron van Oers

Writing a book is an adventure. To begin with, it is a toy and an amusement; then it becomes a mistress, and then it becomes a master, and then a tyrant. The last phase is that just as you are about to be reconciled to your servitude, you kill the monster, and fling him out to the public.

Winston Churchill

In our previous book *The Historic Urban Landscape: Managing Heritage in an Urban Century* (Wiley-Blackwell, 2012), we argued that in spite of the decades long call for interdisciplinary work that was formally codified in the 1975 Amsterdam Declaration on Integrated Conservation, there is currently little integration of professions dealing with the process of heritage conservation and urban development. This leaves the field of urban heritage management seriously compartmentalised, with limited exchanges between the professional 'silos'. This obviously reduces the efficiency and effectiveness of conservation efforts and it creates gaps that can be exploited by the forces that are not interested in the preservation of heritage resources.

In our view, the natural follow-up to the first book was to assemble a range of professional practices and viewpoints related to urban management to broaden the scope and reach of the Historic Urban Landscape as a conceptual framework and operational approach. The Historic Urban Landscape aims to respect and celebrate diversity – of heritage resources *and* cultural traditions – by suggesting a critical *process* (not a model) of identification and analysis to arrive at informed decisions regarding the policies and tools aimed at fostering sustainable urban conservation and management.

While affirming the universal importance of urban heritage, it advocates strongly for local solutions to its management, in the face of rapid urbanisation processes, as well as of the different political, cultural, and economic trajectories of contemporary societies. We argue that urban conservation practices over the past 50 years have been successful in creating a global consciousness of the importance of urban heritage and have allowed the safeguarding of many historic areas and cities. However, we also argue that the time has come to look at urban heritage as a resource for the entire city and for its sustainable development. In our view, this goal can be achieved by advancing the methodology for the implementation of the Historic Urban Landscape approach.

The essays in this book relate to a variety of disciplines and professional practices concerned with urban conservation and management, but they do not cover the entire spectrum. Surely other important tools and practices can be put to the fore, including those from the sociological, philosophical, anthropological, ecological and managerial disciplines. As such, this volume is a contribution that needs to be expanded and continued by other professional disciplines, actors and stakeholders.

The chapters are intersected with additional contributions, in the form of 'case studies' and 'interviews' with prominent professionals and personalities, in order to enlarge the range of opinions and perspectives. The case studies elaborate on particular applications of tools or present relevant examples, while the interviews discuss theoretical issues in relation to cities, urbanisation, communities, and the management of urban heritage in different parts of the world.

The Structure of the Book

Francesco Bandarin's essay 'Urban conservation and the End of Planning' opens the reflection by discussing the situation of the disciplines of conservation and planning, in a world dominated by global processes and social and economic dynamics that have profoundly transformed the approaches to urban management and urban development. The separation between historic areas and the rest of the city that has characterised the twentieth century's experience is seen today as a risk as well as a waste of an important stock of knowledge and experience.

The post-war attempts to reconnect conservation and planning have produced important intellectual results, but have proven inadequate to cope with the emergence of global processes and the *de facto* end of planning as the key urban management tool. In recent years, new methodologies have come to the forefront, based on a landscape approach to urban management that matches the principles expressed by the 2011 UNESCO *Recommendation on the Historic Urban Landscape* and offers a new possibility to reconnect the city management processes, while also valuing the historic city as a resource for the future. The different contributions presented in the book have been organised in two sections, dealing respectively with the disciplinary perspectives on urban heritage conservation and with the development of a tool-kit for the implementation of the Historic Urban Landscape approach.

Section 1: The Layered Dimensions of Urban Conservation

All cities are the product of a gradual layering process that sometimes spans thousands of years of history, like for example in Damascus, Rome and Delhi, and sometimes this lasts just a few decades, as in Brasilia, Chandigarh and Shenzen. Each layer represents a moment in the history of the city, an expression of its culture, of its economic strength, of the ways it adapts to the physical environment, of its innovation capacities and its technological achievements. The layering process is also the result of the interaction, all along history, between human societies and the environment, with the aim to create human settlements adapted to the needs of life, to the changes of population density and size, to the ambition of its inhabitants: in other words an Urban Landscape, which is the expression of the most complex and resilient invention of humankind, the city.

The first part of this section deals with the physical layers of the city, what we can call the 'stratigraphy' of the city. The section starts with an examination of the role of archaeology in interpreting the layers of the urban environment. Tim Williams' essay 'Archaeology: Reading the City through Time' discusses the role archaeology plays in today's urban areas, in the planning process and in the construction of civic identity and sense of place. As much as archaeological remains are fundamental to the understanding of an urban complex, they constitute a presence that needs to be managed and made compatible with modern needs. This interface is multidimensional, and it involves scientific and policy choices that affect the way in which the layers of time can be preserved and exposed. It also needs technical capabilities to make compatible, infrastructure development and preservation, as well as it requires a comprehensive integration of the rationale of archaeology in the processes of civic participation and planning.

As much as cities are a layered built construct, they rest on another layered system, the geological strata formed during Earth's history. This relationship is a fundamental one, albeit often forgotten with dire consequences for urban conservation and for the protection of urban environments from natural hazards.

This dimension is discussed by Margottini and Spizzichino in their chapter on 'How Geology Shapes the Urban Environment', through a number of case studies ranging from the ancient cities of Mesopotamia to more recent urban formations in Italy (Rome, Orvieto), Africa (Lalibela) and South America (Machu Picchu). This essay discusses the way in which the geological setting has determined the ways in which cities were built, their morphology, building materials and building types, as well as the way they were able to adapt to the hydrological and ecological constraints. The relationship of a city to its geological context is not only at the basis of its resilience through time (or of its collapse) but is also the main reason of the continuity of forms and types through the millennia. Obviously, the industrial age has interrupted this continuity, as it has allowed the use of non-local materials and of new building technologies. This 'separation' between the city and its geological context is at the origin of many of today's challenges that are related to urban resilience, sustainability and energy efficiency.

Stefano Bianca's essay on 'Morphology as the Study of City Form and Layering' looks at a discipline that analyses the results of the layering process, Urban Morphology, a powerful tool to understand the city's history and to connect it to the processes of its development and rehabilitation. Urban Morphology analyses the historic urban fabric as a complex cellular micro-system that evolves organically. Because this discipline does not focus only on the outstanding monuments, but on the urban fabric as a whole, it can provide a basis for conservation planning and for renovation and adaptation processes that want to emphasise continuity of the urban form and of urban spaces.

The discussion on urban morphology is enriched by two short contributions. The first is an interview with Architect Wang Shu, who discusses the situation of urban conservation in China and the present trends. The second is a case study by Patrizia Gabellini that presents the evolution of the planning approach to conservation of the historic city of Bologna in Italy, well

known for having been the first to apply the morphological approach to its historic conservation policies.

Finally, within this part dedicated to the urban physical environment, an issue of great importance is discussed, that is the resilience of cities with respect to natural hazards, in particular related to climate change. Anthony Gad Bigio's essay 'Historic Cities and Climate Change' looks at these challenges through a review of the current situation of more than 200 cities inscribed on the World Heritage List. The analysis reveals a high degree of vulnerability, in particular to floods, landslides and climatic events. The essay discusses the policies cities can develop to enhance their resilience and to mitigate the impacts of climate changes in the long term, while preserving their historic character.

Bigio's analysis is complemented by an interview with the physicist Filipe Duarte Santos that discusses the environmental challenges of the 'urban century' and the role urban heritage can play in the future.

The second part of this section deals with what could be termed 'intangible' layers centred on the social dimension of the city. Rohit Jigyasu's essay on 'The Intangible Dimension of Urban Heritage' discusses the nature of the intangible values in historic environments and the process of their representation (or not) and preservation. Furthermore he looks, through the lens of some selected case studies in Asia, at the ways in which intangible values are associated to planning and management processes, and considers the tools that can be used and developed to identify and assess the impact on intangible heritage values.

An interview with Lisa Prosper, an indigenous peoples' intellectual leader, complements this discussion on the role of intangible values as expressions of local identity. In addition, a case study by Feng Han presents the philosophical and cultural basis for the interpretation of Nature in the Chinese tradition that informs directly urban planning and design.

Another fundamental intangible layer of the city is certainly constituted by the planning and management structures that condition and orient its development. The role of this social construct is examined in Francesco Siravo's essay 'Planning and Managing Historic Urban Landscapes', which compares traditional urban planning methods with those needed to preserve the character and social structure of an historic environment. On the basis of an extensive experience in dealing with historic areas in Europe, Africa and Asia, he points to the components that a sensitive planning practice should consider in order to avoid disrupting the physical and social environments, ranging from the analysis of land ownership and tenure to economic activities, infrastructure and services needs, as well as to financial aspects.

Within this discussion the interview with Mohsen Mostafavi on the challenges of urban transformation, brings to our attention the risks of conservation approaches aimed at freezing the historic city, without considering its relationship with the broader urban, regional and ecological contexts.

Finally, the essay of Ken Taylor 'Cities as Cultural Landscapes' rounds up the discussion by presenting, through a number of case studies in Asia and Australia, the possible application of the Historic Urban Landscape approach to complex urban conservation situations. Its main point of concern is the construction of significance in urban places, as well as how urban heritage can become a resource to foster rehabilitation and regeneration of the modern city.

Section 2: Building the Toolkit

Over the course of six years of policy review and best-practice analysis, during the time of discus-
sion and elaboration of the new UNESCO *Recommendation on the Historic Urban Landscape* between 2005 and 2011, the international community of practitioners identified four main types of tools that would be needed to regulate and facilitate heritage management in contemporary and dynamic urban contexts. The four main types include: civic engagement tools; knowledge and planning tools; regulatory systems and financial tools. These tool sets are elaborated respectively by Julian Smith, Jyoti Hosagrahar, Patricia O'Donnell, and Donovan Rypkema, who outline the key issues and the main benefits. These analyses have been supplemented by case studies and interviews aimed to highlight specific issues.

Overall, a great variety of tools for the conservation and management of urban heritage resources can be observed, arguably much more than most practitioners would realise. This should be seen as a reflection of the enormous effort put into urban heritage conservation and management over the last 40 to 50 years,[1] as discussed by Jukka Jokilehto in his essay 'Evolution of the Normative Framework'. This extensive overview clearly reflects his lifelong professional engagement with the topic and serves as a benchmark against which to view the current shifts in thinking about urban conservation practice.

The next important step is to ensure a systematic, integrated and broad use of these tool sets in the management of urban heritage, which is the primary goal of this book. Naturally, not all tools can be easily transferred from one geo-cultural context to another, but the adaptation to one's own context should begin with learning from others' experiences and insights.

---

[1] Bandarin, F. and Van Oers, R. (2012) *The Historic Urban Landscape. Managing Heritage in an Urban Century*, Oxford, Wiley–Blackwell, pp. 41–44.

Civic Engagement Tools[2]

Cultural mapping and community participation in conservation planning and decision-making have become important tools to foster civic engagement. They are based on the recognition that the management of heritage, including its urban expressions, rests in the hands of its creators and custodians, not in the hands of absent caretakers. This is argued in particular by Julian Smith in his essay 'Civic Engagement Tools for Urban Conservation' that deals with experiences of urban revitalisation from within, and the tools needed to make conservation meaningful and successful.

An interview with His Highness the Aga Khan highlights the importance of engaging the communities of beneficiaries from the beginning of an intervention, in order to shape programmes that are responding to the effective local needs, and not to preconceived ideas of decision makers. A case study by Richard Engelhardt presents the Cultural Diversity Lens, a tool developed by UNESCO to recognise and value local cultural specificities.

Knowledge and Planning Tools[3]

Jyoti Hosagrahar in her essay 'Knowledge and Planning Tools' discusses the tools needed to read and interpret the urban landscape, as well as those aimed at its protection and improvement. She points to the fact that as the notion of heritage significance is now extending beyond monuments and architectural ensembles, a broader range of knowledge and planning tools is needed and has to be made available to the urban heritage manager for safeguarding and developing the city's heritage resources. A case study by Hidenobu Jinnai presents a methodology for the investigation of urban structures and meaning that supports an innovative approach to urban conservation and rehabilitation.

Regulatory Systems[4]

In the Western world during the last two to three decades increasingly sophisticated regulatory tools have been conceived and adopted to facilitate a more holistic and integrated approach to heritage landscapes, including historic cities in their wider setting, as summed up in Patricia O'Donnell's essay 'The Role of Regulatory Systems'. Such a comprehensive approach is largely lacking in countries outside the Western context, due to their specific dynamics in the political, economic and social spheres, which have created different legal and institutional frameworks. An interview with Rahul Mehrotra illustrates the importance of these aspects in the

---

[2] The *Recommendation on the Historic Urban Landscape* outlines that: *Civic engagement tools should educate a diverse cross-section of stakeholders and empower them to identify key values in their urban areas, develop visions, set goals, and agree on actions to safeguard their heritage and promote sustainable development. In particular, seeking a strengthening of governance and citizen participation in the reallocation of buildings. These tools that constitute an integral part of urban governance dynamics, should facilitate intercultural dialogue by learning from communities about their histories, traditions, values, needs, and aspirations and by facilitating the mediation and negotiation between conflicting interests and groups.* UNESCO (2011) *Recommendation on the Historic Urban Landscape*, Paris: para. 24a.

[3] Regarding knowledge and planning tools, the 2011 Recommendation further outlines that: *Knowledge and planning tools should help protect the integrity and authenticity of the material attributes of urban heritage. They should also allow for the recognition of cultural significance and diversity, and provide for the monitoring and management of change to improve the quality of life and of urban space. Consideration should be given to the mapping of cultural and natural features, while heritage, social and environmental impact assessments should be used to support sustainability and continuity in planning and design.* UNESCO (2011) *Recommendation on the Historic Urban Landscape*, Paris: para. 24b.

[4] Regarding Regulatory Systems, the 2011 Recommendation outlines that: *Regulatory systems should include special ordinances, acts, codes or decrees to conserve and manage tangible and intangible components of the urban heritage, including their social and environmental values. Traditional and customary systems should be recognised and reinforced as necessary.* UNESCO (2011) *Recommendation on the Historic Urban Landscape*, para. 24c.

construction of cultural significance in non-Western contexts.

Financial Tools[5]

Three authors share their insights and experiences regarding the availability and merits of different financial tools that could be employed in the urban heritage management process. Donovan Rypkema in his essay 'Devising Financial Tools for Urban Conservation' discusses a full range of traditional and innovative schemes that can be introduced to support urban heritage conservation policies within a Historic Urban Landscape approach. Christian Ost and MV Serra in their case studies, emphasise that effective urban heritage management is determined by values and aspirations, constrained by political realities and markets, characterised by shortcomings and externalities, as well as supported by tailor-made schemes involving incentives, regulation and investments. This involves multiple source financing, complex cost recovery mechanisms and subsidies.

The essay by Michael Turner and Rachel Singer 'Researching and Mapping the Historic Urban Landscape' offers a picture of the existing and possible collaborative efforts to expand academic knowledge and operational capacities for the implementation of the Historic Urban Landscape approach. Finally, the interview with Architect Rem Koolhaas provides an outlook on the future of heritage in a metropolitan world and points to the need to expand our view to new types of cultural resources, in response to an increasing demand for urban identity.

Ron van Oers completes the book with the essay 'The Way Forward: An Agenda for Reconnecting the City'. The historic city is not just a fragment of the urban complex, it is a basis for its identity and a fundamental resource for its development. To achieve this, we have to reconnect the different disciplines that operate in the city, in support of a landscape approach which integrates physical and intangible dimensions. The need for new approaches and new instruments to enhance urban conservation policies and the city's management processes is reflected in the Action Plan that accompanies the UNESCO *Recommendation on the Historic Urban Landscape*. The Action Plan, currently being implemented in different parts of the world, is functional to a demonstration of the viability of the Historic Urban Landscape approach. To this purpose, a 20-point agenda is proposed to researchers, planners, administrators and citizens interested in reconnecting the historic environment with the modern city.

---

[5] The fourth set of tools specified in the 2011 Recommendation concerns financial instruments, which: *should aim to improve urban areas while safeguarding their heritage values. They should aim to build capacities and support innovative income generating development rooted in tradition. In addition to government and global funds from international agencies, financial tools should be effectively employed to foster private investments at the local level. Micro credit and other flexible financing to support local enterprise, as well as a variety of models of public-private partnerships, are also central to making the approach financially sustainable.* UNESCO (2011) *Recommendation on the Historic Urban Landscape*, para. 24d.

Contributors

His Highness the Aga Khan is Chairman of the Aga Khan Development Network and 49th hereditary Imam of the Shia Imami Ismaili Muslims, His Highness the Aga Khan is the Founder and Chairman of the Aga Khan Development Network (AKDN), a group of private, non-denominational development agencies dedicated to improving the quality of life of impoverished communities in Asia, Africa, and the Middle East. AKDN's cultural agency, the Aga Khan Trust for Culture, improves the socioeconomic conditions of communities in the Muslim world, through urban regeneration projects and other cultural initiatives. Website: www.akdn.org

Francesco Bandarin is the UNESCO Assistant Director-General for Culture (2010–2014). He has served as Director of the UNESCO World Heritage Centre and Secretary of the World Heritage Convention from 2000 to 2010. He is an Architect and Urban Planner specialising in Urban Conservation, and Professor of Urban Planning at the University Institute of Architecture of Venice. He studied Architecture at the Institute of Architecture of Venice, Italy and City and Regional Planning at the University of California, Berkeley, USA. For over 20 years, he was a consultant for international organisations in the field of urban conservation and development. At UNESCO, he has coordinated the reform processes of the World Heritage Convention and implemented projects in major World Heritage Sites. He has directed the overall process for the drafting and adoption of the UNESCO *Recommendation on the Historic Urban Landscape*. He has coordinated the UNESCO Culture and Development Policy for the Post-2015 UN Millennium Development Goals. He has published extensively on Urban Planning, Conservation and Cultural policies, and has recently co-authored *The Historic Urban Landscape: Managing Heritage in an Urban Century*. Oxford: Wiley-Blackwell, 2012. Email: f.bandarin@gmail.com

Stefano Bianca, Architectural Historian and Urban Planner, studied at the ETH in Zurich, Switzerland, where he gained a PhD in 1972 and was visiting professor in 1978/79. He had a lifelong professional involvement with historic cities in the Islamic world. Since 1975 he has directed many important urban conservation and rehabilitation projects in cities such as Fes, Aleppo, Baghdad, Medina, Cairo, Zanzibar, Hunza, Samarkand and Kabul, some of them as director of the Aga Khan Historic Cities Programme (1992–2006). He is the author of many books and articles dealing with Islamic architecture, cities and gardens, as well as integrated conservation and urban planning. His most recent publications are: 'Hofhaus und Paradiesgarten', Munich, 1991 and 2001; 'Urban Form in the Arab World – Past and Present', Zurich, London and New York, 2000; 'Cairo – Revitalising a Historic Metropolis' (ed.), Torino, 2004; 'Karakoram – Hidden Treasures in the Northern Areas of Pakistan' (ed.), Torino 2005; 'Syria – Medieval Citadels between East and West' (ed.), Torino 2007. Email: stef.bianca@gmail.com

Anthony Gad Bigio is an Urban Advisor with over thirty years of experience of urban development projects across the world. After a 20 years career at the World Bank, he became Adjunct Professor of Urban Resilience and Low-carbon Cities at the George Washington University Graduate Program on Sustainable Urban Planning in Washington DC, USA. He is a Lead Author for the Intergovernmental Panel on Climate Change, contributing to the preparation of the chapter on urban planning and carbon emissions of its 2014 Fifth Assessment Report and an official reviewer of its chapter on urban impacts of climate change (Working Group II). Having graduated as an Architect and Urban Planner *summa cum laude* from the University of Rome, Italy, he has worked on projects aimed at the preservation and rehabilitation of historic cities in the context of social and economic development, primarily in the Middle East and North Africa region. He has published extensively on the topics of urban development, climate change and cultural heritage. Email: agbigio@gwu.edu

Richard A. Engelhardt is the former UNESCO Regional Advisor for Culture in Asia and the Pacific serving in that position from 1994 to 2008. After retiring from UNESCO, he continues to be active in the heritage profession, as consultant in cultural policy and heritage management to governments and as an educator. Mr. Engelhardt is the UNESCO Chair Professor of the Conservation and Management of Historic Towns and Urban Centres at the National College of Art in Lahore, Pakistan, Honorary Professor in the School of Architecture at Southeast University in Nanjing, China, and Visiting Research Professor in the Faculty of Architecture at the University of Hong Kong. During his career, he has received numerous honours and awards in recognition of his services in safeguarding the heritage of the Asia-Pacific region. In 1994, H.M. King Norodom Sihanouk of Cambodia knighted him with the title of *Commandeur de l'Ordre Royal du Cambodge*. Email: richard.a.engelhardt@gmail.com

Patrizia Gabellini is an Architect and Professor of Urban Planning at the Politecnico of Milan, Italy, where she has been Director of the Department of Architecture and Planning. She has been visiting scholar at the IURD of the University of California, Berkeley, USA, and visiting professor in the Academy of Architecture in Mendrisio – University of the Italian Switzerland. She directed the 'Design and Communication' project of the Master Plan of Rome, the Master Plan of Jesi and coordinated the territorial project of the Esino Valley in Marche Region, all in Italy. She was consultant for the Master Plan of Bologna and member of the scientific committee for the Territorial Plan of Emilia Romagna region. She has been Editor of the Journal *Urbanistica* and is the Director of *Planum*, a European Journal of Urbanism online (www.planum.net). She is author of several books and essays, published in Italy and abroad. Since 2011 she is the Deputy Mayor for Planning, Environment, Urban Quality and Historic City of the Municipality of Bologna, Italy. Email: assessoreurbanisticaambiente@comune.bologna.it

Feng Han is Professor at the Department of Landscape Architecture and Assistant Dean of the College of Architecture and Urban Planning, Tongji University (CAUP), Shanghai, China. She represents China in the ICOMOS-IFLA International Scientific Committee on Cultural Landscape and is a member of the World Commission on Protected Areas of IUCN. She works as an expert for international evaluation of World Heritage nomination and heritage related policy making, as expert of the UNESCO program of Conservation and Management of World Heritage Sites in China and as an Advisory Professor for WHITRAP (Shanghai). She has published several papers on heritage landscape conservation and sustainable development. She is the Regional Editor (Asia/Pacific) of the *Journal of Management and Sustainable Development of Cultural Heritage*. She was awarded several prizes including the 2010 Advance Leading 50 Women in Asia from Australia. Email: franhanf@gmail.com

Jyoti Hosagrahar is Director of Sustainable Urbanism International at GSAPP, Columbia University, New York and Bangalore; Chair of the PhD program at Srishti School of Art, Design, and Technology, Bangalore, India. Since 2011, she has served as an expert member of the National Advisory Committee for World Heritage Matters under the aegis of the Ministry of Culture, Government of India and served on the Working Group for revising India's Tentative List 2011–2013. During the past 6 years she has taken part in the UNESCO working groups for development of the Historic Urban Landscapes approach. She is the author of *Indigenous Modernities: Negotiating Architecture and Urbanism*. New York: Routledge, Architext Series, 2005; coauthor of 'Why Development Needs Culture,' in the *Journal of Cultural Heritage Management and Sustainable Development*, 1 (1), 2011; she is the lead author in SUI's 'Integrated Plan for Managing Cultural Heritage and Sustainable Development in the Hoysala Heritage Region of Karnataka, India'. Email: jh2443@columbia.edu

Rohit Jigyasu is a Conservation Architect and Risk Management Consultant, currently working as UNESCO Chair Professor at the Institute for Disaster Mitigation of Urban Cultural Heritage at Ritsumeikan University, Kyoto, Japan and Senior Advisor at the Indian Institute for Human Settlements (IIHS) in Bangalore, Karnataka, India. He is elected member of the Executive Committee of International Council on Monuments and Sites (ICOMOS) and president of ICOMOS International Scientific Committee on Risk Preparedness (ICORP). After undertaking his postgraduate degree in Architectural Conservation from School of Planning and Architecture in Delhi, he obtained his doctoral degree from the Norwegian University of Science and Technology, Trondheim, Norway. He has been teaching as visiting faculty at several national and international academic institutions in India and abroad. He has also been consultant to several national and international organisations for research and training on Cultural Heritage Risk Management and has contributed to several international publications. Email: rohit.jigyasu@gmail.com

Hidenobu Jinnai is Professor of Urban Design at Hosei University in Tokyo, Japan, and since 2004 Director of the Laboratory of Regional Design with Ecology at the Graduate School of Hosei University. He is also the Director of the Historic Museum of Chuo-ward of Tokyo. He graduated in Engineering at the University of Tokyo and studied Architecture at the University Institute of Architecture of Venice, Italy. He specialised in the analysis of historic cities both in Asia and in Europe, with special focus on Japan and Italy. His research interests are in the field of spatial anthropology, urban morphology and conservation of historic cities. He is the author of several books and articles, among which *Turkey: Piligrimage to cities*. Tokyo: Process Architecture, 1990; *Italian Acquascape*. Tokyo: Process Architecture, 1993; *Tokyo: A Spatial Anthropology*. Berkeley: University of California Press, 1995; *Living Places in Southern Italian Cities – Amalfi, Lecce, Sciacca and*

Sardegna. Tokyo: Chuo-Koron Bijutu Shuppan, 2005. Email: jinnai@h-jinnai.jp

Jukka Jokilehto was born in Helsinki, Finland, and graduated as Architect and City Planner at the Polytechnic University of Helsinki; he earned a PhD (DPhil) at the University of York, United Kingdom, in 1986. He worked in Finland as an architect and urban planner in the 1960s. He joined ICCROM in 1972 as the Director of Architectural Conservation and retired with the position of Assistant Director-General in 1998. He has been President of the ICOMOS International Training Committee (1993–2002) and has been ICOMOS World Heritage Advisor from 2000 to 2006. Presently he is Consultant and Lecturer in Architectural and Urban Conservation; Special Advisor to Director-General of ICCROM; Professor at University of Nova Gorica, Slovenia; Honorary Visiting Professor at University of York. Main publications: *A History of Architectural Conservation*. Oxford: Butterworth-Heinemann, 1999; *Management of World Cultural Heritage Sites*, co-authored with Bernard M. Feilden. Rome: ICCROM, 1993; he is the author of several publications on the philosophy of conservation, and on the management of World Heritage properties. Email: j.jokilehto@fastwebnet.it

Rem Koolhaas founded OMA in 1975 together with Elia and Zoe Zenghelis and Madelon Vriesendorp. He graduated from the Architectural Association in London and in 1978 published *Delirious New York: A Retroactive Manifesto for Manhattan* (New York: the Monacelli Press). In 1995, his book: *S,M,L,XL* (New York, the Monacelli Press), summarised the work of OMA in 'a novel about architecture.' He heads the work of both OMA and AMO, the research branch of OMA, operating in areas beyond the realm of architecture such as media, politics, renewable energy and fashion. In 2014, Koolhaas curated the 14th International Architecture Exhibition of the Venice Biennale, under the title, *Fundamentals*. Koolhaas has won several international awards including the Pritzker Architecture Prize in 2000 and the Golden Lion for Lifetime Achievement at the 2010 Venice Biennale. Rem Koolhaas is a Professor at Harvard University where he conducts the Project on the City. Email: park@oma.nl

Claudio Margottini is a Senior Scientist at the Geological Survey of Italy (ISPRA), Vice President of the International Consortium on Landslides at the University of Kyoto, Japan, and acting professor at the Huangzou University in Wuhan, China. His major field of expertise is the development of engineering geological techniques for the conservation and protection of cultural and natural heritage. His activity include projects for the sites of Machu Picchu (Peru), the Buddhas statues of Bamiyan (Afghanistan), the Lalibela Churches (Ethiopia), the Koguryo Tombs (North Korea), the Vardzia caves (Georgia), the Minaret of Jam (Afghanistan), the Stelae Park in Aksum (Ethiopia), the Moai Statues (Easter Island, Chile), the Tiwanaku Pyramid (Bolivia), Petra (Jordan), the Katski Column (Georgia), the Herat Minaret (Afghanistan), the Bayannuur Tomb (Mongolia), the Orongo cliff and village (Easter Island, Chile), the Maaloula cliff and village (Syria), the Zohak Archaeological Fortress (Afghanistan) as well as many Italian sites of high cultural value. He is the author of over 250 publications. Email: claudio.margottini@gmail.com

Rahul Mehrotra is a practising Architect and Educator. He works in Mumbai, India, and teaches at the Graduate School of Design at Harvard University,

Cambridge, Massachusetts, USA, where he is Professor of Urban Design and Planning, and Chair of the Department of Urban Planning and Design as well as a member of the steering committee of Harvard's South Asia Initiative. His practice, RMA Architects, founded in 1990, has executed a range of projects across India. These diverse projects have engaged many issues, multiple constituencies and varying scales, from interior design and architecture to urban design, conservation and planning. As Trustee of the Urban Design Research Institute (UDRI), and Partners for Urban Knowledge Action and Research (PUKAR) both based in Mumbai, he continues to be actively involved as an activist in the civic and urban affairs of the city. Email: rahul@rmaarchitects.com

Mohsen Mostafavi, Architect and Educator, is Alexander and Victoria Wiley Professor of Design and Dean of the Harvard University Graduate School of Design, Cambridge, Massachusetts, USA. His work focuses on modes and processes of urbanisation and on the interface between technology and aesthetics. He has taught at numerous institutions including the University of Pennsylvania, the University of Cambridge, and the Frankfurt Academy of Fine Arts (*Städelschule*). He serves on the steering committee of the Aga Khan Award for Architecture and the board of the Van Alen Institute, and has served on the design committees of the London Development Agency (LDA) and the RIBA Gold Medal. He is a consultant on a number of international architectural and urban projects. His recent publications include *Ecological Urbanism*, Zurich: Lars Müller Publishers, 2010; *Implicate & Explicate*, Lars Müller Publishers, 2011; *Louis Vuitton: Architecture and Interiors*, New York: Rizzoli, 2011; *In the Life of Cities.* Zürich: Lars Müller Publishers, 2012; *Instigations: Engaging Architecture, Landscape and the City:* Zürich, Lars Müller Publishers, 2012; and *Architecture is Life.* Zürich: Lars Müller Pub-

lishers, 2013. Email: mohsen_mostafavi@gsd .harvard.edu

Patricia M. O'Donnell is a Landscape Architect and Planner, and founded in 1987 Heritage Landscapes LLC, Preservation Landscape Architects and Planners, based in Charlotte, Vermont and Norwalk, Connecticut, USA. Her professional firm is dedicated to a vibrant future for communities, territories and cultural landscapes of all types and scales, with some 500 project credits and 66 professional awards. She serves as Global Chair, IFLA Cultural Landscapes Committee, as the US Member of the ICOMOS IFLA Cultural Landscapes International Scientific Committee and as an expert in the UNESCO World Heritage Centre. Since 2004 she has contributed to the development of the UNESCO Historic Urban Landscape process and recommendation as an invited international expert and collaborated with international experts on the final drafting committee. She is currently working on guidance incorporating Historic Urban Landscape constructs and tools for the World Bank on social development projects addressing the sustainable stewardship of heritage villages in Bhutan, as well as Indian heritage cities. Email: odonnel@ heritagelandscapes.com

Christian Ost is an Economist with extensive experience in education institution management and economics of heritage. He holds a PhD in Economics from the Catholic University of Louvain, Belgium, a Master's degree in Economics from Georgetown University, Washington DC, USA, and a Certificate in European Studies from the University of Geneva, Switzerland. He was Dean of the ICHEC Brussels Management School from 2000 to 2008. He has been developing the field of economics applied to cultural heritage since the 1980's, as co-author with Raymond Lemaire of a report on the cultural heritage and economics to the European Commission, as visiting lecturer at the Raymond Lemaire International Conservation Centre in Leuven, the Catholic University of Louvain, and as member of the ICOMOS International Economics Committee, which he chaired from 2000 to 2005. In 2008–2009, he was guest scholar in residence at the Getty Conservation Institute in Los Angeles, with a research entitled '*A Guide for Heritage Economics in Historic Cities: Values, Indicators, Maps and Policies*'. Email: christian.ost@ichec.be

Lisa Prosper is the Director of the Centre for Cultural Landscape at Willowbank, which seeks to further the development of a cultural landscape approach to heritage conservation and community development by emphasising the interrelated physical and sociocultural dimensions of places and regions. She has been contributing to the development of heritage theory and practice for over a decade and is regularly an invited speaker on cultural landscapes and Aboriginal heritage at both Canadian and international forums. She has been a member of numerous expert committees including the working group to develop the federal standards and guidelines for the conservation of cultural landscapes in Canada. She also shares responsibility for the development of the curriculum associated with the Diploma Program at the School of Restoration Arts at Willowbank, where she teaches heritage theory and field studies. Email: lisa.prosper@willowbank.ca

Donovan Rypkema is President of Heritage Strategies International, a firm working at the nexus of heritage conservation and economics. Rypkema has worked in more than 40 countries. International clients include the World Bank, the Inter American Development Bank, the Council of Europe, and others. He is the author of numerous publications including *Feasibility Analysis of Historic Buildings*. Washington, DC: National Trust for Historic Preservation, 2007; 'Public-Private Partnerships and Heritage'. *CreateSpace Independent Publishing Platform*, January 28, 2012; *The Economics of Historic Preservation: A Community Leader's Guide*. New York: The National Trust for Historic Preservation, 2005, that has been translated into Russian and Korean. He holds a Master of Science degree in Historic Preservation from Columbia University, New York, USA. He serves on the Board of Directors of Global Urban Development, and the Senior Advisory Board of the Global Heritage Fund. He teaches a graduate course on the economics of historic preservation at the University of Pennsylvania. In 2012 he received the Louise du Pont Crowninshield Award from the National Trust for Historic Preservation, the highest US award for lifetime contribution to the field of historic preservation. Email: DRypkema@HS-intl.com

Filipe Duarte Santos is Professor of Physics and Environmental Sciences at the University of Lisbon and Director of the Research Centre SIM – Systems, Instrumentation and Modelling in Environmental and Space Sciences and Technology. Mr Santos holds a Master of Sciences in Geophysics by the University of Lisbon and a PhD in Theoretical Physics by the University of London. He published more than 150 scientific papers in the areas of Physics, Environment and Climate Change. He has been visiting researcher or professor at the Universities of Wisconsin, North Carolina, Indiana, Stanford and Harvard in the USA, Munich in Germany, Surrey in the United Kingdom and Vrije University in the Netherlands, among others. Mr. Santos is Vice-President of the UN Commission on the Peaceful Uses of Outer Space and delegate to the

UNFCCC since 1999. He was Coordinator for Sustainable Development, Global Change and Ecosystems in the Ibero-American Program CYTED from 2007 to 2011 and is presently Review Editor for the 5th Assessment Report of the IPCC. Email: fdsantos@oal.ul.pt

MV Serra is an Urban Planner with a 40 years experience in Latin America, South and East Asia, and Africa in the fields of municipal development, housing finance and planning, municipal services and public utilities, urban upgrading, land policies and planning, and urban heritage. He worked for 15 years as Lead Urban Specialist at the World Bank, involved with operations as well as analytical studies and policy analysis. He led the activities of the Bank's Cultural Heritage Unit. Before joining the Bank, he was executive director of Fundação Nacional Pró-Memória, the operational arm of the Brazilian Ministry of Culture for cultural heritage preservation. He currently consults on urban issues and writes extensively on the urban history and problems of Rio de Janeiro. Together with Teresa Serra, he recently published the *Guia de Historia Natural do Rio de Janeiro (Guide of Natural History of RdJ)*. Rio de Janeiro: Editora Cidade Viva, 2013 and published a review article on the impact of the 2014 World Cup and the 2016 Olympics ('Rio Gets Ready', *Arqtextos 17*, 2013: 138–191, Universidade Federal do Rio Grande do Sul). Email: mvserra@jurea.com

Wang Shu is an Architect and Professor at the China Academy of Art in Hangzhou, China. He established the Amateur Architecture Studio with his wife Lu Wenyu in 1997. He has been working and doing research on re-establishment of contemporary Chinese architecture, which reflects in his projects as the Ceramic Houses, New Campus of China Academy of Art in Hangzhou or the Ningbo Historic Museum. The projects have been published widely and exhibited in important venues around the world. He serves as the Head of the School of Architecture in the China Academy of Art in Hangzhou. He has been invited by universities and institutes around the world to give lectures and speeches. He has been awarded the Pritzker Architecture Prize in 2012. He is the winner of the Schelling Architecture Prize 2010 and received the Gold Prize of Architecture from France Architecture Academy 2011. He was the Kenzo Tange Visiting Professor at the Harvard Graduate School of Design in 2011. He has been member of the Jury of the 2013 Aga Khan Award for Architecture. Email: wangshu@caa.edu.cn

Rachel Singer has completed an MA in the Conservation of Material Heritage from Haifa University, and a BA in Geography from the Hebrew University, Israel. She lives in Jerusalem and is currently working towards a graduate degree in Urban Design at the Bezalel Academy of Art and Design. She has previously worked on research projects relating to planning and policy in twentieth century Jerusalem and contemporary issues in the fields of transportation and environment. Currently, she is completing an internship as part of her conservation studies that includes a specialisation in documentation at the Israel Antiquities Authority as well as theoretical research on the Historic Urban Landscapes. She is also a participant in the EU Seventh Framework project 'Designing Safer Urban Spaces'. Email: lambielion20@gmail.com

Francesco Siravo is an Architect specialising in historic preservation and town planning. He received his professional degrees from the University of

Rome, 'La Sapienza', and studied historic preservation at the College of Europe, Bruges and Columbia University, New York. Since 1991, he has worked for the Historic Cities Programme of the Aga Khan Trust for Culture, and has been responsible for planning and building projects in various cities, including Cairo, Lahore, Mopti, Mostar, Samarkand and Zanzibar. Before joining the Historic Cities Programme, he consulted for local municipalities as well as governmental and international organizations, including UNESCO, ICCROM and the World Bank. Previous work includes the preparation of conservation plans for the historical areas of Rome, Lucca, Urbino and Anagni in Italy, and for the old town of Lamu in Kenya. He has been visiting lecturer at ICCROM, the University of Rome and Cassino, and the University of Pennsylvania, and has published extensively on architectural conservation and town planning. Among these are: *Zanzibar: A Plan for the Historic Stone Town* (1996) and *Planning Lamu: Conservation of an East African Seaport* (1986). Email: francesco.siravo@akdn.org

Julian Smith is an Architect, Planner and Educator. He was formerly Chief Architect for Canada's National Historic Sites program. He later set up his own architectural and planning practice, and his projects have involved culturally significant sites in Canada, the U.S., Europe and Asia. In the late 1980s, he established and directed Canada's first English-language graduate program in heritage conservation, at Carleton University in Ottawa. In 2007 he became Executive Director of Willowbank, an independent School of Restoration Arts and Centre for Cultural Landscape near Toronto, Canada. He took part in the drafting process of the 2011 UNESCO *Recommendation on the Historic Urban Landscape*, and is currently President of ICOMOS Canada. He was recently named an honorary member of the Canadian Society of Landscape Architects, in part because of his extensive work on cultural landscape theory and practice. Email: julian.smith@willowbank.ca

Daniele Spizzichino graduated in Engineering at the University La Sapienza in Rome, Italy, in 1999. In 2012, he earned a PhD on Earth System Sciences, Environment, Resources and Cultural Heritage at the University of Modena and Reggio Emilia, Italy. He is an Engineer with an extensive experience in IT applied in different sectors of civil and environmental engineering. From 2000 until today, his work has focused on: natural hazard risk assessment and numerical modelling; sustainable development; environmental impact assessment and natural resource management; conservation plans for Cultural Heritage preservation; monitoring system design and management; design of mitigation works; laboratory tests on rock and soil material. He is the author of several papers and editor of conference proceedings. Presently he works as a researcher at ISPRA, the Geological Survey of Italy. Email: daniele.spizzichino@isprambiente.it

Ken Taylor, Professor Emeritus, is today an Adjunct Professor in the Research School of Humanities, The Australian National University, Canberra, Australia, and a Visiting Professor at Silpakorn University, Bangkok, Thailand. Since 2002 he has been teaching courses on Management of Historic Sites and Cultural Landscapes in the International Program in Architectural Heritage Management and Tourism. He has published nationally and internationally on heritage meanings and values and on cultural landscape conservation. He has also undertaken conservation management plans for historic cultural landscapes. He is a member of ICOMOS International Scientific Committee on Cultural Landscapes and has undertaken work with UNESCO, ICOMOS, and ICCROM. He is an Associate Editor of the Journal *Landscape Research*. His publications include *Canberra: City in the Landscape*. Canberra: Halstead Press, 2006; he is also co-author of *A Contemporary Guide to Cultural Mapping. An ASEAN-Australia Perspective*. ASEAN-COCI and AusHeritage, 2013, and co-editor of *Managing Cultural Landscapes*. London and New York, Routledge, 2012. Most recently he has been working with a group focused on historic towns in Guangzhou, China. Email: k. taylor@anu.edu.au

Michael Turner is an Architect and Professor in the Graduate Program of the Bezalel Academy of Arts and Design in Jerusalem. He is also in charge of the UNESCO Chair in Urban Design and Conservation Studies. He has been involved with municipal urban and environmental issues since 1974 and from 1983 has been in private practice, specialising in the field of urbanism and conservation. Serving on many international professional academic bodies he combines practice and academia contributing many articles and chapters on current design issues. He has been, and is, involved in European research projects, including Partnership for Peace, TEMPUS and the Seventh Framework. Within UNESCO, he has advised on properties around the world and participated in expert meetings on Authenticity, Category 2 Centres and World Heritage for Peace. He was elected in 2005 to the World Heritage Committee and in 2007 he served as vice-president. He has accompanied the debate on Historic Urban Landscapes since its inception. Email: turnerm@013.net

Ron van Oers is an Architect and Urban Planner. He received his PhD from Delft University of Technology in The Netherlands, with a thesis on Dutch Colonial Town Planning (1600–1800). He holds Master's Degrees in Technological Design (MTD) and Urban Planning (MEng). From 2000 to 2012 he worked at UNESCO's World Heritage Centre in Paris in a variety of positions, engaging in project management, programme design and policy development. From 2003 onwards he coordinated the *World Heritage Cities Programme*, under which he spearheaded the international effort to develop new policy guidelines for urban conservation, which were adopted as the *Recommendation on the Historic Urban Landscape* by UNESCO in November 2011. Since October 2012 he has been working as Vice Director of the World Heritage Institute of Training and Research for Asia and the Pacific (WHITRAP) in Shanghai, where he has been appointed also as Researcher at Tongji University to set up a research programme for the application of the Historic Urban Landscape approach in China. He co-authored *The Historic Urban Landscape: Managing Heritage in an Urban Century*. Chichester: Wiley-Blackwell, 2012. Email: whitrap .rvo@gmail.com

Tim Williams is a Senior Lecturer, Institute of Archaeology, University College London. He is Degree Programme coordinator for the MA Managing Archaeological Sites and the MA Urban Archaeology, Director of the Ancient Merv Project (Turkmenistan), Editor-in-chief of *Conservation & Management of Archaeological Sites*, Member of the International Scientific Committee on Archaeological Heritage Management (ICAHM) & Chair of the International Standards Board of

ICAHM. His background is in urban archaeology, especially Roman, Islamic & Central Asian; approaches to complex stratigraphy; earthen architecture conservation; and archaeological site management. He worked for the Department of Urban Archaeology (Museum of London), between 1981–1991 and then was Head of Archaeology Commissions at English Heritage, before joining UCL in 2002. He has undertaken long running projects in Beirut, Lebanon & Merv, Turkmenistan. He is ICOMOS expert member on advisory missions and panels, works on the UNESCO serial transboundary Silk Roads nomination project and undertook the ICOMOS Silk Roads thematic study. Email: tim.d.williams@ucl.ac.uk

About the Companion Website

The book's companion website is at www.wiley.com/go/Bandarin/Historic_Urban
_Landscape and offers invaluable additional resources:

- UNESCO documents on Historic Urban Landscape and New Life for Historic Cities
- Additional illustrations of historic urban landscapes

Introduction
Urban Conservation and the End of Planning

Francesco Bandarin

UNESCO Assistant Director-General for Culture and Professor of Urban Planning at the University Institute of Architecture of Venice, Italy

> *'There is still one of which you never speak.'*
> *Marco Polo bowed his head.*
> *'Venice,' the Khan said.*
> *Marco smiled. 'What else do you believe I have been talking to you about?'*
> *The emperor did not turn a hair. 'And yet I have never heard you mention that name.'*
> *And Polo said: 'Every time I describe a city I am saying something about Venice.'*
>
> Italo Calvino. *Invisible Cities*

It is a paradox that in today's world, where cities have gained a central place in economic, environmental and social policy-making at the global scale, urban planning has declined and *de facto* ended as a unified management system of complex urban processes.

Urban planning, as a rational and comprehensive discipline for the management of urban and territorial development, originated in the last part of the nineteenth century and fully blossomed in the twentieth century, with the aim of governing large-scale urban growth and urban rehabilitation processes.

This functioned effectively (and in some cases it still does) in centralised societies where social change occurred following linear processes: rural-urban migrations, the rise of mass public and private transport systems, as well as planned industrial growth. Nevertheless, it failed with the progressive decentralisation of social decision-making processes, coupled with complex and unpredictable development trends, and the inevitable domination of the global market processes in the economic and social scenes.

Today, urban planning has lost its ability to govern these processes, becoming essentially a socio-economic discipline – one of the many tools for managing cities.

Not all cities in the world are undergoing the same process: some are declining, others expanding, while a few remain stable. But all cities are now interconnected, and the main processes that

Reconnecting the City: The Historic Urban Landscape Approach and the Future of Urban Heritage, First Edition. Edited by Francesco Bandarin and Ron van Oers.

determine their future are of a global nature, due to the shifting of production centres, the exponential increase of communication speed brought about by the Internet, and the accelerated movement of people for work and leisure.

Urban conservation is not immune from these processes, nor is it unaffected by these trends. Urban heritage can no longer be conceived of as a separate reality, a walled precinct protected from the external forces of change by plans and regulations. It simply does not work this way, if it ever did.

In some regions of the world, the relative success of urban conservation as both policy and practice over the past 50 years has created the illusion that sections of the city may be shielded from change and separated from the inevitable evolution of the urban context. This was perhaps possible for monuments, individual buildings and archaeological areas. It proved impossible for a living open system like a city, no matter how historic and protected.

As much as urban planning has ceased to respond to the needs of a mobile, multi-layered, globalised urban society, so urban conservation – as it was shaped in the second half of the twentieth century – has reached its limits, and is losing its ability to deal with the new challenges to the conservation of the urban heritage. This is particularly true in regions of the world where the established principles, mostly of western origin, have been imported and are not embedded in local practices and perceptions.

One of the reasons behind this gradual marginalisation is the 'fracture' imposed by the Modern Movement of the 1920s and 1930s within a disciplinary context that was previously largely inspired by unitary approaches to the management of urban spaces, linked to historical typological research (as seen by Sitte, for instance[1]) or to 'organic' approaches (as those put forward by Geddes[2]).

Modernism rejected the attempts to ensure continuity in urban development, and promoted a radical and revolutionary approach that denied the 'old' city any function in modern life (except for the memory value of a limited number of monuments). The historic city was not considered part of modernity, nor a component to integrate; it was simply and abstractly 'erased' from the urban planning scene.

As a reaction to this approach, an architectural and urban conservation movement came to the forefront, based upon the newly established principles of city conservation and rehabilitation promoted by Gustavo Giovannoni in Italy and enshrined in the documents of the 1931 Athens Conference.[3] The impetus of this movement grew in the post World War II decades, when in the aftermath of the massive destruction of European cities during the war, and of the extensive post-war reconstruction, countries developed legislation and planning practices that fostered the conservation of important sections of the historic urban fabric. This success, however, came at the price of separating two of the main objects of urban management: the historic areas, where special regulations, planning, and subsidy systems were introduced and the development areas, considered non 'historical' or simply new, where urban planning models founded upon the principles of modernism were applied. This was based on medium term projections of the city's physical, economic and demographic development, phased by regulatory tools, such as zoning, or smaller scale plans aimed to regulate building, infrastructure development and use.

[1] Sitte, C.,(1965) *City Planning According to Artistic Principles*, London, Collins. (Originally published in German in 1889, as: *Der Städtebau nach seinen künstlerischen Grudsätzen*).

[2] Geddes, P. (2010) *Cities in Evolution: an Introduction to the Town Planning Movement and the Study of Civics*, Nabu Press. (Originally published in 1915).

[3] Choay, F. (ed.) (2002) *La Conférence d'Athènes sur la conservation artistique et historique des monuments (1931)*, Paris: Les Editions de l'Imprimeur.

This situation generated two different urban management processes, based on different professional approaches and principles. The unified urban planning vision that had characterised development before the modern age came to an end.

Today, we face a twofold challenge. Urban planning, intended as a top-down political and administrative process to regulate urban dynamics, has clearly demonstrated its limits, and is being substituted by a variety of management, participatory and design tools At the same time, urban conservation has also proved unable to ensure the effective and long-term integrity of both the physical and social fabric of historic areas.

These challenges are rendered even more complex by the shifting context of urban management, where issues of sustainability, energy consumption, social inclusion and the transformation of mobility and work patterns have become of paramount importance and will prove critical in the coming decades.

While finding a single direction is unlikely, given the diversity of political, economic and social conditions prevailing in the different contexts, it is clear that traditional 'land management based' planning practices cannot offer viable solutions. Similarly, the traditional 'districting' of historic areas shows conceptual (what is 'historic'?), political (conservation as a way to social exclusion?) and operational (which 'changes' are acceptable?) limits requiring a radical revision of the established paradigms.

After the end of planning, we need to identify which actions will enable us to reflect local conditions, decision-making systems and needs, in order to achieve a higher order of objectives that ensures respect of the principles and frameworks a society wants or is forced to impose upon itself.

These 'limits' and conditions (energy consumption, resource use, the degree of social equality, the production model, the popula-tion mix, etc.) will determine the choices related to urban management, development and rehabilitation.

This new situation, so far mostly addressed by researchers in the domains of Ecological Urbanism and Landscape Urbanism, opens up a new dimension for urban conservation itself. Instead of being a 'marginal' section of the urban complex, the historic city becomes a model, a 'resource' to respond to new needs, to define innovative physical and social patterns, and to value what centuries of experimentation in the design of urban spaces and form has given us. This is the central message of the 2011 UNESCO *Recommendation on the Historic Urban Landscape*, a milestone document aimed at redefining the role of urban heritage in society, and the parameters to be used in managing its conservation, evolution, and integration within the broader urban decision-making process.

Post-War Attempts to Reconnect the City

Following the post-war reconstruction phase, with the limits of the modernist approach increasingly apparent, attempts were made to 'reconnect' the domains of Urban Development and Urban Conservation. Indeed, as demonstrated in greater detail in the previous work on this theme,[4] *reconnecting* the city, in defining the methodological and operational processes that allow the integration, understanding, design and management of the range of urban transformation processes, has been one of the core concerns of modern architects and planners.

Today's urban planners, managers, designers and conservators have at their disposal a vast array of tools and experiences; indeed the past 50 years were crucial in transforming the discourse on the city and in enlarging the scope of the urban disciplines.

---

[4]Bandarin, F. and Van Oers, R. (2012) *The Historic Urban Landscape. Managing Heritage in an Urban Century*, Oxford, Wiley-Blackwell: 23–36.

The foundations of the new approach to urban conservation, as advocated in the UNESCO *Recommendation on the Historic Urban Landscape*, are now an integral part of disciplines involved in urban management, particularly in the area of *civic engagement and participation*, the *analysis of the urban form*, the reflection on the *context and the spirit of place*, the analysis of people's perceptions and memory in the creation of *urban values*, and finally, the understanding of the importance of the *natural processes* in guiding urban development and management and of the *economic roles* of the historic city.

Civic Engagement

The first reactions to the modernist ideology of the Functionalist City announced in Le Corbusier's Athens Charter[5] originated from within the very structure that had propelled Modernism internationally: the CIAM.[6] During the 1950s, a group of younger architects and planners, who subsequently formed 'Team 10', started to critique the official discourse on urbanism. Among many others, the works of Aldo van Eyck[7] and Giancarlo De Carlo[8] in support of a humanistic urbanism based on social participation and on the respect of the values of the urban context, including the historical, social and economic dimensions, constituted an authentic revolution in the traditional planning approach, and paved the way to the development of a new vision of urbanism.

Clearly, the time was ripe for new approaches, even outside the realm of CIAM: the new vision brought about by Jane Jacob's advocacy planning[9] had left a permanent imprint on the work of urban managers worldwide, while in different regions of the world experiments in social participation and urbanism, sensitive to the postcolonial condition took root.[10] The approach to the historic city was bound to change as this was now seen not as a residual dilapidated area for the lower income classes, but, on the contrary, an area rich in social layers and buffers. This perception supported the adoption of specialised legislations in many European countries and provoked a reaction against blind Urban Renewal programmes in the United States.

Analysis of the Urban Form

As this new approach developed, an important methodological contribution arrived in the 1960s and 1970s from disciplines as diverse as geography and planning, with the development of the typo-morphological analysis, a powerful tool for understanding the dynamics and the layering process of the city, which obviously found its greatest applications in the area of urban conservation.[11] The Italian architect Saverio Muratori[12] and his School, pioneered the practical development of this methodology. The archi-

---

[5] Le Corbusier (1957) *La Charte d'Athènes*, Paris, Editions de Minuit. (Originally published in 1943).

[6] The *Congrès internationaux d'architecture moderne* – CIAM (International Congresses of Modern Architecture) was founded in 1928 and disbanded in 1959. Its main objective was the promotion of the principles of the Modern Movement in architecture, urbanism, industrial design.

[7] Lefaivre, L. and Tzonis, A. (1999) *Aldo Van Eyck. Humanist Rebel*, Rotterdam: 010 Publishers.

[8] De Carlo, G. (1972) *An Architecture of Participation*, The Melbourne Architectural Papers, Melbourne: The Royal Australian Institute of Architects. See also: Guccione, M. and Vittorini, A. (2005) *Giancarlo De Carlo: The Reasons of Architecture*, Rome: MAXXI, Catalog of the Exhibition.

[9] Jacobs, J. (1993) *The Death and Life of Great American Cities*, New York: The Modern Library. (Originally published in 1961).

[10] See for instance the work of other Team 10 members, such as Balkrishna Doshi and Charles Correa in India, Kenzo Tange and Fumihiko Maki in Japan, Michel Ecochard and George Candilis in North Africa.

[11] The pioneering work in this field was conducted by the German–British geographer M.R.G. Conzen in the UK. See: Conzen, M.R.G. (2004) *Thinking about Urban Form, 1932–1998*. Bern: Peter Lang.

[12] For a presentation of the approach developed by Muratori, see: Cataldi, G. (2003) From Muratori to Caniggia: the Origins and Development of the Italian School of Design Typology, *Urban Morphology*, 2003, 7 (1):19–34. See also: Caniggia, G.; Maffei, GL, 2001. *Interpreting Basic Building. Architectural Composition and Building Typology*. Firenze: Alinea.

tecture historian and urban planner Leonardo Benevolo later promoted the first significant implementation of this approach in his conservation plan of the historic city of Bologna, which demonstrated the viability of this methodology as an effective tool for the management of urban transformations in historic contexts.

While these experiments proved effective, they nevertheless remained limited to the historic districts of the city. It was the Italian architect Aldo Rossi, developer of the modern theory of city design and management, whose work *The Architecture of the City*[13] was and remains amongst the most influential modern manifestos of urban design. Rossi reaffirms the importance of the historical dimension of the city in guiding architecture and urban design (which he sees as part and parcel of a single process). The city is itself the result of an historical accumulation of human actions, and should be viewed as a living palimpsest of past processes that influence the present and the future. The complexity of the urban sphere is often beyond comprehension, beyond the reach of static rules. Nonetheless rules exist to guide interventions and the design process. The identification of these 'inner rules' entails an understanding of the layering process of the city, of its history and life that is in itself part of the design exercise.

Spirit of Place

Based on a different approach, but with converging results is the contribution of the Norwegian architect Christian Norberg-Schulz, who provided a modern definition of the classical concept of *genius loci*, seen as a result of the relationship of man with the environment.[14]

This dynamic relationship evolves with time, as it engages living human beings. Norberg-Schulz, inspired by the philosopher Heidegger, uses the concept of *Räum-lichkeit*, translated as 'presence', the space of everyday life. In his view, a space changes its meaning from being a *situs* to a *locus*, because life 'takes place' there. In this respect, Norberg-Schulz anticipates many modern views on the role of intangible heritage in the construction of the significance of a place.

An equally important contribution in this direction, that used a landscape perspective rather than an architectural one to examine the city, comes from the work of J.B. Jackson, an interpreter of the meaning of places and of the impact of modernity in their transformation. His analysis of the new forms of urban development is revealing of the problems we face today.[15]

Urban Values

In the post-war period, few urban planners and designers have been able to understand the crisis of the discipline and propose alternative approaches more than Kevin Lynch. His contribution extends to all fields of urban studies and planning, from new city design to urban rehabilitation, to spatial and regional planning to preservation, and is based on the attempt to link the physical space to the perception and the usages of the inhabitants. It is a highly humanistic approach, rendered 'scientific' through the employment of modern notions of psychology, together with a profound understanding of the

---

[13] Rossi, A. (1978) *The Architecture of the City*, New York: Opposition Books. (Originally published in Italian as *L'architettura della Città*, Venezia: Marsilio, 1966).

[14] Norberg-Schulz, C. (1980) *Genius Loci: Towards a Phenomenology of Architecture*, New York: Rizzoli.

[15] 'It is here in the city, not in the open countryside, that the modern road reveals itself. Even a generation ago the old architectural order prevailed: the street was still squeezed between tall and imposing façades, – the urban equivalent of those fences flanking the country road: protecting what lay beyond. But the new road or street is like the eight-hundred-pound gorilla. It goes where it wants to. It is wider than roads were in the past, sinuous in its layout, no longer respectful of the grid, and it devours spaces and structures hitherto though of as sacred. It is creating its own architecture: short-lived, eager to conform to the new type of traffic and to discard its own symbols and any hint of history'. Jackson, J. B. (1994) *A Sense of Place , a Sense of Time*, New Haven: Yale University Press: 9.

spirit and the role of context in modern life. His 'mental mapping' approach to urban design is certainly a precursor of modern cultural mapping processes, and a potent guide to human-centred urban design and conservation proposals. In the field of urban conservation he produced a major reflection, summarised in his book *What Time is This Place?*,[16] an invaluable text that studies the function of preservation in modern society, dissociated from preconceptions and formal schemes, and offering a perspective that allows the interpretation and design of the city, both ancient and modern, in harmony with the laws of time and history, and with a view to the social and cultural needs of modern societies, and rid of dogmatic or elitist approaches.

Natural Processes

The integration of cities and nature was certainly a key aspect of the great planning tradition of the twentieth century, typically in the form of parks and open spaces design.[17] However, contemporary design approaches have, for the most, found their origin and inspiration in the work of the Scottish-American landscape architect and planner Ian McHarg, whose seminal book *Design with Nature*,[18] and research and teachings have helped re-position the idea of landscape and the discipline of landscape architecture at the centre of the urban management process.

Thus, the work of McHarg is linked to the great tradition of environmentally sensitive design rooted in the work of Patrick Geddes[19] and of the great landscape movement in America, as exemplified in the works of Frederick Law Olmsted.[20] However, McHarg had the merit to define an interdisciplinary approach based on scientific methodologies and fully integrating the social and administrative dimensions of urban and territorial planning.

McHarg's 'Environmental Planning' perceives the city within the broader ecological context, reflecting upon the relationships between natural processes and man-made transformations, in order to identify design methods and solutions. At the core of his approach lies an innovative methodology that brings together social and natural processes through a system of analysis and superposition of the layers of values existing in a given area. As the natural processes are linked to geological and hydrological systems, the areas involved in the analysis are subsequently much larger that the city area itself, which leads to a change and an extension of the value system embraced by the plan. This also allows the identification of the intrinsic suitability of the different zones, the compatibility of usages, as well as the economic trade-offs between urban development and conservation. Such an approach is obviously better adapted to a large planning scale at the regional level, as only this scale allows the natural systems to be appropriately understood and managed.

In spite of these important intellectual and institutional advancements, the increasing complexity of urban processes, the difficulty in harnessing market forces as well as ever changing demographic, productive and consumption patterns, reveal how unsatisfactory is the situation of urban planning and design.

The defining moment of the new forms of Urbanism and their veritable point of departure can be said to be *Collage City*, the milestone text of Rowe and Koetter[21] published in 1978. *Collage*

---

[16] Lynch, K. (1972) *What Time is This Place?* Cambridge, Massachusetts: MIT Press.

[17] See: Brantz, D. and Dümpelmann, S. (2011) *Greening the City. Urban Landscapes in the Twentieth Century*, Charlottesville: University of Virginia Press.

[18] McHarg, I. (1969) *Design with Nature*, Philadelphia: The Falcon Press.

[19] Geddes, P. (2010) *Cities in evolution: Evolution: An Introduction to the Town Planning Movement and to the Study of Civics*, Nabu Press. (Originally published in 1915). See also: Welter, V. M. (2002) *Biopolis: Patrick Geddes and the City of Life*, Cambridge, Massachusetts: MIT Press.

[20] Fein, A. (1972) *Frederick Law Olmsted and the American Environmental Tradition*, New York, Braziller.

[21] Rowe, C. and Koetter, F. (1978) *Collage City*, Cambridge, Massachusetts: MIT Press.

City, a manifesto of the modern 'chaotic' urban processes, is where the idea of the city as a continuum is forever abandoned. The city, from antiquity on, is shown as a continuous process; an aggregate of parts built in different eras, of accumulation and overlays. The modern idea of the city as a totalising urban design is seen as an abstract utopia, if not as a dangerous method. As the city grows by small accretions, the focus shifts from the uniform vision of planning to the individual urban design schemes that allow planned or unplanned dynamics. Each part of the city therefore follows its own rules of composition and functioning, and this is what makes the city work and advance, beyond the traditional concepts of order and rational control.

Only after 'chaos' itself becomes a new model we can begin to understand the process whereby the different components of the puzzle establish a process of reciprocal influence and exchange, and how the mutations induced by the economic and social transformations adjust, exalt or condemn the city's parts.

Urban conservation, whether we accept it or not, is part of the urban 'chaos', representing an approach to one of the fragments of the city, the one defined as 'historical' according to a variety of possible processes: political, administrative, academic, community-based or other.

Contemporary Views on Urbanism and Landscape

Chaos as the 'urban condition' has been at the centre of the debate between architects and plan-

ners for over 30 years, and it is out of this discussion that some of the contemporary approaches to urban development and urban conservation[22] have emerged.[23]

On the one hand, in reaction to the increasing complexity of urban functions that the traditional city cannot support, we find responses stemming from large-scale design and urban mega-structures, in reaction to the increasing complexity of urban functions that the traditional city cannot support, or that would require major transformations of existing patterns. This type of Urbanism, partially echoing some of the more provocative proposals of the Modern Movement,[24] aims to define an urban order without having to deal with the underlying urban chaos. Furthermore, it's an approach that allows maintaining the great diversity of the city, including the preservation of its historic parts. Many support this approach, especially as the pace of public and private investment in urban projects has accelerated in past decades. Certainly, its main theoretician and practitioner is Rem Koolhaas, whose proposals[25] and projects have established the models and led the way.[26] Today, many cities promote large-scale projects to address the complexities of urban rehabilitation, the reconversion of terminated industrial areas and of major public services (hospitals, military sites, waterfronts, etc.). These projects thus become both the focal points of the urban development strategy, and the tools that redefine urban development policies in relation to market forces.

In the past half century, vast experiments were conducted throughout the world in the area of

---

[22] For an interesting review of the contemporary approaches, see: Fromonot, F. (2013) Manières de classer l'urbanisme, *Criticat08*: 40–61. (www.criticat.fr)

[23] For a complete review of the evolution of Urban Design in the past half a century, see: Shane, D. G. (2011) *Urban Design Since 1945*, New York: Wiley.

[24] See Le Corbusier's Plans for Rio de Janeiro or the Plan Obus for Algiers: Tsiomis, Y. (2012) *Rio-Alger-Rio*, 1929–1936. Transferts, Fondation Le Corbusier. *Le Corbusier. Visions d'Alger*. Editions de la Villette: 85–101.

[25] Koolhaas, R. (1995) *What Ever Happened to Urbanism?* Koolhaas, Rem; Mau, Bruce: *S,M,L,XL*. New York, The Monicelli Press: 959–971.

[26] Rem Koolhaas has authored several projects addressing the issue of urban large scale recomposition, starting with the project for the Parc de la Villette in 1982 (non selected), to the Toronto Downsview Park Project in 2000 (Competition Prize winner) to the 2010 Hong Kong cultural district (non selected).

urban regeneration, focusing in particular on the transformation of industrial areas or large infrastructures, such as harbour waterfronts, abandoned military zones or railway areas. Very often, these projects, located near the historic centres, have become the motors of urban redevelopment and re-functionalisation, as well as have guided the transformation of urban heritage, in ways more powerful than those possible through traditional urban conservation policies. France is perhaps the country that has implemented the most successful urban regeneration and conservation projects, as shown by the exceptional results achieved in cities like Bordeaux, Lyon, and Strasbourg.[27]

On the other hand, important attempts were made to redefine the parameters of the urban design process through focusing new attention on the extended context of modern urban life, such as the territory and the landscape.

During the past thirty years, McHarg's ideas blossomed, opening the way to significant theoretical and practical developments, both within the landscape architecture and planning disciplines, and at the institutional and normative process. It is largely thanks to this intellectual advancement that the concept of landscape has become today the key for conceptualising and defining urban development process in all regions of the world.

As Charles Waldheim, one of his students, wrote:

> Across a range of disciplines, landscape has become a lens through which the contemporary city is represented and a medium through which it is constructed. These sentiments are evident in the emergent notion of 'Landscape Urbanism'.[28]

The notion of Landscape Urbanism does not point to a specific model or methodological framework of disciplinary nature, but tends to be seen rather as an approach to urban design and management based on an understanding of the natural, physical and social context. In this sense, the lesson of McHarg – who was primarily interested in the role of natural components in the planning process – evolved over the past decades into a variety of approaches, termed 'Urban Landscape', 'Urban Nature', 'Urban Ecology', and 'Green Urbanism',[29] fed by contributions from the work of Carl Troll and the German geographers of his generation.[30]

While McHarg focused mainly on the natural dimension of territorial planning, his followers, both in the USA and Europe, shifted their interest toward the city as an object that could be analysed using the new categories and methodologies, as exemplified by Spirn's groundbreaking work *The Granite City*:

> Nature pervades the city, forging bonds between the city and the air, earth, water, and living organisms within and around it. In themselves, the forces of nature are neither benign nor hostile to humankind. Acknowledged and harnessed, they represent a powerful resource for shaping a beneficial urban habitat ignored or subverted, they magnify problems that have plagued cities for centuries, such as floods and landslides, poisoned air and water. Unfortunately, cities have mostly neglected and rarely exploited the natural forces within them.[31]

The point that is common to the different facets of these approaches is the need to mobilise a variety of disciplines to achieve an understanding of the context within which to operate. The mono-dimensional approaches provided by

---

[27] Tsiomis, Y. and Ziegler, V. (2007) *Anatomie de projects urbains*, Paris: Editions de la Villette.

[28] Waldheim, C. (ed.) (2006) *The Landscape Urbanism Reader*, New York: Princeton Architectural Press:15.

[29] The literature on urban ecological and sustainable management has expanded tremendously in the past decade. See for instance: Newman, P. and Jennings, I. (2008) *Cities as Sustainable Ecosystems. Principles and Practices*, Washington DC, Island Press. In Jenks, M. and Dempsey, N. (eds.) (2005) *Future Forms and Design for Sustainable Cities*, London: Elsevier, Architectural Press. Register, R. (2006) *Ecocities. Rebuilding Cities in Balance with Nature*, Gabriola Islands, BC, Canada: New Society Publishers. In Emelianoff, C. and Stegassy, R. (eds.) (2010) *Les pionniers de la ville durable*, Paris: Editions Autrement.

[30] Troll, C. (1939) *Luftbildplan und ökologische Bodenforschung (Aerial Photography and Ecological Studies of the Earth)*, Zeitschrift der Gesellschaft für Erdkunde, Berlin: 241–298.

[31] Spirn, A W. (1984) *The Granite City*, New York, Basic Books: xi.

'classical' architectural and planning practices were considered insufficient to address the complexity of the challenges posed by the new pace and type of urban development, first in the US and the European scene,[32] and later in many other parts of the emerging world.

Within this framework of action lies a large variety of methodologies and models, from the long-term approach proposed by James Corner[33] to a project-oriented approach proposed, among others, by George Hargreaves[34] in America or Michel Courajoud[35] and Alexandre Chemetoff[36] in France, to name but a few. But all possess – explicitly or implicitly – a common approach based on what Alan Berger[37] termed 'Intelligent Systemic Design', which is on a working method aimed at creating a bridge between different disciplinary fields. In particular, for the concerns of the urban field, the need exists to integrate classical urban design processes with planning approaches, and to associate cultural geography (a discipline that has pioneered the concept of landscape, long before planners and architects[38]), as well as the natural sciences in the landscape plan. Landscape Urbanism is a tool that allows the different disciplines to establish a dialogue, without the need to define a new discipline with its own paradigms and doctrines. This pragmatic approach has allowed Landscape Urbanism to embrace a large variety of situations, from the level of territorial planning to large reconversion projects on the urban scale, to fine-tuned interventions on the built urban environment. It is, in fact, more a process than a product.

As one of today's most prominent theoretical and professional actors in this field, James Corner says:

The emphasis shifts from landscape as a product of culture to landscape as an agent producing and enriching culture. Landscape as noun (as object or scene) is quieted in order to emphasize landscape as verb, as process or activity. Here, it is less the formal characteristics of landscape that are described than it is the formative effect of landscape in time. The focus is upon the agency of landscape (how it works and what it does) rather than upon its simple appearance.[39]

Over recent decades this new approach has allowed an impressive array of experiences in all parts of the world, based on the large-scale consideration of the urban and territorial environment of the city, seen as the framework to orient design choices, densities, and functional mixes down to the detailed choice of materials and of sustainability solutions. Its strong interest in the context of the city and of its territorial dimension, whence its rules are derived, make it an effective tool for some of the most critical issues of contemporary urbanism, based not on urban expansion as such, but on urban rehabilitation and reconversion.[40] In this sense, the historic layers of the city – seen as a much broader context beyond that of the 'historic city' – represent a fundamental guidance for the design, jointly with the understanding of the physical and natural context.

Modern Landscape Urbanists are aware of the risk of transplanting western ideas in other

---

[32] For a review of the European experience in ecological city planning, see: Beatley, T. (2000) *Green Urbanism. Learning from the European City* Washington DC: Island Press.

[33] Corner, J. (ed.) (1999) *Recovering Landscape: Essays in Contemporary Landscape Architecture*, New York: Princeton Architectural Press.

[34] Hargreaves, G. and Czerniak, J. (2009) *Hargreaves: The Alchemy of Landscape Architecture*, London, Thames and Hudson.

[35] Nourisson, D. (2000) *Michel Corajoud, paysagiste*, Paris: Hartman.

[36] Chemetoff, A. (2009) *Visits: Town and Territory: Architecture in Dialogue*, Berlin: Birkhauser Verlag AG.

[37] Berger, A. (2011) *On Landscape Urbanism. A conversation*. In Ferrario, V., Sampieri, A. and Viganò, P. (eds.). *Landscapes of Urbanism*. Roma: Officina Edizioni: 96.

[38] See: Mitchell, D. (2000) *Cultural Geography, a Critical Introduction*, London, Blackwell.

[39] Corner, J. (ed.) (1999) *Recovering Landscape. Introduction*, New York: Princeton University Press: 4.

[40] See: Marot, S. (1999) Reclaiming Sites. In Corner, J. (ed.) (1999) *Recovering Landscape. Introduction*, New York: Princeton University Press: 45–57.

cultural context, with a clear awareness of the disasters generated by western planning as it spread throughout the world in the twentieth century. In fact, the very idea of landscape is not universal or unequivocal: while for western cultures it is connected in its origins to the sense of aesthetics, or of contemplation of nature, for other cultures it is linked to the sacred or it is a constituent of the collective identity through narratives and dreams. For some cultures it is linked to religion and the representation of the otherworldliness, for yet others, it simply has no meaning. However, as the process concerns urban management issues that are nowadays common to many different contexts, and because the idea is fundamentally pragmatic and operational, and sustained by technical protocols, it is conceivable that a similar methodology could support different cultural visions of the city. As Charles Waldheim writes:

> In this context, the discourse surrounding landscape urbanism can be read as a disciplinary realignment in which landscape supplants architecture's historic role as the basic building block of urban design. Across a range of disciplines, many authors have articulated this newfound relevance of landscape in describing the temporal mutability and horizontal extensivity of the contemporary city.[41]

Since its appearance a decade ago, the approach proposed by Landscape Urbanism has not been accepted without critique and opposition,[42] predominantly pointing to the lack of sufficient 'sustainability' and to the excessive focus on the rural-suburban dimension of the city, and less to the city as an existing built environment.

As global concern for the environment expanded in the last two decades, following the surge of China and other countries as major industrial powers, and the entry into the global market for energy, food and raw materials of hundreds of millions of new consumers, the attention of urban thinkers and planners has, in fact, moved toward the issue of sustainability and resource management, and paying greater attention to the existing building stock and its relationships with energy flows.

This has prompted a specific dimension of urbanism that has been termed 'Ecological Urbanism', which may also be considered a derivation of the Ecological Planning School of the 1960s. While supporting principles similar to Landscape Urbanism, Ecological Urbanism places more emphasis on the ecological, economic and social condition of the modern metropolis, distancing itself from the cultural orientation that is a structural feature of the landscape approach. The present financial and economic crisis affecting the industrialised countries, together with the upcoming grave climatic change perspectives, have shifted the attention of urban thinkers and managers to the future impact on cities, and therefore to urban ecology as a tool for sustainability and resilience. As Mohsen Mostafavi writes:

> The recent financial crisis, with all its ramifications, suggests the on-going need for a methodological reconceptualization of our contemporary cosmopolitan condition. In this context, it is now up to us to develop the aesthetic means – the projects – that propose alternative, inspiring, and ductile sensibilities for our ethico-political interactions with the environment. These projects will also provide the stage for the messiness, the unpredictability, and the instability of the urban, and in turn, for more just as well as more pleasurable futures. This is both the challenge and the promise of ecological urbanism.[43]

Despite their different emphases, Landscape Urbanism and Ecological Urbanism offer, for the

---

[41] Waldheim, C. (2006) *Landscape as Urbanism*. In Waldheim, C. (ed.) *The Landscape Urbanism Reader*. New York: Princeton Architectural Press: 37–53.
[42] Duany, A. and Talen, E. (eds.) (2013) *Landscape Urbanism and its Discontents. Dissimulating the Sustainable City*, Gabriola Island, Canada: New Society Publishers.
[43] See: Mostafavi, M. (ed.) (2010) *Ecological Urbanism*, Harvard University, Graduate School of Design. Baden, Switzerland: Lars Muller Publishers.

first time in a century, the possibility to observe, understand and manage cities in a unitary manner. Clearly, the presence of the natural dimension is a strong unifier of the different parts of the urban 'collage'. What appears, from an architectural viewpoint, separated and different in terms of structure, form and function, is seen as part of the same system when perceived from the point of view of nature and ecology. Furthermore, the modern global condition has rendered the urban, social and economic structures quite fluid, and not necessarily matching the originally intended forms. This opens the way to innovation in the way cities are used, lived and transformed. Only an open-ended, flexible, even undetermined approach, can match the speed and unpredictability of such changes.

Landscape planning has the potential to address all urban scales, but it is particularly apt to manage the very large metropolitan scale that has become the main challenge of contemporary societies. A recent planning exercise launched by French President Sarkozy in 2007 for the *Grand Paris*[44] allowed the comparison of different contemporary approaches to large scale metropolitan planning. While most of the proposals hinged on traditional infrastructure development, one project in particular addressed the issue through an innovative landscape-based approach. The project, by the Italian urban planners Secchi and Viganò, named 'The Porous City'[45] addressed the critical issues of most contemporary conurbations, that is inclusion, mobility, and sustainability, through an approach based on the analysis of the specificity of the different landscapes of the city and of its existing networks. The key concepts used for this analysis are those of *porosity* as a measure of the space available for movement; *connectivity* as a measure of the degree of mobility; *permeability* as a measure of the ease of movement; and *accessibility* as a measure of the ability to move from one point to another. All these measures require an understanding of specific landscape features (artificial and natural) of the different places.

Under a Landscape approach, the 'reconnection' of the city becomes a possibility, albeit not in the classical sense of a system unified by a single vision and planning process. On the contrary, it is in other dimensions of the urban complex, namely ecology, resilience, sustainability, porosity and resource management, that we find the unifying factors. Furthermore, we need to reinterpret the relationship between the social structure and the built environment to better understand how urban spaces and environment are used and transformed by people.[46]

Repositioning Urban Conservation, Reconnecting the City

As argued earlier, the idea of the Historic Urban Landscape is part of a broader reflection on the evolution of urbanism, as a response to the increasingly complex challenges brought by global processes. The historic city is not an island, and all global social, economic and physical transformation processes affect both it and its spaces. The normative 'barriers' created by special legislation and programmes aimed at its protection are unable to shield it – if this was ever possible or intended – from the rest of the

---

[44] The initiative was announced on 17 September 2007 during the inauguration of 'La Cité de l'architecture et du patrimoine', when Sarkozy declared his intent to create a 'new comprehensive development project for Greater Paris'. In 2008 an international urban and architectural competition for the future development of metropolitan Paris was launched. The architects leading the ten multi-disciplinary teams were: Jean Nouvel, Christian de Portzamparc, Antoine Grumbach, Roland Castro, Yves Lion, Djamel Klouche, Richard Rogers, Bernardo Secchi, Paola Viganò, Finn Geipel, Giulia Andi, and Winy Maas.

[45] Secchi, B. and Viganò, P. (2011) *La ville poreuse. Un projet pour le Grand Paris et la métropole de l'après-Kyoto*, Geneva, Metis Presses.

[46] See also in this direction: Shane, D. G. (2005) *Recombinant Urbanism. Conceptual Modeling in Architecture, Urban Design, and City Theory*, New York: Wiley.

city. In as much as classical planning has proved incapable of mastering contemporary urban processes, largely dominated by market forces and by increasingly swift and fluid social changes, classical urban conservation schemes are likewise demonstrating their own limits.

Protecting the historic city requires not only a special status, but also public investment in urban infrastructure and direct or indirect subsidies to private owners to sustain the cost of maintenance. Inevitably, this has an impact on land and housing values, and subsequently on social access to the historic city. Gentrification is not just a market process; it is often the result of public policies.

As the historic city embodies a higher urban quality and identity, it attracts tourists and visitors, shops and services. This higher quality of space and density of functions – often replicated and pursued in modern waterfront rehabilitation programmes – tends to transform the historic city into something closer to a shopping centre or a theme park,[47] where the original values linked to history, artistic forms, memory functions and social atmosphere are converted into commercial opportunities for a transient population.

This situation is obviously not applicable to all historic cities – as in fact many of them still need major interventions to be preserved. Also, in many situations, cities have been able to strike a balance between conservation and development and have been able therefore to retain their character.

In this scenario, there is a compelling need to reposition urban conservation within the overall urban management process, and to redefine some of its basic operational principles.

Some of the questions that need to be raised are related to the definition and role of heritage in contemporary and future society. Others address the relationships of 'heritage' areas with other parts of the city and the territorial dimension. And yet others tackle the ways and means of preserving urban values in a changing world.

For those interested in urban conservation, the key issue is clearly *what* is to be preserved. Until now, the issue has been solved, as discussed earlier, by placing a distinction separating what is 'historic' (to be preserved) and what is 'modern' (that can be changed). All the modern urban conservation policies follow – more or less – this dichotomy, with the related tool-kit of 'preservation' areas and districts, special rules and procedures, supervision etc. This model of urban heritage conservation has been enshrined in national legislations and in international systems, in particular in the World Heritage Convention.[48]

Today, it is becoming clear to policymakers and conservationists throughout the world that

---

[47] See the forward-looking analysis of these processes in: Ashworth, G. J. and Tunbridge, J. E. (1990) *The Tourist-Historic City*, London: Belhaven Press.

[48] As of today, historic cities are defined as 'Groups of buildings' in the Operational Guidelines of the World Heritage Convention. For the inscription in the World Heritage List, an urban area has to be listed with a perimeter (the 'property') and a 'buffer zone'. Following the adoption of the UNESCO *Recommendation on the Historic Urban Landscape* in 2011, the World Heritage Committee has launched a process for the revision of the definition of historic cities. A workshop was organised in Rio de Janeiro in November 2013. The meeting recommended that "*in the future it would be preferable for historic cities, towns, and urban areas to be nominated as "sites" rather than "groups of buildings" within the definition of cultural heritage provided in Article 1 of the Convention. The meeting expressed the idea that, as all urban areas are works of humans or the combined works of nature and humans, the category of sites is a more appropriate way of expressing the layering and attributes as laid out in the Recommendation on the Historic Urban Landscape. It is felt that the definition of groups of buildings is much more limiting as it refers only to the physical attributes of the group and emphasises homogeneity rather than the complexity and diversity found in most urban areas*". The meeting also recommended to: "*change the name of the existing category (Historic Towns and Town centers) to become "Urban Heritage" to better reflect HUL approach.*" See: UNESCO World Heritage Centre (2013) *Report on the International World Heritage Expert Meeting on the Mainstreaming of the Methodological Approach Related to the Recommendation on the Historic Urban Landscape in the Operational Guidelines*, Rio de Janeiro, Brazil, 3–5 September 2013. UNESCO World Heritage Centre, unpublished document.

this traditional approach is no longer valid: it springs from a historicist tradition rooted in nineteenth century ideology; fails to provide a convincing definition of what historic values are appreciated by modern societies; excludes communities in the definition of heritage, and, above all, it does not allow the understanding and management of change.

Urban conservation policies have so far achieved important results, as they have allowed preserving historic areas that would have otherwise lost their character. While this is undisputed, there is today a rising consciousness of some of the dangers of historic preservation that far-seeing planners such as Kevin Lynch spotted long ago:

Rather than simply save things I emphasize the use of saved things to say something. Money gained by forgoing preservation would be spent on education. Preservation rules ought to be simpler and more flexible and yet also more widely applied. In now concentrating our historical anxieties on a few sacred places, where new construction is taboo, we encounter multiple dilemmas: everyday activities progressively decamp, leaving behind a graveyard of artefacts; tourist volume swells, making it impossible to maintain the site 'the way it was'; what is saved is so self-contained in time as to be only peculiar or quaint. A sense of the stream of time is more valuable and more poignant and engaging than a formal knowledge of the remote periods. New things must be created, and others allowed to be forgotten.[49]

This type of critique is a persistent feature in the architectural and urban design debate, as epitomised by the provocative exhibition *Cronocaos* presented by Rem Koolhaas at the 2010 Venice Architecture Biennale and later in New York.[50] Today's urban conservators are confronted with these issues in a more direct manner than in the past. In particular, two questions dominate today's discussion: 1) What needs to be preserved? 2) What is the 'tolerance' for change within protected areas?

Some important innovations in the way historic cities are defined and protected have indeed appeared in the past decade, heralding a new approach to heritage management, largely based on the building of consensus between planners, administrators and the inhabitants on the identification of areas of historic value. While examples abound, the case of Bologna, a city that pioneered urban conservation processes since the 1960s,[51] stands out for its innovative approach. In remaking its urban conservation Plan in 2007, the planners proposed a transition from the traditional concept of *Historic Centre* to the new idea of *Historic City* that includes many areas external to the traditional medieval centre, some of which are, in fact, the result of modern public housing programmes.[52] This fundamental change of policy is opening new perspectives for urban conservation and is part of the new approach proposed by the UNESCO *Recommendation on the Historic Urban Landscape*.

---

[49] Lynch, K. (1972) *What Time is this Place?* Cambridge, Massachusetts, The MIT Press: 237.

[50] "*Has preservation become a dangerous epidemic? Is it destroying our cities? That's the conclusion you may come to after seeing "Cronocaos" at the New Museum. Organized by Rem Koolhaas and Shohei Shigematsu, a partner in Mr. Koolhaas's Office for Metropolitan Architecture, the show draws on ideas that have been floating around architectural circles for several years now – particularly the view among many academics that preservation movements around the world, working hand in hand with governments and developers, have become a force for gentrification and social displacement, driving out the poor to make room for wealthy homeowners and tourists. Mr. Koolhaas's vision is even more apocalyptic. A skilled provocateur, he paints a picture of an army of well-meaning but clueless preservationists who, in their zeal to protect the world's architectural legacies, end up debasing them by creating tasteful scenery for docile consumers while airbrushing out the most difficult chapters of history. The result, he argues, is a new form of historical amnesia, one that, perversely, only further alienates us from the past.*" Ouroussoff, N. (2011) An Architect's Fear That Preservation Distorts, *The New York Times*, May 24, 2011, page C1.

[51] Bandarin, F. (1977) The Bologna Experience. Planning and Historic Renovation in a Communist City. In Appleyard, D. (ed.). *The Conservation of the European City*. Cambridge, Massachusetts: The MIT Press: 178–202.

[52] See: Patrizia Gabellini's essay: 'Bologna: from Urban Restoration to Urban Rehabilitation' in this book.

The issue of the 'limits' or 'tolerance' for change has been at the core of an important discussion led by ICOMOS in recent years.[53] While this discussion has been important in breaking new ground of reflection for conservators, it has yet to produce clear operational outcomes, especially with reference to urban conservation. The derivation of many conservation principles from the practice of monument restoration represents a clear limit to the development of innovative approaches within the conservation profession.[54]

In this respect, the *Recommendation on the Historic Urban Landscape* opens important avenues for rethinking the urban conservation paradigm. In fact, the Recommendation moves away from the traditional concept of 'historic area/centre/city' and puts at the core of preservation policies the concept of 'urban heritage'. This allows a much more flexible, open-ended and people-driven approach to conservation.

Should this approach be mainstreamed in the conservation profession, an important paradigm shift will be possible in the future, with important consequences for the role of heritage in the urban development processes. In fact, linking heritage conservation and sustainability has become increasingly necessary to ensure social stability and long-term flows of resources in areas that have the potential of becoming poles of growth and attractiveness for the entire city.

As traditional policies based on the transfer of public resources to historic areas ended up penalising the weaker social groups and have now become increasingly unsustainable, there is a need to reshape the approach to urban heritage conservation and to better integrate it within the urban development processes.

This new, integrated vision of the urban conservation process is matched by the other important principle put forth by the Historic Urban Landscape Recommendation: the need to integrate the management of natural heritage processes and urban conservation – a relationship that has been consistently underplayed by a tradition largely inspired by architectural and monument conservation principles. Linking conservation and nature management requires an understanding of processes that have, by definition, a territorial scale, and that cannot be restricted to districts. This requires a new approach and a new disciplinary integration. And this is precisely why the landscape and ecological urbanism approaches that emerged in urban management in recent years are so important for heritage conservation.

Another dimension underemphasised by traditional urban conservation, yet that plays, on the contrary, a central role in the landscape approach, is the identification and preservation of urban intangible values. The growing awareness of the importance of living traditions, memory and spirit of place in the creation of values in the urban space is a reaction to the dramatic losses to the authenticity of many urban heritage areas, brought about by global economic processes. Today, many historic places that have maintained their architectural appearances are turned into empty shells, tourist supermarkets and theme parks, no different in substance (although possibly different in appearance) from other historic or pseudo-historic models presented to the public as heritage places, or even theme parks built in imitation of historic areas.[55]

As the urban thinkers of the previous generation, Lynch, Norberg-Schultz, Jacobs, van Eyck,

---

[53] ICOMOS (2011) *Paradigm Shift in Heritage Protection? Tolerance for Change – Limits of Change.* VIth Conference of the ICOMOS International Scientific Committee on Theory and Philosophy of Conservation and Restoration, Florence, Italy – March 4th–6th, 2011.

[54] For a presentation of this issue, see: Bandarin, F. and Van Oers, R. (2012) *The Historic Urban Landscape. Managing Heritage in an Urban Century*,. Chichester: Wiley-Blackwell. In particular see: Chapter 1. Urban Conservation: Short History of a Modern Idea: 1–36.

[55] Many historic districts in Europe, North America and Asia and have become specialised tourist commercial areas. In China, the market pressures led to complete reconstruction of heritage zones in pseudo-historical form, like for instance in the case the redevelopment of the area of Qianmen south of Tien An Men Square, turned into a pedestrian shopping mall.

De Carlo and others demonstrated very clearly, a city is not just architecture or a monument. It is, most of all, a living space, where the meaning of the built environment has to be understood in relation to the living society, its needs for the preservation of memory as part of its culture and life, its sense of beauty, its use of places and its changing processes. The values of the city cannot be understood without accurate cultural mapping, without the participation of the people living, using and shaping the space.[56] This is what is proposed by the landscape approach. Conservation is a powerful tool that helps in building the collective memory and should not be allowed to perform the opposite task, as Arjun Appadurai recently observed.[57]

All these aspects of urban heritage: the physical space, its natural environment, the value system, the social and economic dimensions –

must be fully integrated to allow a full understanding and sustainable management of the urban heritage. The methodologies initially proposed by McHarg and the powerful analytical and representation tools available today[58] enable urban managers and designers to address the issue of complexity and integration of urban process in ways that also allow public participation in the creation of alternative scenarios.

While an adequate tool kit still needs to be formally produced and tested, many examples exist of methodologies that enable the implementation of a landscape approach to urban conservation. The experiments that preceded the adoption of the European Landscape Convention, such as the English Historic Landscape Characterisation programme,[59] or the Belvedere Memorandum in the Netherlands,[60] for instance, are of great interest, as well as other proposals

---

[56] For an analysis of a grass-root approach to urban management, see: Rosa, M. L. and Weiland, U. E. (2013) *Handmade Urbanism* Berlin: Jovis Verlag GmbH.

[57] "*These reflections about architecture raise a deep question about conservation and the related matters of heritage in the great cities of India. The standard critiques of conservationist ideologies is that they are elitist and expensive, that they take resources away from bigger projects of social housing and urban planning for India's exploding urban masses. There is the related critique of nostalgia, which is seen as out of place in an environment of fast-forward development, utopian urbanism, and nationalist modernism. There is something to these charges. But from the point of view of the argument about architecture and amnesia that I have presented here, conservation, especially of heritage sites, could present another sort of opportunity for recovery and recollection, though not of the sites themselves (valuable though that may be). Rather, conservation could enable the recovery of the pedagogical purpose of the debates that lay behind these heritage sites and the possibility that these sites themselves foreclosed as they entered the history of the built environment and made it harder to imagine the possibilities of the unbuilt. Thus conservation, usually seen as the most important tool of remembering, could actually be seen as running the risk of a second forgetting, since it restores the history of the built environment as the only possible history. Were conservation also to develop an interest in the unbuilt, the unremembered, and of abandoned options, it could bring alive the archive of architectural possibilities that always lie around us and behind us.*" Appadurai, A. (2013) Architecture and Amnesia in Indian Modernity. In Mostafavi, M. (ed.) (2012) *In the Life of Cities.* Zürich: Lars Müller Publishers: 331.

[58] For instance, Carl Steinitz has developed interesting computer methodologies that enable managing complexity in an integrated process. Steinitz, C. (2012) *A Framework for Geodesign*, Redlands, California: Esri Press.

[59] An interesting example is The Historic Landscape Characterisation (HLC) program defined and launched by English Heritage in the 1990's, in order to support conservation of historical dimensions within a sustainability framework. The HLC supports the principle that conservation should be based on 'management of change' and in the integration of plans and processes of all stakeholders. This program allowed the creation, for the first time in England, of a detailed view of the archaeological, historical and cultural interest of the landscape. Fairclough, G. (2001) *Cultural Landscape, Sustainability, and Living with Change?* Teutonico, Jeanne Marie; Matero, F. (2003) *Managing Change: Sustainable Approaches to the Conservation of the Built Environment*, Los Angeles: The Getty Conservation Institute: 23–46.

[60] In 1999, the government of the Netherlands adopted the Belvedere Memorandum, a policy document aimed at promoting culture-oriented sustainable development programs. The objective of the Belvedere strategy is to promote a respectful approach in regard to cultural and historic values within spatial development. This is to be accomplished neither by vetoing change nor burying the past, but by seeking effective ways to create win-win situations: to use space in such a way that an object of cultural and/or historic importance is given a place and will contribute to the quality of its newly created surroundings. Netherlands State Government (1999) *The Belvedere Memorandum: a Policy Document Examining the Relationship Between Cultural History and Spatial Planning*, The Hague.

such as the ones developed in recent years in Kyoto[61] and Cape Town.[62]

The implementation of the Historic Urban Landscape approach, however, could not bring about a substantial change in the management of historic areas if restricted to the traditional 'historic' districts. Its main value is indeed in the proposal to 'reconnect' the so-called 'historic' and 'modern' city, in order to enable a full under-standing of the significance of urban heritage, its changing dimensions and its power to inspire and guide contemporary urban rehabilitation processes and urban development.

Reconnecting the city does not signify operat-ing a unitary management process, but the rec-ognition of the diversity of the urban experience. What is required is a more holistic understand-ing of urban processes, in order to define a unified intent that respects different outcomes. This 'meta-planning'[63] can then be used to define how the different parts of the city express their nature and become resources for the rest. In this sense the exchanges between the historic, modern and contemporary cities are not unidi-rectional; they operate following the needs of the changing environment and population, and within the available resources.

Urban heritage conservation has the potential of becoming a leading process in the sustainable management of cities in the future, provided its proponents learn to address modernity and manage change, and do not lock themselves inside indefensible trenches.

Let's read again and reflect on the forward-looking words of a great urban historian, Spiro Kostof:

The urge to preserve certain cities, or certain build-ings and streets within them, has something in it of the instinct to preserve family records; something of the compulsion to protect a work of art. We can all rejoice that medieval Rome did not scrap the remains of antiquity. But we must not be innocent of urban process as a principle. With pretentions of historical purity to one side, and talk of a scientific approach to urban conservation kept modest and conditional, we can regain the central direction in assessing cities. They are live, changing things – not hard artefacts in need of prettification and calculated revision. Cities are never still; they resist efforts to make neat sense of them. We need to respect their rhythms and to recognize that the life of city form must lie loosely somewhere between total control and total freedom of action. Between conservation and process, process must have the final word. In the end, urban truth is in the flow.[64]

---

[61] City of Kyoto (2007) *Kyoto City Landscape Policy. Forming Timeless and Radiant Kyoto Landscapes.*
[62] City of Cape Town (2005) *Integrated Metropolitan Environmental Policy (IMEP). Cultural heritage strategy for the City of Cape Town*, Cape Town: Environmental Management Branch, Heritage Resources Section.
[63] For a definition, see: Wilensky, R. (1981) Meta-Planning: Representing and Using Knowledge About Planning in Problem Solving and Natural Language Understanding, Computer Science Division, Department of EECS, University of California, Berkeley. *Cognitive Science* 5: 197–233.
[64] Kostof, S. (1992) *The City Assembled. The Elements of Urban Form Through History*, London, Thames and Hudson: 305. Professor Spiro Kostof died untimely in 1991 while completing this book. These are his last words.

SECTION 1
The Layered Dimension of Urban Conservation

1 Archaeology: Reading the City through Time[1]

Tim Williams

Senior Lecturer, Institute of Archaeology, University College London, UK

> *Memory is not an instrument for exploring the past but its theatre. It is the medium of past experience, as the ground is the medium in which dead cities lie interred.*
>
> Walter Benjamin

Introduction

Cities are undergoing massive transformations, creating irreversible demands on limited archaeological resources. During these processes, it is often difficult to reconcile the interests of development with archaeological research and heritage conservation, which is a particularly pressing issue in the context of urban growth and regeneration. In a world where *'the future of humanity is irrevocably linked to the city'*[2] we might argue that it has been so for millennia – there is a need for urban archaeology and development to be seen as complementary strands of an approach to creating vibrant twenty first century urban communities. Heritage, and specifically archaeology, has a crucial role to play in helping to produce resilient cities, capable of sustaining and developing their inhabitants.

Advances in promoting dialogue between government agencies, planners, development companies, heritage professionals and international agencies currently are threatened by a combination of rapid urban growth, financial crises and decentralised decision-making. Whilst these problems are differently expressed in different parts of the world, there are many areas of common interest to those concerned with the study and care of the historic fabric of the world's cities.

Almost all cities are the result of complex processes of layering through time. These processes have both contributed to the shaping of the physical landscape inhabited today and also, much more subtly, created an atmosphere of use, a demarcation of physical and social space, and an experience of the sense of the city.[3] Archaeology offers a unique source of informa-

---

[1] I am grateful to the UCL Research Group 'Managing archaeology in the new urban context', and specifically Hana Koriech, Joe Flatman and Dominic Perring, for their stimulating input to discussions and their enthusiasm for the issues.
[2] *Cities of Asia*. Available at: http://whc.unesco.org/en/activities/498/ (Accessed September 2013).

Reconnecting the City: The Historic Urban Landscape Approach and the Future of Urban Heritage, First Edition. Edited by Francesco Bandarin and Ron van Oers.

tion on how urban societies have been conceived and sustained. Whilst the study of towns cannot be divorced from the study of wider settlement landscapes, urban archaeology has a distinct identity: it involves both the study of past urban systems and the practice of field research within modern cities. The main contribution that archaeology makes to the study of towns is through its description of spatial and temporal change. Archaeological research has developed a greater awareness of the social and temporal dimensions of space,[4] and the potential of morphological analyses[5], but the study of how urban spaces were navigated and experienced requires much greater attention.

Against this backdrop, the concept of Historic Urban Landscape seeks to recognise the layering of values present in any historic city.[6] UNESCO's Historic Urban Landscape initiative led to the *Recommendation on the Historic Urban Landscape*.[7] This specifically highlighted the 'time dimension' to managing historic cities (indeed, the fact that most of the world's cities

have historic antecedents), and also implicitly recognised that archaeological methods are a primary source of knowledge that needs to be integrated with management practice. The development of a landscape approach reflects the growing emphasis upon holistic approaches to heritage management.[8] These have sought to move archaeological resource management away from reactive interventions, fighting over the future of individual sites when they are threatened, towards a long-term engagement with urban policies and practices, and integrating archaeological issues with urban planning, conservation and development processes. By understanding and exploring the archaeology of our cities we can contribute so much more. This also raises questions about the balance between preserving archaeological remains *in situ* and excavating them to enable the knowledge gained from the process to contribute to contemporary interpretation.

This paper explores these issues with the aim of defining research areas and tools that will

---

[3] See in particular: Lynch, K., (1960) *The Image of the City*, London: MIT Press. Lynch, K. (1976) *Managing the Sense of a Region*, Cambridge: MIT Press. In Banerjee, T. and Southworth, M. (eds) (1990) *City Sense and City Design. Writings and Projects of Kevin Lynch*, Cambridge: MIT Press. Hall, P. (1998) *Cities in Civilization: Culture, Innovation and Urban Order*, London: Weidenfield & Nicolson. Ouf, A. (2001) Authenticity and the Sense of Place in Urban Design. *Journal of Urban Design* 6 (1): 73–87. In Schofield, J. and Szymanski, R. (eds) (2011) *Local Heritage, Global Context: Cultural Perspectives on Sense of Place*. Farnham, Surrey: Ashgate.

[4] See in particular: Laurence, R. (1994) *Roman Pompeii: Space and Society*, London: Routledge. In Laurence, R. and Newsome, D. (eds) (2011) *Rome, Ostia, Pompeii: Movement and Space*, Oxford: Oxford University Press.

[5] See for instance: Hillier, B. and Hanson, J. (1984) *The Social Logic of Space*, Cambridge: Cambridge University Press. Batty, M. and Longley, P. (1994) *Fractal Cities: A Geometry of Form and Function*, London: Academic Press. Batty, M. (2005) *Cities and Complexity: Understanding Cities with Cellular Automata, Agent-based Models, and Fractals*, Cambridge, Massachusetts: MIT Press.

[6] Bandarin, F. and Van Oers, R. (2012) *The Historic Urban Landscape. Managing Heritage in an Urban Century*, Chichester: Wiley-Blackwell. In Van Oers, R. and Haraguchi, S. (eds) (2010) *Managing Historic Cities*, Paris: UNESCO. Van Oers, R. and Roders, A. P. (2013) Road Map for the Application of the Historic Urban Landscape Approach in China. *Journal of Cultural Heritage Management and Sustainable Development* 3(1): 4–17. Rodwell, D. (2007) *Conservation and Sustainability in Historic Cities*, Oxford: Wiley-Blackwell. Araoz, G. (2008) World-Heritage Historic Urban Landscapes: Defining and Protecting Authenticity, *APT Bulletin* 39 (2/3): 33–37.

[7] UNESCO (2011) *Recommendation on the Historic Urban Landscape*. Please see: http://whc.unesco.org/en/activities/638.

[8] See for example: Cleere, H. (ed) (2000) *Archaeological Heritage Management in the Modern World*, London: Routledge. Teutonico, J. M. and Matero, F. (eds) (2003) *Managing Change: Sustainable Approaches to the Conservation of the Built Environment*, Los Angeles: The Getty Conservation Institute. McManamon, F. and Hatton, A. (eds) (2000) *Cultural Resource Management in Contemporary Society: Perspectives on Managing and Presenting the Past*, London: Routledge. Aplin, G. (2002) *Heritage: Identification, Conservation, and Management*. South Melbourne: Oxford University Press. English Heritage (2000) *Power of Place: the Future of the Historic Environment*, London: English Heritage. Hall, C. M. and McArthur, S. (1996) Strategic Planning. In Hall, C. M. and McArthur, S. (eds) *Heritage Management in Australia and New Zealand*, Oxford: Oxford University Press: 22–36. Mason, R. (2002) Assessing Values in Conservation Planning: Methodological Issues and Choices. In De la Torre, M. (ed) *Assessing the Values of Cultural Heritage*, Los Angeles: The Getty Conservation Institute: 5–30.

assist in the integration of archaeological think-ing into contemporary urban management.

Problems and Issues

Over-simplistic Dichotomy between Preservation and Development

> *With the world turning into a global village, urban encroachment is one of the major factors endanger-ing historic cities, and the pressure of economic devel-opment is seen as one of the underlying causes of this daunting threat.*[9]

This common portrayal places urban heritage, and particularly buried archaeological resources, as being in opposition to the needs of twenty-first century communities. Archaeology is per-ceived not as an asset, but rather as an obstruction or hindrance. There will often be conflicting values placed upon any given space: the archaeo-logical knowledge of buried or exposed remains, versus the economic value of the space for reuse, being an obvious example. And very real ten-sions do exist: as McGill states '*there has rarely been a time like the present when new develop-ment has been so necessary*'[10] while '*conservation-ists, on the other hand, have vociferously argued that the archaeological heritage is a finite resource that is rapidly diminishing due to development*'.[11] It is also portrayed as rare that these elements can work together: '*municipalities ... are more focused on urban development to support eco-nomic growth and job creation*'.[12] They may be, but again by over-simplifying the tensions we do little to explore the solutions.

Values are much more complex, and looking for a more holistic approach, beyond the appar-ently oppositional elements, can reveal more common ground between stakeholders: the desire to create a sense of place, the attempt to create a distinctive impact on the built urban environment, the aim to draw people into navi-gating and engaging urban space, etc., all facets to which archaeology and historic fabric can, and should, make a significant contribution. Archaeology also needs to be bolder with its contribution – we have the ability to engage with powerful narratives of place, and through these with community engagement.

Heritage conservation is often criticised as 'monument-centric', concentrating on individual historic buildings to the exclusion of their context. This is often because the connection between buildings and their urban landscape is poorly understood or articulated.[13] This lack of integration is exemplified in the *Recommendation on the Historic Urban Landscape* which only mentions archaeology twice (and both of these in the glossary). In the *Edinburgh World Heritage Site Management Plan* there are references to archaeology, giving some idea of the city's archaeological potential, and espousing that buried archaeology is an integral and vital part of the World Heritage Site and that its conserva-tion, promotion and interpretation are objec-tives.[14] But this is not actually followed through: there is no mention of archaeology under threats/ risks, sustainability, measuring the state of con-servation, or implementation. This is fairly typical: archaeology is recognised as a character-istic of place, but not really as a contributor to it, and certainly not as something to be actively used to create a sense of place (see the case study in Box 1.1).

[9] CyArk blog site: http://archive.cyark.org/heritage-at-risk-urban-encroachment-blog [Accessed October 2013].

[10] McGill, G. (1995) *Building on the Past: a Guide to the Archaeology and Development Process*, London: Spon: xvii.

[11] Skeates, R. (2000) *Debating the Archaeological Heritage*, London: Duckworth: 58.

[12] Global Heritage Fund (2010) *Saving our Vanishing Heritage*, Palo Alto: Global Heritage Fund: 34.

[13] Menon, A. G. K. (2005) Heritage conservation and urban development: beyond the monument, *Heritage Conservation and Urban Development*. New Delhi: INTACH. Sinha, A. and Sharma, Y. (2009) Urban Design as a Frame for Site Readings of Heritage Landscapes: A Case Study of Champaner-Pavagadh, Gujarat, India. *Journal of Urban Design* 14 (2): 203–221.

[14] City of Edinburgh Council (ed) (2011) *The Old and New Towns of Edinburgh World Heritage Site. Management Plan 2011–2016*, Edinburgh: City of Edinburgh Council.

Box 1.1 Beirut, Lebanon

Beirut has witnessed the contestation of preservation versus development, with major arguments between conservationists and developers, and archaeologists seen as 'colluding' with the developers to remove extensive areas of remains to facilitate the development process. There is no doubt this is true, one function of archaeological recording is to document archaeological evidence before it is destroyed, and if the wider 'will of society' (or at least those with power in the decision-making process) is not to retain *in situ*, then urban archaeological methods provide a powerful tool to document and communicate significance (Figure 1.1). However, in Beirut not all the archaeology was removed in advance of redevelopment: the development company Solidere had a strong agenda of creating a sense of historical reference in the post-war city and were keen to integrate archaeological remains and historic fabric, along with historical motifs, into the urban design process (Figure 1.2). In some instances these were token gestures, and there is no doubting the scale of archaeological resources removed in the overall process, but nevertheless it was also evident that archaeology did play a significant role in urban design, and specifically an attempt to integrate it into a sense of place, with historically rooted identity.

These engagements have also created a longer process, as now *in situ* preservation and display of archaeological remains is considered an important strategy in urban planning in the city (Figure 1.3), with a strong emphasis on creating distinctive locations and settings in the urban landscape. However, the debate over destruction continues, and the issues are far from resolved.

Figure 1.1 Extensive excavations in advance of development in Beirut, Lebanon.

Figure 1.2 A reconstruction of excavated mosaics in the modern shopping area of the Beirut Souks, Lebanon.

Sources:

Alkantar, B. (2013) Minister of Culture 'Dismantles' Beirut's Roman Hippodrome. *Al Akhbar Newsletter*. Available at: http://english.al-akhbar.com/node/5155 (accessed 28-10-2013).

Battah, H. (2013) Activists Fight to Preserve Beirut's Roman Heritage, *BBC website*. Available at: http://www.bbc.co.uk/news/world-middle-east-24222755 (accessed 28-10-2013).

Carver, M. (2009) *Archaeological Investigation*. London: Routledge.

Gavin, A. (1996) *Beirut Reborn: the Restoration and Development of the Central District*. London: Academy Editions.

Lawler, A. (2011) Rebuilding Beirut, *Archaeogy (Archaeological Institute of America)* 64 (4).

Naccache, A. (1998) Beirut's Memorycide. Hear no Evil, see no Evil. Meskell, L. (ed) *Archaeology under Fire. Nationalism, Politics and Heritage in the Eastern Mediterranean and Middle East.* 140–158. London: Routledge.

Perring, D. (1983) *Manuale di archeologia urbana*. Milan: Gruppi archeologici Nord Italia.

Perring, D (2009) Archaeology and the Post-war Reconstruction of Beirut, *Conservation and Management of Archaeological Sites* 11(3–4): 296–314.

Roskams, S. (2001) *Excavation*. Cambridge: Cambridge University Press.

(*Continued*)

Figure 1.3 Display *in situ* of part of the rock cut city ditch within the new shopping complex of the Beirut Souks, Lebanon.

Sandes, C. (2010) *Archaeology, Conservation and the City: Post-conflict Redevelopment in London, Berlin and Beirut*. Oxford: Archaeopress.
Seif, A. (2009) Conceiving the Past: Fluctuations in a Multi-value System, *Conservation and Management of Archaeological Sites* 11: 282–295.
Spence, C. (1990) *Archaeological Site Manual*. (2nd ed) London: Department of Urban Archaeology, Museum of London.

There is still a conceptual separation for many people between below ground archaeological remains, whether buried or exposed, and above ground fabric. The latter tends to be divided between ruins and habitable structures, with the former perhaps more often seen as archaeological and the latter classified as historic buildings. However, most archaeologists do not see such a distinction: all standing structures have below-ground elements (Figure 1.4) and buried structures do not cease to be historic buildings (e.g. Pompeii). There are obviously different challenges when faced with the active reuse of historic structures, but the separation of these elements, and the discussion of the contribution archaeology makes to the historic urban landscape, is often hampered by such distinctions. These narrow definitions within the historic

Figure 1.4 The Duomo in Syracuse, Sicily, Italy, showing elements of a classical structure incorporated into later builds of the medieval and post-medieval periods: there is no distinction between the complexity of the built fabric and its below-ground elements.

environment constitute a major obstacle to a discourse on heritage and the city.

Challenges to Presenting Archaeological Sites in Modern Urban Landscapes

There are a number of very practical issues that make the presentation of archaeological sites within modern urban settings a particularly challenging activity. The most obvious is the basic physical separation of archaeological remains from the modern street level. In the pre-mechanised era, urban redevelopment (on whatever scale) seldom was able to mobilise or commit the scale of resources needed to transport large quantities of building debris away from the urban environs. A new street was simply laid on top of an old surface, rather than as would often happen today, where the old surface would be removed and the debris transported out of town; the result was street surfaces

Figure 1.5 Parco Archeologico della Neapolis, Syracuse, Sicily. A general air of abandonment and the lack of interpretation does little to integrate this area into the modern urban landscape.

rising up in comparison to the floor levels of adjoining buildings. When the latter were redeveloped, the building rubble from the preexisting structure was often used to raise up the building plot to the new street level. This build-up of archaeological deposits over many centuries, with the superimposition of one building or street over another, has led to modern urban street levels often being considerably higher than the buried archaeological remains: in London, for example, Roman floor levels commonly lie between 3 and 6 meters below the contemporary ground surface. The result of this is that most archaeological remains exposed within modern urban settings appear to be 'down holes' – physically separated from the modern streetscape. This physical separation often makes it difficult to integrate those remains into modern contexts,

such as parks or open spaces. Neglected holes often become traps for rubbish, either deliberately discarded or simply blown in. Access for maintenance can be difficult and the combination of rubbish and vegetation often creates the impression that the spaces are neglected and unused (Figures 1.5 and 1.6). Indeed, this perception of a barrier between the present and the past does little to actually improve the integration of archaeological remains into a contemporary intellectual landscape of urban life. The remains appear as an echo of the past rather than as an integral part of the shaping of the landscape they inhabit.

Another key challenge in the presentation of urban archaeological remains is the issue of their legibility. Archaeologists in an urban environment seldom have the opportunity to select the

Figure 1.6 Parco Archeologico della Neapolis, Syracuse, Sicily. Vegetation growth makes the interpretation of the site impossible, but also suggests that it is no longer making a contribution to the present.

areas for archaeological excavation: these areas are nearly always the consequence of wider factors of urban development. While this can have strengths, in terms of examining the complexity of past urban landscapes, it also means that archaeologists do not have control over the extent to which specific building plans are exposed. The discovery, for example, of part of the Roman amphitheatre in London did not lead to the exposure of the entire structure but rather the presentation of the relatively small part that fell within the development footprint.[15] This can work adequately when the fragment of the structure within the development has an understandable relationship to the entire

complex: so in the case of the Roman amphitheatre even a relatively small fragment can act as a catalyst to articulate the interpretation of the wider building and its landscape setting (Figure 1.7). However, many archaeological remains provide a small window on much larger structures or built landscapes; a few walls from a large urban monastic complex, for example, are challenging to use to engage the visitor or resident in a meaningful understanding of the past landscape (Figure 1.8). In addition, *'while these sites tell interesting stories in themselves, it is sometimes difficult to connect them conceptually or physically to one another, or to envisage the urban or rural landscapes in which they once existed'*,

[15] Bateman, N. (2001) *Gladiators at the Guildhall: the Story of London's Roman Amphitheatre and Medieval Guildhall*, London: Museum of London Archaeology Service.

Figure 1.7 Part of the Roman amphitheatre on display beneath the Guildhall Art Gallery in London, UK. Although only a relatively small part of the overall monument, because of the nature of the structure the remains are easily understood by visitors.

Figure 1.8 Impressive remains of the thirteenth century Winchester Palace in Southwark, London. While visually striking, most visitors find it difficult to understand the original context and setting of the huge medieval complex, nor can relate these remains to the development of the modern urban landscape.

Figure 1.9 Partially excavated and conserved remains, covered by a shelter, in the centre of Bukhara, Uzbekistan. This fragment of a wider landscape is currently under-interpreted, and the design of the shelter visually divorces the remains from the contemporary landscape: indeed, most visitors pass it by without recognising that archaeology is on display, nor how these remains contribute to their sense of place, either past or present.

and this 'can also obscure evidence of broader historic urban planning principles that are still evident (archaeologically or otherwise) in street grids, views, setbacks, open space and density of development'.[16] Again, archaeologists need to demonstrate and articulate the relevance of material remains to the understanding of the development of the urban landscape.

The below-ground nature of exposed remains also poses challenges for their conservation. It can often be difficult to ensure adequate drainage, with sites becoming damp and prone to problems with vegetation. As with all archaeological sites, in whatever context, the process of excavation and exposure creates new challenges for the previously buried fabric, but the tensions in an urban context of trying to provide visual access to remains whilst also ensuring their conservation are perhaps particularly challenging. Conventional approaches to sheltering do not easily adapt themselves to modern urban settings (Figure 1.9). However, the incorporation of archaeological remains into new buildings offers an interesting approach to this dilemma and also

[16] Allen, C. (2005) Archaeology and Urban Planning: Using the Past in Design for the Future, ICOMOS (ed) *15th ICOMOS General Assembly and International Symposium: Monuments and Sites in their Setting. Conserving Cultural Heritage in Changing Townscapes and Landscapes.* 17–21 Oct 2005, Xi'an, China. ICOMOS. Available at: http://openarchive.icomos .org/357/.

potentially to some of the tensions between the economic pressure to utilise urban space for modern development and the desire to retain archaeological remains *in situ* and on display. It is evident, however, that in the majority of cases the pressure for conservation of archaeological remains and new build has often been addressed by a process of reburial rather than conservation and display.

Preservation *in situ* and Mitigation Strategies

Arising from the period in the 1970s and 1980s of rapid urban development and concomitant struggles to adequately document the archaeology being destroyed, the move to a preventive approach seemed both logical and essential.[17] This movement ultimately led to the *Valletta Convention*,[18] which required that signatories of the treaty should implement measures for the physical protection of archaeological heritage *in situ* (see Articles 4.2 and 5.4).

Preserving archaeological remains *in situ* can also offer advantages to the developer, especially in enabling a clear strategy to be developed at a relatively early stage in the planning process,

thus reducing or removing archaeological risk. In addition, given the concomitant concept that developers often have to pay for the excavation of archaeological deposits, which the development will destroy (the 'polluter-pays' principle), developers can avoid archaeological excavation if it is cheaper to preserve *in situ*. These principles have been widely disseminated and implemented in many parts of the world[19] and this prioritisation of preservation *in situ* has become the new orthodoxy.

However, this approach is not without its critics: it has become the '*central dogma of western archaeological heritage management … [which] … while surely useful and important in some situations, preservation in situ is too problematic in several ways to be acceptable as an ethical principle with broad validity*'.[20] Indeed, it has often developed, in practice, into a rigid and inflexible approach by decision-makers, supported by the mantra of 'preserving the past for the future', which is used to justify preservation policies. But all heritage is about ascribed values[21] and decision-making prioritising one value, preservation for future generations, over all others certainly risks the accusation of dogma. In the first instance there is seldom clear agreement regarding which archaeological resources are considered significant enough to preserve,

[17] Hodges, H. (ed) (1987) *In Situ Archaeological Conservation*. Mexico: The Getty Conservation Institute/INAH. Stanley Price, N. (ed) (1995) *Conservation on Archaeological Excavations, with Particular Reference to the Mediterranean Area*. Rome: ICCROM. Biddle, M., Hudson, D. and Heighway, C. (1973) *The Future of London's Past: a Survey of the Archaeological Implications of Planning and Development in the Nation's Capital*. Worcester: Rescue. ICCROM (1986) *Preventive Measures during Excavation and Site Protection*. *Conference, Ghent, 6–8 November 1985*. Rome: ICCROM. Mathewson, C. and Gonzalez, T. (1988) Protection and Preservation of Archaeological Sites Through Burial. Marinos, P. and Koukis, G. (eds) *The Engineering Geology of Ancient Works, Monuments and Historical sites. Preservation and Protection. Volume 1*. 519–526, Rotterdam: A.A. Balkema. Olsen, O. (1980) Rabies archaeologorum, *Antiquity* 64: 15–20. Thorne, R. (1989) *Intentional Site Burial: a Technique to Protect against Natural or Mechanical Loss*. Washington, DC: US Department of the Interior, National Park Service.

[18] Council of Europe (1992) *European Convention on the Protection of the Archaeological Heritage*, Strasbourg: Council of Europe.

[19] Naffé, B., Lanfranchi, R. and Schlanger, N. (eds) (2008) *L'archéologie préventive en Afrique: enjeux et perspectives*. Actes du colloque de Nouakchott, 1–3 février 2007, Saint-Maur-des-Fossés: Sépia.

[20] Willems, W. (2012) Problems with preservation *in situ*, *Analecta Praehistorica Leidensia* 43/44: 1.

[21] De la Torre, M. (ed) (2002) *Assessing the Values of Cultural Heritage*, Los Angeles: The Getty Conservation Institute. In Mathers, C., Darvill, T. and Little, B. (eds) (2005) *Heritage of Value, Archaeology of Renown: Reshaping Archaeological Assessment and Significance*, Gainesville: University Press of Florida. Smith, G., Messenger, P. and Soderland, H. (eds) (2010) *Heritage Values in Contemporary Society*. Walnut Creek, Calif.: Left Coast Press. Spennemann, D. (2011) Beyond "Preserving the Past for the Future", Contemporary Relevance and Historic Preservation, *CRM: the Journal of Heritage Stewardship* 8 (1–2): 7–22.

and very few attempts have been made to be explicit about the criteria or values used in such decision-making. There is an evident bias towards the monumental and the elite: large public buildings of the Roman era, constructed of stone or brick, such as amphitheatres, theatres, bath houses, fora, etc., litter our modern cities, but how many clay and timber domestic buildings have been prioritised for *in situ* conservation (let alone display)? So is the archaeological evidence of the urban life of the majority of the population less valuable than the expression of power? And what of the potential to prioritise other values in decision-making, such as research, public engagement, education, capacity building, developing a sense of place, constructing meaningful urban destinations, etc.? As Tunbridge and Ashworth stated: '*the present selects an inheritance from an imagined past for current use and decides what should be passed on as useful to an imagined future*'.[22] An *a priori* decision to preserve *in situ* suggests that we have yet to balance preservation with other goals.

The implementation of preservation *in situ* strategies is also seldom as clear cut as we may wish. There is a danger that heritage management, under pressure to reach compromises with the economic values of development, will allow mitigation strategies that involve invasive measures (peripheral trenching, bored piles, etc.) that result in partially excavated (but poorly understood) sites, with the remaining portions '*preserved in situ in awful conditions … with virtually no chance of survival until a very hypothetical future research excavation*'.[23]

The efficacy of preservation *in situ* has also been questioned: while there have been considerable advances in approaches to the preservation of archaeological remains *in situ*, and for strategies to mitigate the impact of new build on buried remains, research suggests that archaeological materials in many circumstances continue to decay and the corrosion rates of buried metal artefacts have been a particular concern.[24] Wider concerns regarding fluctuating water tables, de-watering, compression from overlying structures,[25] demonstrate that archaeological deposits are not always static and that reburied sides, in some cases, are actively degrading. This concern, of course, exists outside the urban field.[26] There has certainly been too little research into the impact of reburial and the effective monitoring of sites in the long-term (see case study in Box 1.2).

Nevertheless, all of these reservations do not actually detract from the need to preserve *in situ*: there is no doubt that archaeological deposits in our urban centres are a finite resource. However, what is required is a robust policy towards decision-making, which encompasses other values and recognises that local context is vital.

Civic Engagement and Urban Community

Civic engagement contributes to the vitality of urban communities, reflecting on ambitions for a higher quality of life, promoting different forms of representation, and offering an outlet for expressing different identities and cultural

[22] Tunbridge, J. and Ashworth, G. (1995) *Dissonant Heritage: the Management of the Past as a Resource in Conflict*, Chichester: Wiley-Blackwell: 6.

[23] Willems, W. (2012) Problems with Preservation *In Situ*, *Analecta Praehistorica Leidensia* 43/44: 4.

[24] Gerwin, W. and Baumhauer, R. (2000) Effect of Soil Parameters on the Corrosion of Archaeological Metal Finds, *Geoderma* 96 (1): 63–80.

[25] See, for example, 'Theme 1: Degradation of Archaeological Remains'. Gregory, D. and Matthiesen, H. (eds) (2012) *The 4th International Conference on Preserving Archaeological Remains In Situ (PARIS4): 23–26 May 2011, the National Museum of Denmark, Copenhagen*. Special Edition of Conservation and Management of Archaeological Sites 14 (1–4). Martens, V. V. and Vorenhout, M. (2013) Guidelines for *In Situ* Preserved Archaeological Sites and Areas, *The European Archaeologist* 38 (Winter 2012/13): 61–62. Willems, W. (2012) Problems with Preservation *In Situ*, *Analecta Praehistorica Leidensia* 43/44: 2.

[26] In marine contexts: Manders, M. (2009) *In Situ* Preservation: the Preferred Option, *Museum International* 60(4): 31–41.

Box 1.2 UK

In 1990 the UK Government published *Planning Policy Guidance 16 Planning and Archaeology* (PPG16). PPG16 established a process for dealing with archaeological remains affected by development, in which the key element was *'where nationally important archaeological remains, whether scheduled or not, and their settings, are affected by proposed development there should be a presumption in favour of their physical preservation'* (PPG 16 paragraph 8). Issues arise, of course, regarding 'significance': what constitutes nationally important remains and who decides? Perhaps rather more importantly, there was no presumption in the PPG to study, articulate, present or interpret the archaeological remains; although not specifically stated in the PPG, the practice of reburial has become the preferred option.[27] This has been hugely damaging in establishing archaeology as a viable contributor to modern urban design: with nothing visible to form any linkage between the archaeological past and the contemporary urban landscape it is little wonder that archaeology can be perceived as an obstruction, not an asset.

Source: Department of Environment 1990. *Policy Planning Guidance 16: Archaeology and planning.* London: HMSO.

marginalised groups, in a manner that encourages education and a sense of place, while acknowledging the contribution of the wide array of identities already formed.

Archaeological knowledge, including the tangible material culture and eco-factual data generated from excavations, along with *in situ* sites, can all play an important role in framing local identities, and provide a resource for civic and community engagement. However, there are a number of issues about the speed with which archaeologists are able to disseminate the results of their work. A major obstacle for well-recorded urban archaeological sequences is that they produce vast quantities of stratigraphic, material culture, and bio-archaeological data. These require considerable time and effort to effectively analyse and present, often meaning that the results of complex archaeological work are unavailable for some years after the excavations. While this is an inevitable process, it has the end result of divorcing the development process and the outputs of archaeological research. The immediacy of discovery is often reflected in the local media, but this long analytical process often means it is difficult to build upon the sense of momentum and interest at the time of discovery. This is also reflected in funding difficulties: while excavations are on-going developers, and the development process, recognise the need to resource archaeological activities; once the excavation is over and the archaeology 'passes out of sight' it is often much more difficult to sustain funding streams to enable the archaeology to be adequately analysed, archived, and perhaps most importantly presented/disseminated. Once again, new urban archaeologies need to convince funders and policy makers of the contribution archaeology can make to wider societal processes, and develop processes that enable archaeological data to rapidly and effectively make those contributions.

values. Such issues are of great concern to cities, addressing the problems and opportunities of high population densities, immigration, unregulated development, housing shortages (and homelessness), and gentrification. Planning issues build on intangible issues such as conflicting senses of place, identity and diversity. While city landscapes fix themselves into neighbourhoods with particular 'identities', space can also be divided along economic and social lines. Transient populations add to the problems of maintaining social cohesion and nurturing participatory civic and local identities. City administrations are increasingly occupied with how to provide for youth, diverse communities, and

---

[27] English Heritage's professional advice website specifically states: 'The key element is the presumption in favour of reburial' (http://www.english-heritage.org.uk/professional/advice/advice-by-topic/heritage-science/archaeological-science/preservation-*in situ*/) (Accessed October 2013).

The irony, of course, is that less rigorous work, with less attention to detail, is easier to publish. There is a major issue here about quality control in archaeological work: the principle that the developer funds the archaeological process (the 'polluter-pays') falls down if they are not engaged with the quality of the process – which in many cases they are not; they are engaged with enabling the site to be developed. If the developer has been engaged in the outputs of the archaeology, in terms of creating a destination, a sense of place, building a sense of community, etc., then they may be more concerned with the product of the archaeological research; but in reality this is still a rarity. Archaeology needs robust quality control,[28] but it also needs to engage planners and other decision-makers with the quality and contribution archaeology has to make to civic issues.

Designation

In many countries areas are legally designated for their historical and/or archaeological value.[29] In these areas, special regulations apply and sometimes special administrations have responsibility. However, these measures often only concern a small part of the historic urban fabric and most of the significant changes have been framed in response to public opposition: for example, the Rose Theatre in the United Kingdom.[30] The emphasis of many legislative frameworks, therefore, has been with resolving past conflicts rather than anticipating future needs. The growing emphasis on localism and decentralisation, accentuated by the neo-liberal emphasis on deregulation, suggests that local interpretations of legislative mechanisms and approaches will become ever more divergent. The relaxation of planning controls to encourage entrepreneurial development in many countries may also lead to widespread destruction and lost research opportunities. Archaeologists will need to be more proactive in the process.

Integration into Strategic Planning

The integration of archaeology into strategic planning, especially spatial planning, has been much more successful in rural landscapes; as is evidenced by the development of cultural landscape approaches significantly earlier in this area,[31] in contrast with its much later adoption within 'historic urban landscapes'.[32]

Some attempts have been made to develop a stronger linkage between archaeological data and urban planning processes, for example the *Urban Archaeological Database* programme in

---

[28] Willems, W. and Van der Dries, M. (2007) *Quality Management in Archaeology*, Oxford: Oxbow Books.

[29] Aplin, G. (2002) *Heritage: Identification, Conservation, and Management.* South Melbourne: Oxford University Press. Carman, J., (1996) *Valuing Ancient Things: Archaeology and the Law*, Leicester: Leicester University Press. Cookson, N. (2000) *Archaeological Heritage Law*, Chichester: Barry Rose Publishers.

[30] Corfield, M. (2004) Saving the Rose Theatre: England's First Managed and Monitored Reburial, *Conservation and Management of Archaeological Sites* 6: 305–314. Corfield, M. (2012) The Rose Theatre: Twenty Years of Continuous Monitoring, Lessons, and Legacy, *Conservation and Management of Archaeological Sites* 14(1–4): 384–396. Sidell, J. (2012) Paris, London: One Hundred and Fifty Years of Site Preservation, *Conservation and Management of Archaeological Sites* 14(1–4): 372–383.

[31] Fairclough, G. and Rippon, S. (eds) (2002) *Europe's Cultural Landscape: Archaeologists and the Management of Change*, Brussels, Belgium: Europae Archaeologiae Consilium. Clark, J., Darlington, J. and Fairclough, G. (eds) (2003) *Pathways to Europe's Landscape: European Pathways to the Cultural Landscape 2000–2003*, Heide: Council of Europe. Fairclough, G. (2003) Cultural Landscape, Sustainability, and Living with Change? Teutonico, J. M. and Matero, F. (eds) *Managing Change: Sustainable Approaches to the Conservation of the Built Environment*, 23–46. Los Angeles: The Getty Conservation Institute. Rossler, M. (2006) World Heritage Cultural Landscapes: A UNESCO Flagship Programme 1992–2006, *Landscape Research* 31 (4): 333–353.

[32] Bandarin, F. and Van Oers, R. (2012) *The Historic Urban Landscape. Managing Heritage in an Urban Century.* Chichester: Wiley-Blackwell.

the UK;[33] unfortunately most of these are not fully embedded within local planning processes. Most urban archaeological information systems, if they exist at all, are not well structured for strategic planning: most tend to be focused on individual monuments/structures (as excavated and observed), or on surviving scales of resources (depths of deposits, waterlogging, etc.). All of these are useful, but seldom provide a wider analysis of values and significance that would help to underpin their role in strategic urban planning. An example is Lincoln in the United Kingdom[34] where deposit and monument based data were analysed to develop significance and vulnerability maps for the local authority.

The declining number of archaeologists engaged in local civic administration urban planning, in the face of the global financial crisis, puts further pressure on this integration. A recent report in England[35] has highlighted the issue, as well as an associated press release from the *Institute of Historic Building Conservation* stressed that: '*This massive loss in conservation knowledge and specialist advice equates to the reduction of one in three conservation officer posts in local government, threatening the proper care of heritage as well as the huge investment of public monies into England's historic environment.*'[36]

City Services, Infrastructure and Archaeological Impact

One of the main characteristics of an urban landscape is its developing infrastructure, designed to service ever-increasing and changing populations. Cities need to develop advanced city services and infrastructure: public open space; recreation facilities; complex transportation systems; reliable communication networks (including fiberoptics, telephone services, etc.); public services such as water and sewage.

The archaeological implications of such work are considerable, but are often neglected in the face of the political and economic needs of regeneration, and within the context of the pace of the development processes. There is little doubt that archaeology is perceived as an agency that slows or hampers development; either through the time needed for archaeological research and documentation, or through the demands of *in situ* preservation. Civic responses to the problems of infrastructure development, and different approaches to the process of assessing and mitigating impacts, provide a telling measure of the perceived role and status of heritage resources.

Sustainable Development

The importance of sustainability, the buzzword of the twenty-first century, should not be underestimated because of tokenistic overuse. Economic and political responses to the perceived environmental impacts of urbanisation are bound to condition the way in which we engage with the archaeological past and manage archaeological resources. Similarly most heritage professionals recognise the need to describe their work in terms of its sustainability, but the perceived desirability of promoting activities as 'sustainable' can obscure important distinctions between what is environmentally, economically, socially and physically sustainable. This is particularly the case in the area of heritage tourism, where investment in the study and interpretation of archaeological landscapes is often geared towards increasing visitor numbers as a way of promoting economic regeneration.

---

[33] Thomas, R. (2004) Urban Characterisation: Improving Methodologies, *Conservation Bulletin* 47: 11–17.

[34] Jones, M., Stocker, D. and Vince, A. (2003) *The City by the Pool: Assessing the Archaeology of the City of Lincoln*, Oxford: Oxbow.

[35] English Heritage, Association of Local Government Archaeological Officers, and Institute of Historic Building Conservation 2013. A fifth report on Local Authority Staff Resources. *English Heritage*. Available at: http://www.ihbc.org.uk/skills/resources/5th-rep-LAStaff.pdf [accessed 9-11-2013].

[36] Institute of Historic Building Conservation: http://ihbconline.co.uk/newsachive/?p=6410 (Accessed October 2013).

Approaches and Potential

Strategic Planning

Most cities and towns have short, medium and long-term strategic plans to inform future development and policies. For archaeology to make a significant contribution to the contemporary urban environment and its planning it needs to engage with this strategic planning process. To do this archaeology needs to develop solid information platforms and strong strategic visions.

Geographic Information Systems (GIS) offer great potential for integrating earlier individual monument/site/find data, with below ground deposit modelling, to develop a solid platform for future decision-making. It is crucial that this data is well documented, but also that the interpretations of the data, in terms of significance and values, are effectively communicated to all the potential users. Archaeological priorities need to be clearly established and expressed, to create a proactive, not reactive, environment: this is not about depth of deposits, for example, but why they are important.

All of this requires a landscape approach. The development of historic landscape approaches[37] recognised the significance of urban landscapes. This emphasised the need to move from a site or monument based approach to one that considers the contribution that archaeology makes to the understanding of the development of the whole urban landscape. This has been a major shift in the approach to archaeological heritage manage-ment; which is ironic, considering that most early archaeological research into the city was very cognisant of the importance of seeing each archaeological observation as simply a window into a broader landscape.

Integrating the Physical Archaeological Remains into the Contemporary Urban Environment

The integration of archaeological remains into the urban environment is a complex process, which faces challenges from shifting attitudes and values, questionable standards of practice, poor legislative frameworks, and the on-going struggle of establishing a beneficial relationship between archaeology, development and the community. The process of integrating archaeo-logical remains into the urban environment touches on many of the obstacles and challenges discussed above, but these are magnified by the condensed, densely-populated and complex characteristics of the urban landscape.

Numerous strategies have been adopted in urban areas to display archaeological remains. Archaeological parks, or large areas of exposed archaeological remains, are relatively rare inside modern urban centres. Athens[38] and Rome[39] (Figure 1.10) are perhaps the two of the best examples of large areas of archaeological remains within functioning modern urban centres. They create a sense of destination, especially for tour-ists, and a strong sense of place in those areas around the exposed sites, with restaurants, cafes

---

[37] Fairclough, G. and Rippon, S. (eds) (2002) *Europe's Cultural Landscape: Archaeologists and the Management of Change*, Brussels, Belgium: Europae Archaeologiae Consilium. Lozny, L. (ed) (2006) *Landscapes under Pressure: Theory and Practice of Cultural Heritage Research and Preservation*. New York: Springer. Bloemers, T., Kars, H., Van der Valk, A. and Wijnen, M. (eds) (2010) *The Cultural Landscape & Heritage Paradox: Protection and Development of the Dutch Archaeological-historical Landscape and its European Dimension*, Amsterdam: Amsterdam University Press. Turner, S. (2006) Historic Landscape Characterisation: A Landscape Archaeology for Research, Management and Planning, *Landscape Research* 31(4): 385–398.

[38] Papageorgiou, L. (2000) The Unification of Archaeological Sites of Athens. The Birth of an Archaeological Park? *Conservation and Management of Archaeological Sites* 4: 176–184. Parlama, L. and Stampolidis, N. (2000) *Athens: the City beneath the City: Antiquities from the Metropolitan Railway Excavations*. Athens: Greek Ministry of Culture, Museum of Cycladic Art.

[39] Ricci, A. (2011) Archaeology and Today's Cities: the Case of Rome, *Heritage Reinvents Europe. A Critical Approach to Values in Archaeology, the Built Environment and Cultural Landscape, 12th EAC Heritage Management Symposium Ename, Belgium, 17–19 March 2011*, Europae Archaeologiae Consilium.

Figure 1.10 Trajan's Market and the forum, Rome, Italy.

and other amenities vying for space within the visual catchment of these landscapes.

But such large areas are unusual within modern urban landscapes: they are difficult to create, certainly today, and difficult to maintain against the pressures on urban space and the impacts of urban life.[40] It is much more common to encounter projects to display specific monuments (or more often parts of monuments), juxtaposed with modern buildings, public spaces and streetscapes. There are two principal strategies: exposed remains in open spaces, and covered remains contained within (or more accurately under) modern structures. The former tended to be the preferred option, not least because of engineering constraints. The latter has become increasingly popular, as modern construction systems have enabled larger areas to be spanned without intervening supports, and the possibility of retaining the built space while sacrificing the basement area to the archaeology has sometimes seemed an attractive option for developers.

Displaying remains inside modern buildings (Figure 1.11) has numerous advantages for the archaeology: the potential for controlled access and security, conservation gains (such as a climate controlled space, lighting levels, and protection from weathering), and interpretative space (for example, the ability to create space for different forms of interpretative media that would not be feasible in an open air context, or the opportunity to integrate material culture from the excavations within the site display).[41]

[40] In Rome – see Day, M. 10 November 2013, 'Rome Wasn't Rebuilt in a Day: Mayor Faces Long and Costly Fight to Make its Ancient Treasures Traffic-free'. *The Independent*.
[41] Keily, J. (2008) Taking the Site to the People: Displays of Archaeological Material in Non-Museum Locations, *Conservation and Management of Archaeological Sites* 10 (1): 20–40.

Figure 1.11 Archaeological remains displayed under a modern building in Ravenna, Italy. Accessed via the Church of Sant'Eufemia.

This approach can also enable the continued use of urban space, for example with the excellent Museu d'Història de la Ciutat-Plaça del Rei in Barcelona, Spain (Figure 1.12), where the display of the archaeological remains, integrated with an on-site museum, take place beneath the functioning urban public square. But with display inside modern structures there is also the danger that this isolates the archaeological remains from the contemporary urban landscape and makes it difficult for it to contribute to a sense of place.

Open sites are problematic (as discussed earlier), with issues over access, conservation and interpretation, but they can also provide strong visual links and can create a sense of place and impact upon residents and visitors. The strengths of visual cues within the landscape should not be under-rated.

A possible middle ground is archaeological remains under purpose built shelters, combining the strengths of a managed and controlled environment with greater opportunities to design linkage between the remains and the city (Figure 1.13).

Archaeological Knowledge and Its Potential Impact on Urban Communities

We have noted the challenge of communicating archaeological information rapidly and effectively from the time of discovery through to the culmination of analytical work. This is seldom achieved with complex urban sequences. Perhaps one of the best examples was the excavation at

Figure 1.12 The Museu d'Història de la Ciutat-Plaça del Rei, Barcelona, Spain. (Photograph from APPEAR project).

Figure 1.13 Roman Saragossa, Spain: 'Roman Walk' links monuments and the city wall, integrating them in the urban setting and with other cultural walks, and with linked signposts in the town (Photograph by © Tony Rotondas (under Creative Commons Attribution-Share Alike 3.0 Unported)).

No. 1 Poultry in the City of London. A popular publication, *Heart of the City*,[42] reached tens of thousands of people while the excavations were still fresh in their minds, and this was followed by an impressive reconstruction of the domestic buildings/shops in a Museum of London exhibition *High Street Londinium*,[43] arguably one of the best exhibitions of urban archaeology ever mounted. All of this preceded the detailed presentation of the analytical archaeological programme much later.[44] In more recent work at Bloomberg Place, Museum of London Archaeology (MoLA) have very effectively used hoardings around the site (physical access was too difficult in terms of health and safety) to communicate discoveries, and QR codes have linked these visual panels to more in-depth web-based resources (Figure 1.14).

There is also a need to think through the landscape connections between individual elements of archaeological evidence (our small windows on to the past) and the wider historic landscape. Rather than articulating each individual element, the aim should be to develop approaches, which link *in situ* evidence, archaeological knowledge, and the contemporary landscape. Urban landscapes offer rich grounds for exploring visual connections and a sense of place (see also Box 1.3). An example of this is the *Archaeological Landscape Management Strategy* (ALMS) prepared for Parramatta in Sydney, Australia.[45] Here opportunities for understanding and conserving important archaeological sites were developed on a landscape basis, using planning, interpretation and architectural tools to build an interpretation of the historic development of the urban centre.

Box 1.3 APPEAR

An important European Commission project, *APPEAR: Managing archaeological remains in towns and cities: from discovery to sustainable display*, was undertaken to provide advice and guidance for a wide range of potential stakeholders: public authorities, communities, developers, museum curators, archaeologists, conservators, etc. The final report appeared in 2007, and the crucial supporting data and case studies were made available on a website, but unfortunately this has subsequently disappeared. Nevertheless, the report focused on the concept that archaeological remains can benefit:

- Today's inhabitants: by helping them to understand their historical roots.
- Urban design: by integrating ancient remains with the modern environment.
- The urban economy: by providing jobs.
- Visitors: by enriching their experience of the town.
- Students: by direct contact with historical evidence.

Source: European Commission (2007) APPEAR: Managing Archaeological Remains in Towns & Cities: from Discovery to Sustainable Display, http://www.*in situ*.be/guide_en.pdf.

Urban museums also have a crucial role in utilising the knowledge and material culture gained through archaeological tools to reconceptualise the urban landscape for residents and visitors. These are powerful places to create a distinctive destination and provide the opportunity to engage diverse audiences. City museums can 'change lives' by contributing to strong and resilient communities.[46]

---

[42] Rowsome, P. (2000) *Heart of the City: Roman, Medieval and Modern London Revealed by Archaeology at 1 Poultry*, London: Museum of London Archaeology Service.

[43] Hall, J. and Swain, H. (2000) *High Street Londinium: Reconstructing Roman London*, London: Museum of London.

[44] Hill, J. and Rowsome, P. (2011) *Roman London and the Walbrook Stream Crossing: Excavations at 1 Poultry and Vicinity, City of London*, London: Museum of London Archaeology.

[45] Allen, C. (2005) Archaeology and Urban Planning: Using the Past in Design for the Future. ICOMOS (ed) *15th ICOMOS General Assembly and International Symposium: 'Monuments and Sites in their Setting - Conserving Cultural Heritage in Changing Townscapes and Landscapes'*, 17–21 Oct 2005, Xi'an, China. ICOMOS. Available at: http://openarchive.icomos.org/357/.

[46] Museums Association (2013) *Museums Change Lives*, London: Museums Association: 5.

Figure 1.14 (a,b) Hoarding around the MoLA Bloomberg excavation site in the City of London, UK, with imagery linked to QR codes for further information.

Communication technologies (and widespread social networks) are increasingly at the forefront of how we communicate our results and engage with different/wider stakeholders. Many of these approaches offer the potential to go far beyond the transfer of information: the site that links to the web via a QR code is a powerful tool, but beyond that there is the opportunity to engage people and enable them to participate in both the process and its interpretation (e.g. History Pin[47]).

Preservation In situ and Mitigation Strategies

The environments of buried sites are affected by anthropogenic or natural changes. Issues regarding the nature of the ground environment, how archaeological evidence changes through time, and the impacts of short- and long-term change, are key areas of research. Considerable work has been undertaken on the complexity of preservation and mitigation strategies,[48] but there are still concerns that the 'results remain limited because of the complexity of degradation processes'.[49] (Specifically there has been much discussion of saturated urban deposits and the concerns of de-watering.)[50]

There have been a number of significant technological advances: for example, approaches to mitigation strategies and piling[51] and specific design solutions (e.g. a variety of structural solutions used at Vienne, France[52]). There is also a need to emphasise monitoring: Jim Williams review of 30 years of monitoring made recommendations to help improve future monitoring projects, including that 'more work is needed on assessing the state of preservation of a site before monitoring is considered; that a proper project design needs to be developed at the outset of the work; and that more thought should go into deciding why monitoring is needed for a given site, including identifying mitigation options that can be initiated if monitoring data suggest optimum conditions for survival are no longer being maintained'.[53] All of this also requires a significant change in the education and training of archaeological conservators.[54]

New Urban Archaeologies

The future of urban archaeology lies in demonstrating it has relevance to twenty first century urban communities. In general, archaeology has enormous potential to create narratives that help to develop a sense of place and a sense of purpose.

---

[47] History Pin website: www.historypin.com (Accessed October 2013).

[48] Corfield, M. (ed) (1998) *Preserving Archaeological Remains In Situ: Proceedings of the Conference of 1st-3rd April 1996*, London: Museum of London Archaeology Service. Nixon, T. (ed) (2004) *Preserving Archaeological Remains In Situ? Proceedings of the 2nd Conference, 12–14th September 2001*, London: Museum of London Archaeology Service. Kars, H. and van Heeringen, R. (eds) (2008) *Preserving Archaeological Remains In Situ: Proceedings of the 3rd Conference, 7–9 December 2006, Amsterdam*. Amsterdam: Institute for Geo and Bioarchaeology. Gregory, D. and Matthiesen, H. (eds) (2012) *The 4th International Conference on Preserving Archaeological Remains In Situ (PARIS4): 23–26 May 2011, the National Museum of Denmark, Copenhagen*. Special edition of Conservation and Management of Archaeological Sites 14 (1–4).

[49] Willems, W. (2012) Problems with Preservation *in Situ*, *Analecta Praehistorica Leidensia* 43/44: 3.

[50] Caple, C. forthcoming, *Preservation of Archaeological Remains in Situ*, London: Routledge.

[51] Davis, M. (ed) (2004) *Mitigation of Construction Impact on Archaeological Remains* London: Museum of London Archaeology Service. English Heritage (2007) *Piling and Archaeology*. Swindon: English Heritage.

[52] Example of approaches using 'ribbed, reinforced frames, on restored and compacted soil' and 'steel-reinforced piles driven in and connected by horizontal beams or longitudinal girders': see http://www.culture.gouv.fr/culture/arcnat/vienne/en/annexe62.htm (Accessed October 2013).

[53] Williams, J. (2012) Thirty Years of Monitoring in England - What Have We Learnt? *Conservation and management of archaeological sites* 14(1–4): 442–457.

[54] Caple, C. (2008) Preservation *in Situ*: the Future for Archaeological Conservators? *Conservation and Access: Contributions to the 2008 IIC Congress, London*. International Institute for Conservation of Historic and Artistic Works.

To achieve this we need to ensure the quality of the process: high quality excavation[55] and properly funded research, clear and transparent decision-making on *in situ* preservation, creative strategies for on-site presentation, and valuing and developing urban museums. It is also about enabling complex narratives to be developed that explore the historic urban landscape, not isolate fragments of it. *in situ* archaeological remains, and archaeological excavations, will, by their very nature, be fragmentary windows into past landscapes. What is urgently needed is more attention to transforming these into historic urban landscapes; building research, and from that narratives, that enable communities, planners and developers to understand the significance of the time depth of their cities and weave these into their contemporary mental maps.

Preservation and conservation agendas have confused a need to preserve, for an unknown future, with actually engaging with significance. Other issues, such as legislative frameworks, design solutions, etc., all follow on from this.

Some of the major issues that the discipline needs to grasp include:

- **Engineering practices:** There is much work to do to develop engineering practices that will conserve archaeological strata in the long term, and monitoring systems that will support this.
- **Urban archaeological recording systems:** Developing recording systems that improve on-site interpretation and enhance research output.
- **Research and outreach strategies:** We need to develop rapid approaches that sustain the process from the excitement from the time of discovery, through to an understanding of relevance and impact that shapes development and reaches the community. This in no way compromises the long-term goal of rigorous archaeological analysis and timely academic publication, but this process recognises the value of archaeological information to wider community planners, developers, residents and visitors, and sees this as a vital and dynamic part of the archaeological process, not as an optional extra.

- **Synthesis:** We need to build upon synthetic archaeological research into past urban societies, exploring the complexity of urban systems from broad morphology, such as road networks and public spaces, through to the complexities of urban life, such as rubbish disposal, water supply, and feeding the city. These broader narratives of urban development and life are powerful tools in communicating the complexity of urban change and are often the most effective mechanisms for engaging in the broader debates about a sense of place and community.
- **Communication in the urban environment:** Museums, as well as new technologies, have a significant role in the transmission of information about archaeological heritage to as large an audience as possible. Wider communication and interpretation strategies, and the integration of city museums into education and interpretation, are vital if we are to bring a realisation of the value of archaeological material to the contemporary city.
- **Urban archives:** There is a key challenge in how we choose to archive archaeological material and research in an environment that is finding storage space increasingly expensive. We must ensure the physical enation of this research material, especially if we are to do more than nod at the concept of 'preservation by record' that helped to justify its removal from the ground in the first place. But perhaps even more significantly we need to ensure that this material is used, particularly to create dynamic relationships between

---

[55] Excavation is not a bad thing; it enables us to engage communities and planners in the relevance of archaeological knowledge to contemporary society.

residents and their historical past and enable meaningful urban experiences.

- **Supporting urban planning:** Archaeology must actively engage in strategic planning: '*urban designers and city administrators [are] concerned with providing a historical urban identity as much as an authentic urban identity ... through selectivity in deciding the locations to be conserved*'.[56] Archaeology must be proactive in participating in these debates. This is not just about preservation strategies, either city-wide or site specific, but about the contribution that archaeology can make to urban planning: sense of place, identity, physical and visual references, etc.
- **Develop linkages:** There is a need to build on the links between decision-makers and urban communities. Archaeology needs to develop an outward looking research culture for the archaeological study of urbanism that embraces both field practice and academic inquiry, and develops inter-disciplinary working.
- **Creating and sustaining a sense of place:** Archaeology needs to actively research the ways in which it can contribute to the wider debate on how the historic environment plays a role in communities' sense of place (Figure 1.15). This needs to stop paying patronising lip-service to the idea that simply having fragments of the archaeological past in the urban landscape makes a significant contribution and needs to really engage with the concept of using archaeological sites, buried deposits and the results of excavations to truly engage people with these issues.

Mental mapping, developed by Kevin Lynch,[57] explores the representation of time in place and the visual perception of urban form, and is based on the theory that people experience landscapes as places, not artefacts. As such, it provides a useful tool to explore how people understand and reference their urban landscape, and as an element of this, how archaeological evidence and *in situ* sites impact upon the sense of place for urban communities. It is the author's opinion that we would be depressed by the results, but what such research does show is the potential for archaeology, through better interpretation, education and engagement, to make a significant impact.[58]

This understanding can then help to integrate archaeology into urban design. For example, at Champaner-Pavagadh in Gujarat, India:

Urban design interventions can provide a framework for thoughtful and imaginative site reading and interpretation. The interventions use a different medium of expression than reproducing historical precedent – the attempt is not to mimic the past but to evoke it through a visual and spatial vocabulary of design.[59]

Conclusion

We need to develop a much more nuanced approach to articulating the complex values archaeological remains and evidence have in contemporary urban societies. We must move beyond a simplistic preservation for preservations sake; as William Lipe suggested, preservation is only a means, not an end:

In sum, what should drive archaeological preservation is the social benefit that archaeology can provide to society over the long run. That benefit is primarily the contribution of knowledge about the past derived from systematic study of the archaeological record. In situ preservation of archaeological resources is a tool for optimizing that benefit. (…..…) Long-term, frugal consumption of the archaeological record by well-justified research—both problem-oriented and

---

[56] Ouf, A. (2001) Authenticity and the Sense of Place in Urban Design, *Journal of Urban Design* 6 (1): 73–87.

[57] Lynch, K. (1960) *The Image of the City*, London: MIT Press.

[58] Dhanjal, S., forthcoming, *Is There a Role for Archaeology in Diverse Urban Communities?* PhD thesis.

[59] Sinha, A. and Sharma, Y. (2009) Urban Design as a Frame for Site Readings of Heritage Landscapes: A Case Study of Champaner-Pavagadh, Gujarat, India,. *Journal of Urban Design* 14 (2): 203–221.

Figure 1.15 Part of the 'Topography of Terror' in Berlin, Germany, where archaeological evidence of the Gestapo buildings and Berlin Wall are juxtaposed with graphic and textural interpretations to engage the visitors with an understanding of the evolution of the landscape of Berlin through these processes.

mitigation-driven—must be an accepted and integrated part of the preservation program. If the research doesn't get done, or if it gets done and we don't learn anything from it, or if only scholars learn from it and the public is shut out, then preservation will have been in vain, because its goals will have not been achieved.[60]

Thus we need to achieve greater articulation of archaeological narratives, to promote an appreciation of the potential benefits of integrating archaeological knowledge and fabric into the urban landscape. This needs to be coupled with a much more sophisticated appreciation of what urban planners, designers and developers are trying to achieve. The contribution that archaeology can make to concepts such as resilient and sustainable cities are immense, but we have yet to carry that debate forward with conviction.

---

[60] Lipe, W. (1996) In Defence of Digging. Archaeological Preservation as a Means, Not an End, *CRM* 19 (7): 23–27.

2 How Geology Shapes Human Settlements

Claudio Margottini[1] and Daniele Spizzichino[2]

[1]Senior Scientist at the Geological Survey of Italy (ISPRA)
[2]Researcher at ISPRA, the Geological Survey of Italy

Cities were once rural and one day they will turn to be rural again.

Mehmet Murat Ildan

Introduction

For almost 3 million years, the different lineages of humans survived by carrying out two basic activities: hunting (or fishing) and gathering edible items of all kinds (from fruit to insects). We are unusual among animals in combining the two functions, and we have been greatly helped in this by the development of language. But basically, as hunter-gatherers, we have lived using what nature provided.

A radical change came about 10 000 years ago, after the last glacial age, when humans first learned to cultivate crops and to domesticate animals, in what can certainly be considered the most significant development in human history. This process took place during the Stone Age,

when tools were still made of flint rather than metal. However, there is a dividing line, which separates the old Stone Age (Palaeolithic) from the new Stone Age (Neolithic). This passage has been rightly called the Neolithic Revolution. The causes of this development are probably linked to the end of the most recent cold phase of the last ice age (22 ± Kyears cal. BP).[1] This created new temperate regions, in which humans could live comfortably, especially after the de-glaciation which ended in the Holocene climatic optimum (8.5 ± Kyears, cal. BP).[2]

It is not hard to imagine, in these circumstances, a strong human impulse to abandon the hunt of wild animals and to settle, instead, in regions where edible plants were growing in sufficient quantity to be worth cultivating and

---

[1] Dates determined using radiocarbon dating come in two kinds: uncalibrated (also called Libby or raw) and calibrated (also called Cambridge) dates. Uncalibrated radiocarbon dates consider a constant ratio 14C/12C during time. It has been demonstrated that this assumption is not correct and there is the need to modify uncalibrated radiocarbon age, using corrections based on local independent techniques, such as tree ring, varves, coral, and other studies. Calibration curves are now available for most parts of the world. Source: Wikipedia.

[2] Margottini, C, and Vai, G. M. C. (eds) (2004) *Litho-Palaeoenvironmental Map of Italy During the Last Two Climatic Extremes.* Bologna: Litografia Artistico Cartografica s.r.l.

Reconnecting the City: The Historic Urban Landscape Approach and the Future of Urban Heritage, First Edition. Edited by Francesco Bandarin and Ron van Oers.
© 2015 John Wiley & Sons, Ltd. Published 2015 by John Wiley & Sons, Ltd.

protecting (by weeding around them, for example). Some human groups adapted to a new way of life, although the adaptation was quite slow.

If the impulse was to settle, there was also a strong incentive to ensure that animals remained nearby as a supply of food. This would support attempts to herd them, to pen them in enclosures, or to entice them near the settlements by laying out fodder.

Starting with the Age of Agriculture, however, humans began to prosper, and the population began to grow (one of the basic facts of ecology is that when a group has surplus food and available habitat, its number increases). Farming produced a lot more food than had ever been available previously, and the population grew in response. More people needed more food, so production was increased, allowing the population to grow even further.

The conversion to agriculture, the development of permanent settlements, and the rapid growth of the population brought a number of significant changes that we identify with the emergence of human civilisation.

One of the most significant changes concerned the process of planning, designing and building structures for human settlements. Architectural works, in the material form of buildings, are often perceived as cultural symbols and as works of art. Historical civilisations are often identified with their architectural achievements. However, architectural works are not related to construction only. They represent the synthesis of a complex system that, in all ages, was guided by human genius, but depended on the availability and types of construction materials (natural geological resources). Their form is determined by social and economic conditions, as well as by local morphological situations (e.g. defensive settlements on top of cliff), and it is influenced by local meteo-climatic conditions.

Before the present time, characterised by a heavy impact of humans on the environment at a global level, by a high concentration of population in metropolises and by the introduction of construction techniques that are similar all over

the world,[3] most urban centres were the outcome of processes linked to local natural and social conditions and, as a consequence, they constituted unique urban landscapes.

The Historic Urban Landscape approach carries a strong interest for the processes of ancient city building and for their lessons for modern planning and sustainable urban development. The following figure summarises the conceptual framework for the Historic Urban Landscape we have just discussed (Figure 2.1).

For all the reasons described, and primarily for the importance of the factors linked to the availability of natural resources and to the morphology of the places, earth sciences play a fundamental role in urban conservation. They show how geology and geomorphology have been the determinants of urban evolution before the industrial era, and how they have shaped human settlements.

This tight connection ended in the nineteenth and twentieth centuries, when steel and reinforced concrete ('the liquid stone', according to the Italian Engineer Pier Luigi Nervi) became the universal building materials, their introduction being accompanied by the development of a new Science of construction.

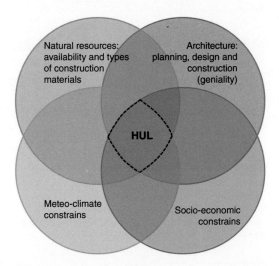

Figure 2.1 A Conceptual Framework for the Historic Urban Landscape.

---

[3] Gregotti, V. (2001) *Architettura e postmetropoli*, Torino: Passaggi Einaudi.

Table 2.1 An attempt to classify the geo-diversity of the Historic Urban Landscape.

Clay based urban landscapes		Soft rock-based urban landscapes		Hard rock-based urban landscapes		Complex environment	
Sun-dried bricks buildings	Burnt bricks buildings	Buildings realised with soft rock	Historic settlements directly built up into soft rock	Squared blocks	Rounded natural blocks	Mixed materials	Continuous evolution in time

Some case studies highlighting the relationship between geology, architecture, climate and socio-economic factors in shaping the urban landscape will be presented in this essay. A classification of the different types of urban landscapes will also be proposed, based on the relationship between building materials and the form of human settlements (Table 2.1).

Clay-Based Human Settlements

Clay and sand are among the most common materials found on the Earth's surface. They originate through mechanical weathering of igneous rock material and its accumulation in river and marine deposits. Clay and sand, together with water and some kind of fibrous or organic material (sticks, straw, and/or manure) can be shaped into bricks (using frames) and then dried in the sun.

The common brick, a truly 'artificial stone', is one of the greatest inventions of humanity. Brick making transforms low-strength mud into a strong material that endures for centuries when properly maintained. The following table shows the different type of bricks and their characteristics.[4] (Table 2.2)

The earliest bricks, made in areas with a warm climate, were simply sun-dried for hardening. Sun-dried bricks, which were used extensively in ancient times, especially in Egypt, were made of clay mixed with straw. However, since the early civilisations, bricks were also fired in ovens with the use of a fuel. These bricks were made of clay mixed with straw to give them added strength during sun-drying and baking in ovens. The Bible contains the earliest written record of the production of bricks, which were made by the Israelites under their Egyptian masters.[5] The Bible also records that fired brick was used in the building of the Tower of Babel.[6] The Greek historian Herodotus, in the fifth century BC, stated that fired bricks were used to build the walls of Babylon. In addition to the Egyptians, the ancient Assyrians, Chinese, and Romans also used bricks. In China, bricks were used to build several sections of the Great Wall, which was started in the third century BC. The Romans made wide use of sun-dried bricks and fired bricks; they used sun-dried clay bricks until the time of Augustus (63 BC–14 AD), but after that time they generally used bricks baked in kilns.

Sun-Dried Bricks for Building

Sun-dried bricks, also called adobe, were quite common in West Asia, North and West Africa, South America, South-western North America, Spain (usually in the *Mudéjar* style), Eastern Europe and East Anglia, particularly Norfolk, where they are known as 'clay lump'.

The Afghan fortress of Shahr-i-Zohak is located in the Shapol area, in the eastern end of the Bamiyan Valley in central Afghanistan, 15 km east of the provincial capital of Bamiyan. The fortress is situated on the hilltop at the confluence of a triangular mountain massif generated by the activities of the Bamiyan and the

---

[4] http://www.fao.org/docrep/s1250e/S1250E07.htm
[5] Exodus 1:14; 5:4–19.
[6] Genesis 11:1–9.

Table 2.2 Different types of bricks and related masonry units.

	Cost	Compressive strength	Resistance to moisture	Resistance to erosion	Uniformity of shape	Room comfort	Speed of production	Skilled to make	Labor requirement	Ease of transport	Energy requirement	Remarks
Sun dried bricks (Adobe bricks)	1	1	1	1	1–3	4	3	1	3	3	1	Most easily produced locally, much improved with stabilisers, least stable and durable.
Manually rammed stabilised soil blocks	2	1–2	1–2	1–2	2–3	5	1	2	5	2	2	A little more strength, better quality and stability.
Mechanically pressed stabilised soil blocks	2	2	2	2	4	5	2	2	4	3	2	Stronger, more durable
Hydraulically pressed stabilised soil blocks	3	3	3	3	4	5	3	3	3	3	2	Even stronger and more durable
Locally made burnt clay bricks	3–4	2–3	3	3–4	1–3	4	2	4	4	3–4	4	Labour insensitive production. Bricks are generally low quality
Factory produced burnt clay bricks	5	5	5	5	5	4	5	5	1–3	5	3	Commercial production is common. The plant requires large investment
Concrete blocks (sand & cement)	5	5	4	5	5	3	3–4	4	2–3	4	5	Long life, strong but heavy. Local production is generally more labour intensive than commercial
Building stones	4	5	5	5	1–4	3	1	5	5	4	2	When locally available, a strong, durable and attractive material.

Rate 1 – 5: 1 = lowest, 5 = highest

Kalu rivers. The steep cliff surrounding the site provides a natural form of protection.

Although the exact occupation periods of the site still remain unclear, the establishment of Shahr-i-Zohak probably goes back the Buddhist period (fourth to fifth centuries AD). In fact, the main building work at Shahr-i Zohak seems to belong to the period of the Yabghu Kingdom, when it appears to have functioned as a fortified residence for the rulers of Bamiyan. This function continued, albeit sporadically, down to the Timurids,[7] though with a break in occupation in the Ghurid[8] period.

The hill is composed of a heterogeneous complex of red conglomerates and marl, relicts of large mudflows, reaching a thickness of about 1000 metres and dating back to the Eocene, about 54 million years ago.[9]

Important faults and tectonic deformations are clearly evident in the site where the fortress is located, the most evident part of the outcropping materials being composed of red marl. The best clayed soil was used to manufacture sundried bricks that were then used to build the walls and structure of the fortress.

In order to limit water flows and to protect them against soil erosion, the foundations of buildings and walls were built with rock blocks from the conglomerate part of this formation. For the protection against weathering, all the walls and structures were also covered with a mud plaster (Figures 2.2 and 2.3).

Figure 2.2 The red town and fortress of Shahr-i-Zohak in Afghanistan.

---

[7] The Timurid dynasty was a Persianate, Sunni Muslim dynasty of Turco-Mongol lineage which ruled over modern-day Iran, Afghanistan, much of Central Asia, as well as parts of contemporary Pakistan, India, Mesopotamia, Anatolia and the Caucasus. The dynasty was founded by Timur (*Tamerlane*) in the fourteenth century. The Timurids lost control of most of Persia to the Safavid dynasty in 1501, but members of the dynasty continued to rule parts of Central Asia and parts of India, sometimes known as the Timurid Emirates. Source: Wikipedia.

[8] The Ghurids or Ghorids were a native Sunni Muslim dynasty of Eastern Iranian, possibly of Tajik origin, which at their zenith established rule over parts of modern day Afghanistan, Iran, India, Pakistan and Tajikistan. Their rule lasted from 1011 to 1215. The dynasty succeeded the Ghaznavid Empire. Their empire was centered in Ghor Province or Mandesh, in the center of what is now Afghanistan. Source: Wikipedia.

[9] Reineke, T. (2006) *Environmental Assessment of the Bamiyan Valley in the Central Highlands of Afghanistan*. Bamiyan Masterplan Campaign 2005. Aachen: University of Aachen.

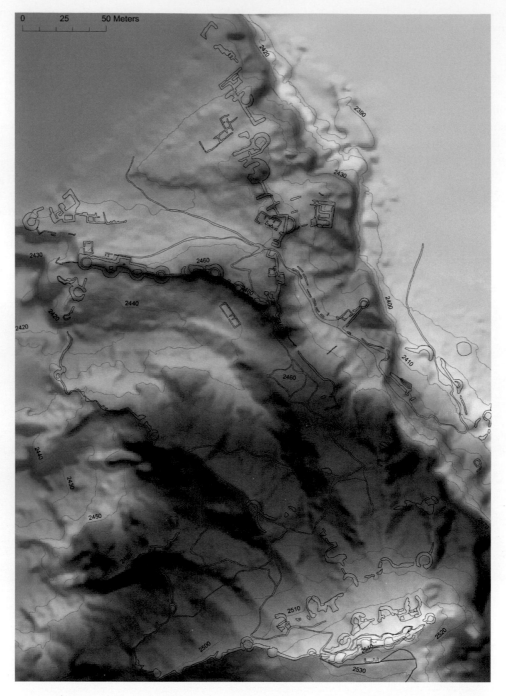

Figure 2.3 Location of the ruins of Shar-i-Zohak (topographic map courtesy of ICONEM, elaboration ISPRA).

Figure 2.4 Aerial photo of the upper part of the Fortress (courtesy of ICONEM).

The historic buildings of the red town of Shahr-i-Zohak and the surrounding areas are therefore fused in an urban landscape where the red coloured marl and clay and bricks are composing a homogeneous scenery (Figure 2.4).

The fortress is currently affected by continuous weathering, and the mud bricks are decaying under the effects of water infiltrations (Figure 2.5). Similarly, weathering of the constructions has caused the loss of some parts of the retaining walls, and that in turn has allowed rainwater to flow freely, producing important gully erosion all around the cliff. The prevention of these erosive processes has now become quite urgent.

Many villages in sundried bricks (adobe) were constructed all over the world. One of the most famous is the Bam Citadel in Iran (Figure 2.6), which was almost totally destroyed (up to 80%) by an earthquake on 26 December 2003. Other large adobe structures are the Huaca del Sol in Peru, made of about 100 million signed bricks, the *ciudellas* of Chan Chan and Tambo Colorado, also in Peru.

Typical adobe buildings arranged close to each other can be found in the Sub-Saharan region (Figure 2.7). The need to offer protection from direct sun radiation produced a unique architecture that is well integrated in the environment, and is also very comfortable and sustainable from a climatic point of view.

Fired Bricks for Building

Fired clay bricks have a good resistance to moisture, insects and erosion. They have medium to high compressive strength. Bricks can be made with sophisticated factory methods, different

Figure 2.5 Mud bricks construction over rock foundation.

Figure 2.6 The ancient citadel of Bam before the 2003 earthquake.

Figure 2.7 Typical adobe construction in Southern Morocco, with buildings close to each other for protection against direct sunlight.

types of mechanical techniques or even simple labour-intensive methods. The labour-intensive production methods are most suitable for rural areas where the demand for bricks is relatively small. The bricks produced by hand have normally a relatively lower quality, especially as regards their compressive strength, and tend to have irregular dimensions. However, they are economical, require little capital investment and have a low transportation cost. Bricks made in this manner have been used in buildings that lasted for centuries. Their durability depends on the quality of the ingredients, the skill of the artisans and the weather to which they are exposed.

It is however during firing that the bricks acquire their strength. At high temperatures, the alkalis in the clay, together with some iron oxides and other metals, link in a chemical bond with the alumina and silica in the clay to form a dense and durable mass.

Environment, materials and crafting skills have all had an important role in the development of so called 'Nepali' architecture, also known as *Newar*-style, in the Kathmandu Valley.[10]

The architecture of the Valley typically consists of a combination of a structural wooden frame and masonry or brick walls (Figure 2.8). *Kasthamandapa* is an example of a temple built

[10] Bonapace, C. and Sestini, V. (2003) *Traditional Materials and Construction Technologies Used in the Kathmandu Valley.* Paris: UNESCO.

Figure 2.8 Durbar Square in Kathmandu with its traditional architecture.

using this system. *Kasthamandapa*, meaning the wooden house, was originally built during the Licchavi period[11] and is considered to be the oldest known wooden building still standing in Nepal. The wooden structural system is not always apparent as the wooden components have also a decorative function and do not look structural. This architecture is characterised by intricate carvings on the pillars, brackets, struts, beam frames of the peristyles, window and door frames, that are full of symbolic, often coloured figures. The overall impression is as though sculptural works have been inserted into the walls rather than being an integral part of the structural support. This style is very different from the one found in neighbouring Asian countries with similar cultures, traditions and religions.

A typical feature of this style is the way in which the two principal materials, wood and brick, are combined to create a distinctive style of architecture. The material that unifies architecture with its environment is the clay soil of the Valley. The soil dug from a dry lake, whose waters previously reached a gorge in the south, as the *Manjusree* legend goes, is still used for the fabrication of bricks. The widespread use of fairfaced bricks with their intense colour visually unifies and gives character to the urban spaces of Nepal (Figures 2.9 and 2.10). Bricks are used for the construction of palaces and temple walls, roofs, pavements of narrow lanes, streets and

[11] Licchavi (also *Lichchhavi*, *Lichavi*) was an ancient kingdom in Nepal, which existed in the Kathmandu Valley from approximately 400–750 AD. The language of Licchavi inscriptions is Sanskrit, and the particular script used is closely related to official Gupta scripts, suggesting that the other major kingdoms of the Classical Period to the south were a significant cultural influence. Source: Wikipedia.

Figure 2.9 Bakhtapur in Kathmandu Valley.

squares. Bricks are also used for different parts of the building to create different patterns (Figure 2.11). Their extensive use gives a visual continuity to Nepalese architecture.

There are many types of clay found in the Kathmandu Valley, with different colours ranging from white to black, and from yellow, red to brown. It can be observed that the clays are used in different parts of the buildings according to their properties and quality. For common bricks and tiles grey and black clays are most generally used. Grey and brown clays are best for quality plaster. Grey clays are also used to make mortar, while red clay is used to make pavements. White clays (pure kaolin) are used for internal and external paintings, yellow clays for plastering of joints and for the mortar used to join trapezoidal-cut bricks.

Research carried out to investigate the quality of clay soil used in the Valley looked at the mineralogical properties of clay samples taken in different places and for different uses. Results showed the presence of quartz (sand) and fedspati and a variety of clay substances (silicate of

Figure 2.10 Bakhtapur in Kathmandu Valley and the traditional craft makers.

Figure 2.11 Traditional construction in Patan, in Kathmandu Valley, with brick, stone and wood.

alluminia, in particular Illite and kaolinite) ranging from 42% (soil for unfired bricks) to 77% (red clay for paintings) of the total. Some clays are dug from river beds, while others are dug directly from agricultural terraces on hill slopes where a temporary nearby kiln is usually found. Most kilns remain *in situ* only for a short period of time, after which they are dismantled, allowing the soil to return to its original state for agricultural uses. Before firing, the bricks are dried in the open air. They are turned from time to time to ensure they are evenly dried so as to avoid cracks from uneven shrinkage. Bricks are then fired in traditional kilns. There are two types of kiln: the so-called Hoffmann kiln, which has a series of chambers,[12] and the intermittent kiln, which has a single chamber. The use of these materials, together with exceptional architectural skills linked to religious beliefs has produced a unique urban landscape not found in other parts of the world.

Soft Rock-Based Human Settlements

Introduction and Definition

The basic geological description of rock masses (BGD), officially proposed by the International Society for Rock Mechanics (ISRM), provides a standardised description with the purpose to define the quality of rock masses in relation to

---

[12] A Hoffmann kiln consists of a main fire passage surrounded on each side by several small rooms. Each room contains a pallet of bricks. In the main fire passage there is a fire wagon that holds a fire that burns continuously. Each room is fired for a specific time, until the bricks are vitrified properly, and thereafter the fire wagon is rolled to the next room to be fired. Source: Wikipedia.

Figure 2.12 Buildings made with local soft rock material, volcanic tuff in Civita di Bagnoregio, Italy.

their mechanical behaviour. In the technical and scientific literature,[13] classification methods for rock masses agree in defining a rock as 'weak' (*sensu lato*) with respect to the Uniaxial Compressive Strength (UCS),[14] when the reference values are lower than 20 MPa (Megapascal).[15] This definition includes the quartzite, very porous rocks (weathered sandstone and tuff), most of the clay rocks, salt rocks and any other type of weathered and degraded rocks.

Human Settlements Made up of Soft Rocks

Availability of materials that are 'soft' or 'weak', and therefore suitable for the construction of buildings and structures, is the main reason of the origin and survival of some wonderful historic urban landscapes in different parts of the world. Like in the case of hard rock, the use of soft rock as building material is directly influencing the structural and morphological processes. For construction with soft materials, two distinct categories can be proposed:

1. Urban landscapes and structures built with soft rock (Figures 2.12 and 2.13);
2. Urban landscapes directly built into soft rock (Figures 2.14, 2.15 and 2.16).

Concerning the first category, the development of human settlements is strictly dependent on the availability of this type of construction

---

[13] Deere, D. U. and Miller, R. P. (1966) *Engineering Classification and Index Properties for Intact Rock.* Tech. Rept. No. AFWL-TR-65-116, Air Force Weapon Lab., Kirtland Air Force Base, New Mexico.

[14] By definition, the ultimate compressive strength of a material (generally defined Uniaxial Compressive Strength) is that value of vertical compressive stress that a standard rock sample can sustain, without lateral support, before the material fails completely. The compressive strength is usually obtained experimentally by means of a compressive test or with other indirect techniques. Source: Wikipedia.

[15] Anon. (1979) Classification of Rocks and Soils for Engineering Geological Mapping. Part 1: Rock and Soil Materials. IAEG Commission of Engineering Geological Mapping. *Bulletin of the International Association of Engineering Geology,* 19: 364–371.

Figure 2.13 Buildings made with local soft rock material, volcanic tuff in Sorano, Italy.

Figure 2.14 Varzia, Georgia, example of a settlement directly built into a volcanic tuff strata (soft rock).

Figure 2.15 Rock cave houses on typical soft rock formation (volcanic pumice) in Cappadocia, Turkey.

Figure 2.16 Cave dwellings, typical peasant homes in the Sassi of Matera, Italy.

material in the area. Very often the historic centre is located on top of a soft rock plateau that is used both as supporting platform and as a quarry for the extraction of the material used to build the city. Small villages over soft rock plateau cliffs, as well as large historic centres like Naples, are examples of urban landscapes of soft rock. The historic centre of Orvieto also belongs to this category (Figure 2.17).

Orvieto is located within the Tiber River valley, placed between the steep margins of a volcanic plateau to the west and the Apennine Mountain Range to the east. All along the margin of the plateau, several ancient hamlets and towns overlook the valley, from the top of *buttes*,[16] *mesas*[17] and *inselbergs*[18] that are isolated by erosion.

These isolated hills are generally formed of soft rock masses and lavas erupted by the four main volcanic groups of Northern Latium since the Early Pleistocene.[19] Erosion is accelerated and morphologies are sharper where the

---

[16] A *butte* is an isolated hill with steep, often vertical sides and a small, relatively flat top. Source: Wikipedia.

[17] A *mesa* (Spanish and Portuguese for table) is the American English term for table and, an elevated area of land with a flat top and sides that are usually steep cliffs. It takes its name from its characteristic table-top shape. Source: Wikipedia.

[18] An *inselberg* or monadnock is an isolated rock hill, knob, ridge, or small mountain that rises abruptly from a gently sloping or virtually level surrounding plain. Source: Wikipedia.

[19] The Early Pleistocene (also known as the Lower Pleistocene) is a subepoch in the international geologic timescale or a subseries in chronostratigraphy, being the earliest or lowest subdivision of the Quaternary period/system and Pleistocene epoch/series. It spans the time between 2.588 ± 0.005 Ma (million years ago) and 0.781 ± 0.005 Ma. The Early Pleistocene consists of the Gelasian and the Calabrianages. Source: Wikipedia.

Figure 2.17 General view of Orvieto: most of the building material is from quarries located under the town.

Plio-Pleistocene[20] clay formation, underlying the volcanic plateau, emerges. This occurs at the margins of the plateau and along the deepest and narrowest valleys of the tributaries of the Tiber River.

Orvieto is a major historic town endowed with an inestimable artistic heritage dating back to the Etruscan period. The ancient town was built on the top of an elliptical slab (1500 by 700 metres) delimited by vertical cliffs up to 60 metres high, which is formed by the 'Orvieto and Bagnoregio Tuff', a soft rock formation.[21]

Particularly interesting is the underground of Orvieto. It consists of the combination of more than 1200 cavities excavated in the tuff for different purposes: old aqueducts; channels; tanks; quarries for the extraction of tuff blocks; and ancient cult places. The peculiar geology of the cliff on which the town is built allowed, over the course of Orvieto's 3000 years of history, the inhabitants to excavate a system of caves overlapping one another under the urban fabric. The caves bear a close connection with the urban development above ground, because of their complementary service functions with the city. The present historic urban landscape is the final result of construction activities conducted in all historic periods, showing how the history of the town is strictly connected to its geological setting.

Human Settlements Directly Built up into the Soft Rock Material

Compared to the first category, the second is often linked to a pre-existing morphological setting (e.g. karst, cave, or morphogenetic processes), which provides the initial natural setting, then modified by human action and transformed into an urban landscape.

An extraordinary example of this type of urban landscape, directly 'sculpted' into the soft rock masses, is the Holy City of Lalibela in Ethiopia. Lalibela is located in the northern-central part of Ethiopia, approximately 600 km north of Addis Ababa. The town, which has about 12 000 inhabitants, is located at an altitude of 2500 metres. In its centre a unique complex of 11 rock-hewn Christian Orthodox churches can be found, cut out of the soft rock (compact and weathered basalt) some 800 years ago, the construction being attributed to King Lalibela (1167–1207). The churches are still used daily for religious practices and ceremonies. In 1978, Lalibela was included in UNESCO's World

---

[20] The term Plio-Pleistocene refers to the geological period more recent than circa five million years (Ma) ago, incorporating both the formally defined epochs of the Pliocene and the Pleistocene. Strictly, this would exclude the Holocene – the last 12 000 years – but informally most Earth Scientists would probably consider the term to incorporate this period. Source: Wikipedia.

[21] The site of this settlement (Etruscan period, eighth–ninth centuries BC) has in fact been identified with the Etruscan centre of Velzna (Volsinii in Latin), a city that began flourishing at the beginning of the sixth century BC. Its economic prosperity was based mainly on the production of ceramics and bronze work. From a political point of view, Velzna was at the forefront of the struggle against Roman expansionism, but in 254 BC it was conquered by the Roman armies and then razed to the ground.

Following the destruction of the city its inhabitants were dispersed, and for the most part were forced to relocate in the highlands overlooking the lake of Bolsena. At the time of the 'barbarian' invasions, Orvieto was occupied by Alaric the Goth and by Odovacar. Vitige took advantage of its strategic natural position to create a defensive stronghold in the war against the Byzantines.

In the eleventh century Orvieto became a Commune or city-state. Noblemen of the area, who had relocated back to the city, began to build the towers and palaces of the city. The institution of the Commune is documented beginning in 1137 and 20 years later a treaty was signed with Pope Adriano IV, which increased papal influence in the city. The vitality of Orvieto can be seen in the construction activity of the period. It is at this time that the churches of San Lorenzo degli Arari, of San Francesco, of San Domenico, of Santa Maria dei Servi were built, as well as many public buildings such as the Communal Palace, the Palazzo del Capitano del Popolo, and the Papal Palace. And in 1290, the building of the Cathedral began.

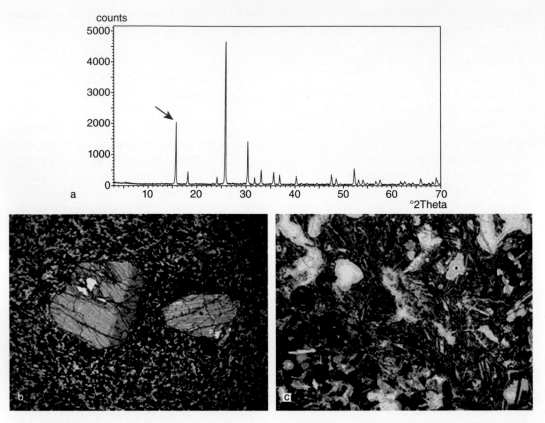

Figure 2.18 (a) X-ray diffractogram from a sample of weathered basalt in Lalibela. The red arrow shows the peak interpreted as *montmorillonite* (smectite family), (b) thin section of intact basalt, and (c) thin section of weathered basalt.

Heritage List. Some of the churches are sculpted out of solid hydrothermally-altered and nearly aphyric vesicular basalts with incipient lateritisation.[22] Some are quarried enlargements of natural caves. They are almost all connected by long underground tunnels and passages that show a wide variety of architectural styles.

Over the years, the churches have been exposed to wind, rainfall, thermal changes and human activities. Laboratory tests have highlighted the large concentration of montmorillonite (smectite family) in the weathered vesicular basalts, that is a largely hygroscopic clay mineral,

exhibiting a changing volume up to 10–20 times during the wet and dry cycle (Figure 2.18 a–c). Montmorillonite is a weathering product from silica-poor rocks, like basalts, when permeated by hot water. As a consequence of such alteration, the basalt becomes easy to work and to carve, allowing the construction of churches and monuments that constitute a unique 'built' landscape.

This is a good example of a construction material easy to carve, albeit with a very low degree of durability, due to the hygroscopic effect of montmorillonite. Because of the outstanding

[22] Renzulli, A., Antonelli, F., Margottini, C., Santi, P. and Fratini, F. (2011) What Kind of Volcanite the Rock-hewn Churches of the Lalibela UNESCO's World Heritage Site Are Made of? *Journal of Cultural Heritage*, Elsevier, 12: 227–235.

Figure 2.19 The rock-hewn Church of Biet Giorgis in Lalibela, Ethiopia.

value and the nature of the monuments, the conservation of the Lalibela churches (Figure 2.19) attracted long ago the interest of the international community. Presently, almost all churches are protected from rain and humidity by shelters, although these shelters have some adverse impacts on the monuments. Furthermore, centuries of rainfall have produced a decay of the mechanical strength of the original basalt, even if partially weathered. Nowadays some churches (Figures 2.20 and 2.21 a–c) are at risk of collapse due to the loss of bearing capacity of the rock-wall.

Hard Rock-based Human Settlements

Introduction and Definition

In the scientific literature, classification methods for rock masses define a rock as 'hard' (*sensu lato*) with respect to the Uniaxial Compressive Strength (UCS), when the reference value is not lower than 50–70 MPa.[23] Included in this definition are the quartzite, diabase basalts (>200 MPa), most of the igneous rocks, well-cemented sandstone, and most massive limestone and dolomites. A rock mass is usually comprised of rock

---

[23] Anon. (1979) Classification of Rocks and Soils for Engineering Geological Mapping. Part 1: Rock and Soil Materials. IAEG Commission of Engineering Geological Mapping *Bulletin of the International Association of Engineering Geology*, 19: 364–371.

Figure 2.20 The Church of Biet Emmanuel in Lalibela, in critical condition due to the loss of bearing capacity of the rock-wall.

blocks separated by joints or discontinuities, and its mechanical behaviour thus depends both on the mechanical properties of the rock material and the characteristics of the discontinuities.

Human Settlements Made up with Hard Rocks

Availability of a relatively hard material, suitable for building adequate protections against military attacks, animals and bad weather, while at the same time easy to maintain, provided the best conditions for ancient settlements. For these reasons, and because of the labour cost of excavation of hard rock, their use as building material depended on the presence of some diagenetic, structural and morphological processes. These processes facilitate the formation of hard rock construction material that is easy to manage and work, with a size and geometry that is suitable for building construction.

A classification of this material into two distinct categories can be proposed with respect to size and geometry:

1. Square blocks of hard rock (big and small size), as in Figure 2.22;
2. Rounded natural blocks of hard rock (small size), as in Figure 2.23 (a,b).

Large Size Square Blocks of Hard Rock

Concerning the first category (square blocks of hard rock), the forms of the urban landscapes

Figure 2.21 (a) Weathering processes in Lalibela, (b) the roof of Biet Medane Alhem, and (c) Biet Macorewos; damage to internal reliefs in Biet Mikael Golgota.

Figure 2.21 (*Continued*)

Figure 2.22 Some examples of square blocks of hard rock: the top is Machu Picchu, Peru, and above the fortress of Saksaywaman in Cusco, Peru.

Figure 2.23 (a,b) Example of constructions with small sized, rounded blocks in Ficulle, Italy.

depend on the original geo-structural conditions (e.g. spacing, fracturing, tectonic conditions, type and degree of metamorphism, sin-genetic and post-genetic conditions). The archaeological city of Machu Picchu, built with large size squared blocks of hard rock clearly belongs to this category (Figures 2.22 and 2.24).

Recent studies have proven that the choice of the site for this magnificent and mysterious city, symbol of the heyday of the Inca empire, is linked to the specific geomorphologic processes at the site, that made granitic material for construction easily available (Figure 2.25).

This material could also have been formed by paleo-landslide(s) affecting the area. One of the main evidences of this interpretation is the recent discovery (September 2004) of a large landslide, on which the citadel lies (Figure 2.26). Depending on the particular topographical and morphological situation, the area was presumably used as a quarry for construction materials or as the foundation of entire neighbourhoods of the city (Figure 2.27).

Such hypothesis assumes that, like in other relevant archaeological sites, a strong connection exists between the foundation, evolution and prosperity of a city and the original morphological setting of the chosen area. Machu Picchu's citadel illustrates the causal link between the geomorphologic process, the availability of building material and the urban and spatial evolution of an urban landscape from its origin to its decline.

Another clear example of the results of the availability of squared hard rock blocks (of small size, as a result of the rapid cooling of lava flows) can be found in the Orongo village, built in the Huri-Moai period (from about the sixteenth to the eighteenth century) on Easter Island, Chile (Figure 2.28).[24]

[24] Mulrooney, M., Ladefoged, T., Stevenson, C. and Haoa, S. (2009) The Myth of A.D. 1680: New Evidence from Hanga Hoʻonu, Rapa Nui (Easter Island). *Rapa Nui Journal*, Vol. 23, No. 2, October: 94–105.

Figure 2.24 General view of Machu Picchu.

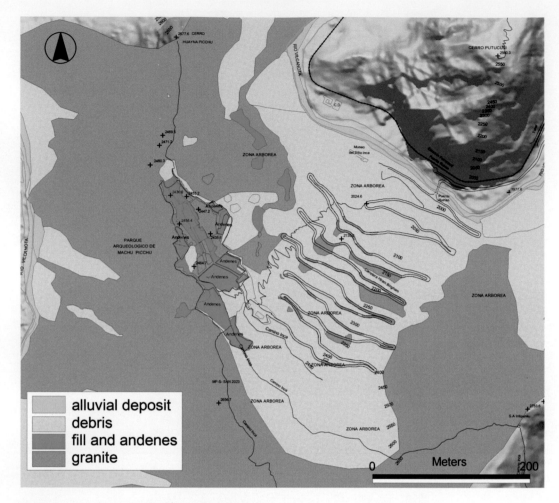

Figure 2.25 Machu Picchu detailed geological map.

The ceremonial village of Orongo is consti-tuted of 53 stone masonry houses made of easily available highly fractured basalt rock. The houses were built with local stones placed one over the other. Rocks are fractured from the cooling process of volcanic lava in the surrounding area and do not need to be re-shaped (Figure 2.29).

Rounded Natural Blocks of Hard Rock

Compared with squared rock blocks, the rounded ones are more difficult to use, and require the use of an important quantity of mortar. The strength characteristics of the struc-ture are therefore poorer, especially under dynamic stress. This type of material has had a great diffusion among historic towns and com-munities located near rivers and alluvial deposit areas. In these locations, this type of material can be collected and used in a very economical way. It does not require special processing (sizing) and can be put in place with dry-stone tech-niques and also with different types of mortar. Wonderful examples of historic villages built up with this construction material can be found for instance in the high valley of the river Tiber in the Umbria region in Italy. The villages of Ficulle

Figure 2.26 (a) Possible quarry/paleo landslide deposit of the archaeological site of Machu Picchu and (b) technique for squared hard rock block production.

Figure 2.27 Location of the stone quarry used in Machu Picchu.

Figure 2.28 Type of construction in the Orongo Village, Easter Island, Chile.

Figure 2.29 Type of construction in the Orongo Village, Easter Island, Chile.

and Parrano in Central Italy are a typical example of historic urban landscapes related to type and availability of this material (Figure 2.30 a,b). The limitation of this construction technique is its high vulnerability to earthquakes, even with a low vibratory character and a rapid ageing of the structure that requires frequent replacement of the mortar between the blocks.

Time Variability and Complex Urban Environments

The cases described above represent settlements built in a given time period and characterised by a relative uniformity of building materials. The period normally coincides with the more extensive urban development phase.

An opposite example is the town of Dubrovnik, where the old town remained unchanged until the beginning of the twentieth century, but witnessed a rapid growth process during the last century. The aerial photo of the area clearly shows the two stages of urban development (Figure 2.31 a,b).

Important historic towns generally differ from the case of Dubrovnik, where urban development takes place in a specific time period. For them, evolution is a continuous process, and they show a layering of the urban landscape from the oldest to the most recent times. Rome is a paramount example of this type of layering (Figure 2.32).

During the so-called 'Kings period' (753–509 BC) at the beginning of the history of Rome, there was an important Etruscan influence with a prevalent use of different types of tuff coming from the surrounding regions. These materials were used mainly for aqueducts, wastewater drainage channels (e.g. *cloaca maxima*), such as the first paving of the Roman Forum done by King Tarquinio Prisco in seventh century BC, and the first surrounding wall (*Murus Servii Tullii* or *Mura Serviane*), a defensive barrier built in the early fourth century BC. The wall was built from large blocks of tuff (a volcanic rock made from ash and rock fragments ejected during an eruption) quarried from the Grotta Oscura quarry near Rome's rival city of Veius.

After the Kings period, during the Republican age (509–27 BC), a large population growth occurred as a consequence of Roman expansion. The town changed and extensive use was made of tuff, travertine, marble and the *cementizio*, a precursor of modern concrete, made of sand, *pozzolana* (volcanic ash), lime, water and stone or rock fragments. A classical example of this mix of techniques and materials can be found in one of the most significant places of the Republican period, Augustus' Forum (Figure 2.33).

The peak of the development of Roman architecture and of its population was reached during the Imperial Period (first – fifth century AD, but mainly between the first and the second centuries AD). During this period, the availability of building materials, jointly with the spread of advanced construction techniques made possible the creation of an urban complex that is still visible and preserved today.

The use of travertine blocks and tuff for the foundations and, for the first time, the large use of fired bricks and cement (e.g. *opus reticulatum*, arch and vaults), allowed the creation of monuments such as the Colosseum, the Pantheon, the thermal baths (Caracalla, Diocletianus), Marcello's Theatre, the Basilica of Maxentius and many others.

As an example, the Colosseum, the world's largest amphitheatre (Vespasian, 72 AD), has foundations, pillars and the main structure in travertine blocks, the internal walls in fired bricks (e.g. *praecinctiones* or *baltei*), covered by marble and even some parts in *opus quadratum* in tuff. It is important to notice the appearance of fired bricks in this period, and of a quarry inside the city. A late eighteenth century drawing by F. Becker shows an ancient clay stone quarry within the Monte Mario formation along the south flank of Saint Peter's Basilica.

Similarly, the Theatre of Marcellus (44–11 AD), reflects the Roman builders' careful selections of tuff and travertine for construction stone and volcanic aggregates for the preparation of

Figure 2.30 (a,b) Street views of Parrano, Italy: rounded stones from river alluvial deposits are the main construction material, very often cut in two or more pieces to allow a better stabilisation of the wall.

Figure 2.31 (a,b) Dubrovnik's historic centre surrounded by walls and the large expansion of modern constructions, not present at the beginning of the twentieth century.

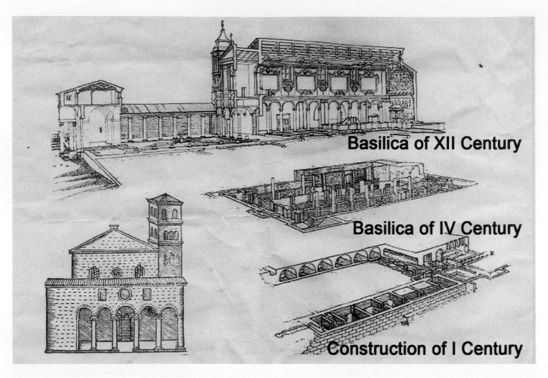

Figure 2.32 The progressive development of the San Clemente Basilica in Rome, from the initial construction in the first century AD to the Basilica in the twelfth century.

Figure 2.33 Example of the variety of building materials in the Forum of Augustus in Rome.

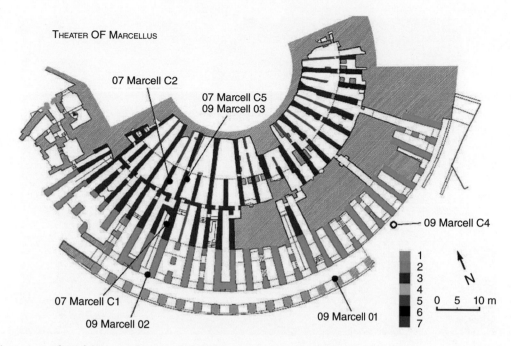

Figure 2.34 Plan of the ground level of the Theatre of Marcellus. The following numbers refer to the colour grid in the figure: 1. Travertine dimension stone; 2. *Tufo Lionato* dimension stone; 3. Augustan age concrete with *Tufo Lionato opus reticulatum* facing; 4. Augustan age concrete with *opus testaceum* brick coating; 5. Later Julio-Claudian age concrete with Tufo Giallo della Via Tiberina *opus reticulatum* coating; 6. Julio-Claudian age conglomeratic concretes; 7. Structures of Late Antiquity.

'pozzolanic' concrete (Figure 2.34). Sophisticated insertions of travertine stone reinforce the tuff masonry, which is complemented with durable concrete walls and barrel vaults (Figure 2.35). Concrete walls are coated with *Tufo Lionato opus reticulatum* and *tufelli*, and *opus testaceum* of fired, greyish-yellow brick.[25]

With the recognition of the Christian religion at the time of Emperor Constantine (272–337 AD) begins the great period of the early Christian basilicas (e.g. St John Lateran, Saint Peter). However, major architectural developments were not possible throughout the Middle Ages (fifth – fifteenth centuries AD).

The first part of the medieval period was characterised by the traditional use of 'Roman' brick, although a number of buildings were still made of natural stone. Only during the second part of the medieval period the use of stone increased. Indeed, in this period, new techniques were introduced by the Gothic builders, with the prevalent use of carved stone, instead of brick. It was only during the Renaissance that, with the translations of the ancient texts of Pliny the Elder and Vitruvius, the Roman construction techniques were rediscovered.

During this period the use of brick was never abandoned, and even regained the upper hand at some point, limiting the use of stone exclusively to sculptural elements, due to its high cost. Stone lost its role as a structural element and was increasingly considered a material for completion and finishing (coating or decoration), and, as an aggregate, for road works. Concrete was used for all types of masonry up to the nineteenth century. The twentieth century was finally

[25] Jackson, M. D., Ciancio Rossetto, P., Kosso, C.K., Buonfiglio, M. And Marra, F. (2011) Building Materials of the Theatre of Marcellus in Rome. *Archaeometry* 53, (4): 728–742.

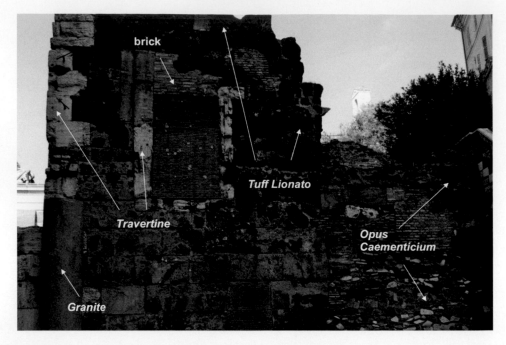

Figure 2.35 Types of building material in the Theatre of Marcellus in Rome.

characterised by the 'new science' of construction based on mathematical models. Steel and reinforced concrete became the globalised building materials, forever changing and standardising urban landscapes all over the world.

Conclusions

The Historic Urban Landscape approach aims at re-establishing the connection between management of the historic environment, contemporary urban development and the geological context. Doing so would result in a higher degree of sustainability and risk control, as well as harmony and continuity in urban forms, building structures and materials. For this, a stronger integration between sectors and disciplines is required, as well as the development of *ad hoc* knowledge and management tools.

The present work has investigated the role of geology, with respect to the availability of materials and to the morphological, climatic and natural hazard constraints, in shaping the architecture and urban planning of ancient villages and town.

The analysis of the historic settlements allows, from a geological point of view, the identification of the four different typologies of urban landscapes. The proposed approach can also constitute the basic methodological framework for a new tool aimed to guide the definition of conservation boundaries and/or buffer zones of archaeological, architectural and urban heritage areas:

1. **Clay based:** Characterised by constructions made of bricks, both sun-dried (e.g. adobe) and fired;
2. **Soft rock:** Characterised by constructions made with easy-to-carve soft rock or by settlements directly built into the soft rock;
3. **Hard rock:** With constructions made of stone blocks of large or small sizes, rounded or squared;
4. **Complex situations:** Mainly typical of important towns, showing a layering over time, with the use of various materials in the different historical periods, and with the use of a mix of materials in the last century.

3 Morphology as the Study of City Form and Layering

Stefano Bianca

Architectural Historian and Urban Planner, Zurich, Switzerland

The chief function of the city is to convert power into form, energy into culture, dead matter into the living symbols of art, biological reproduction into social creativity.

Lewis Mumford

Introduction

Over the past 200 years the understanding of urban systems and urban values has undergone radical changes, mainly due to the impact of the Scientific Age, the Industrial Revolution and, more recently, the Modern Movement. Wholesale demolition of pre-existing historic structures and *creatio ex nihilo* of complete new towns have become prevalent, due to rapid demographic growth and corresponding new urbanisation and transportation modes. The conception of the city as a mechanical system geared towards (existing or invented) material needs and to industrial production processes has led to new architectural concepts and corresponding planning techniques that are less rooted in overarching cultural frameworks.

In the course of this historic shift, a radically new notion of 'development' was born, as an aggressive, rationally produced and externally controlled activity that is alien to internally driven, evolutionary growth processes steered by local communities, and ignores traditional environmental and spiritual concerns. Abstract planning schemes based on the isolation of previously interacting social components, have often destroyed the sense of urban integrity. Large-scale segregation of formerly linked land uses have called for artificial reconnection through ever increasing streams of vehicular traffic that dissect the city and destroy the surrounding landscape, giving birth to mushrooming suburban extensions. As we experience today, the result has been an increasing loss of human scale and sense of place that were natural ingredients of historic cities. Urban transformation (in both time and space) has ceased to be an incremental process, embedded as it was in meaningful cultural patterns; and the emerging modern townscape often no longer corresponds with given human ways of emotional and sensorial perception.

It is against the background of such recent urban development trends and their disregard for historic structures that Urban Morphology

Reconnecting the City: The Historic Urban Landscape Approach and the Future of Urban Heritage, First Edition. Edited by Francesco Bandarin and Ron van Oers.
© 2015 John Wiley & Sons, Ltd. Published 2015 by John Wiley & Sons, Ltd.

needs to be appreciated. Its approach can provide an antidote, as it were, to the shortcomings of disruptive modern planning concepts; for Urban Morphology is based on the recognition of the historic urban fabric as a complex cellular micro-system that shows strong internal coherence and evolves in ways not dissimilar to natural, that is organic growth and transformation processes. Through its particular procedures, Urban Morphology is able to un-riddle the inner rationale of a given urban structure:

- to grasp the formative powers active within its deep tissue;
- to identify the formal archetypes that have governed the constitution of the urban fabric;
- to read the range of variegation of such archetypes and interpret their various connective patterns; and
- to detect the successive layers of urban growth that have accumulated or superseded each other throughout time.

Going through such a process of 'cultural anamnesis' must not be seen as an obsession with the past, but as the best way to prepare the ground for the future. For it will allow the management of on-going urban renewal in a continuous and coherent manner, and, where necessary, help heal damage inflicted upon the historic fabric.

Origins and Implications of the Term Morphology

Before addressing the more practical implications of Urban Morphology, its analytical tools and the specific modes of intervention to be derived from it, it may be useful to dwell for a moment on the significance of the actual term, which is composed of the Greek words *logos* and *morphe*. Logos is a key notion in Greek philosophy, and its content can oscillate between 'word' (as the creative principle), 'discourse', 'reason' or 'meaning', while morphe stands for the visible shape or form (one could say *Gestalt*) of individual types of living phenomena, seen as an external manifestation of their intrinsic constitution. 'Morphology' could thus be interpreted as the knowledge of how to deal with the formal appearance of living structures.

Perhaps the most representative propagator of the morphological approach was Johann Wolfgang von Goethe (1749–1832) not only a great poet, but also a universal mind who ventured from the arts into philosophy and the natural sciences.[1] While living in the age of Enlightenment and Rationalism, he rejected contemporary attempts to look at nature as a mere mechanical system and, instead, saw the creation as an on-going transformation and adaptation process (*metamorphosis*) based on the continuous evolution of specific archetypes (*Urphänomene*) that have their roots in the spiritual world but manifest and incorporate themselves in a great variety of material forms. The underlying archetypes provide inner unity within external variety, and their unfolding is ruled by *entelechy*, that is their need is to comply as far as possible with their inherent spiritual imprint, while their actual shape is conditioned by the environmental circumstances and constraints of the material medium, through which they come into existence and pursue their path of growth. At the end of his poem 'Primal Words – Orphic', Goethe has summarised this view in two lines:

No time, no power will decompose
Minted form that through life arose.[2]

---

[1] Goethe's main morphological writings are contained in: *Goethes Werke*. Hamburger Ausgabe, Munich 1981, Vol. 13: 53–250.
[2] The original lines in Goethe's German poem "Urworte – Orphisch" read:

Und keine Zeit und keine Macht zerstückelt
Geprägte Form, die lebend sich entwickelt.

It should be noted that such ideas of multi-form unfolding of non-material seeds in material reality, of 'animated' formal structures and of everlasting cyclical renewal of manifested phenomena of nature are contrary to the mainstream of modern thinking. Modernity usually does away with transcendent analogies that would root mundane realities in spiritual archetypes. It has given preference to more 'objective' perspectives, where the products of creation (whether of natural or human origin) are alienated, commodified and standardised – which makes them become soulless, rigid and uniform. Meanwhile, human history, in its modern form invented in the nineteenth century, no longer unrolls as a continuous and repeated process of manifestation, but as a series of critical revolutions, each new stage of development discrediting and obliterating the previous one.

The Scope of Urban Morphology

Whereas the above approach to nature and its organic phenomena may not be fully applicable to the study of urban form and historic cities, it provides interesting lessons for understanding the relevance of architectural archetypes and typologies in the context of any given cultural matrix. It also sheds light on the importance of internally driven cybernetic processes that have guided the evolution of many historic cities. If historic cities are to survive as living entities, that is beyond mere material conservation of their physical shell, such self-supported internal revitalisation (that could be compared to the natural principle of *auto-poiesis*, such as self-reproduction) must be strengthened from within, rather than being undermined by hostile external forces.

In this sense, far from introducing alien tools and methodologies, Urban Morphology provides for the greatest possible affinity in dealing with pre-industrial historic cities and their intricate urban fabric. What it actually does, is to render more explicit the formerly implicit rules and modes of internal 'organic' evolution.[3] It proceeds by careful in-depth analysis of the cellular urban tissue, both at micro-scale, (considering its smallest units and components) and at macro-scale (considering topographic constraints), historic layers, existing articulations and possible disjunctions due to external interventions. In doing so, it can define the rules for on-going renewal of the historic fabric according to its own premises. It can simulate (and stimulate) traditional evolutionary processes by making use of existing 'genetic codes', as it were, and adapting them to contemporary circumstances and needs. As a result, the past can be projected into the future without radical disruption, and a bridge can be built between tradition and modernity.

Thus Urban Morphology is an appropriate instrument to investigate the formation and gradual transformation of the physical structure of human settlements. It attempts to detect 'from within' the live forces and processes that have conditioned (and continue to shape) a specific urban fabric and are responsible for its specific spatial qualities. In doing so, it enables planners and architects to penetrate the secret inner ordering principles of a given urban structure – and also to project and extend them into the future, thus ensuring natural continuity in urban evolution. During the pre-industrial past (and until recently in many Third World countries), this underlying order has unfolded mostly in an unconscious manner. For it was rooted in collectively applied building customs and archetypes that reflected implicit beliefs and social codes of conduct, rather than being subject to explicit decision-making of planners and architects. As a present-day approach, Urban Morphology takes into account that such informal traditional cultural patterns may no longer

---

[3] A ground-breaking modern contribution for understanding the morphologic structure of living ('organic') beings can be found in: Capra, F. (2002) *The Hidden Connections*. New York, Anchor Books.

exert the same cohesive power and therefore need to be reinforced in more formal manner, so that they can be harnessed for consistent and creative reproduction – without necessarily taking recourse to literal replication of architectural forms.

Indeed, the structural archetypes that form the basic units of the typological and morphological approach are virtual (or ideal) forms in the Platonic sense. Hence their physical manifestation is not subject to stylistic features representative of a specific period, but determined by intangible structural laws. Their various 'incarnations' therefore offer an inbuilt flexibility for variation (*metamorphosis*), as they rely on the interplay between constant qualitative archetypes and variable material manifestations. And even in the case that the social and cultural parameters should change, or new functional needs occur, inherited architectural shells may well lend themselves to adaptive re-use, thus preserving the integrity of the existing urban fabric. Examples of such creative recycling of inherited architectural 'containers' are abundant throughout architectural and urban history. One may think of extreme cases, such as, for instance, of the re-use of former Roman amphitheatres as a basis of housing compounds in several Italian cities, of the re-use of ancient palaces for modern museums, or of the conversion of old industrial compounds into lofts, shops and public meeting spaces.

As an applied theory, Urban Morphology and the related survey and intervention methods are a relatively young 'discovery' that was first introduced and experimented with, in Italy in the late 1950s, in a country blessed with an extreme wealth of historic city centres and at a time when actual or potential destruction (be it by war damage or by the impact of Modern Movement ideologies) became evident. Without going into a full account of this current, suffice it to mention here a few representative professionals, such as Saverio Muratori, Carlo Aymonino and, to some extent, Aldo Rossi.[4] The new movement then spread to other European countries such as France and England,[5] and became particularly successful once combined with social rehabilitation measures capable of supporting and stabilising resident communities. Moreover, the morphological and typological approach eventually transcended the limits of urban conservation and urban rehabilitation, for it became a way of thinking that could be used in contemporary design – as demonstrated, for instance, by the brothers Leon and Rob Krier, or, with a different accent, in Christopher Alexander's 'Timeless Way of Building'.[6]

Methodology and Procedures

In the following part of this essay, we shall concentrate on the practical implications of Urban Morphology, as applied to relatively homogeneous historic districts. The premise of this approach is, again, that any intervention must be based on a thorough analysis of the urban fabric both at micro-level and at macro-level, in order to detect the rules that have governed the constitution (and the subsequent evolution) of the deep urban structure. Particular importance must be attached to grasping the interrelations that exist between single components and the overarching spatial system. As in any organic cellular tissue, small components already contain virtual prefigura-

---

[4] See for instance: Muratori, S. (1959) *Studi per una operante storia urbana di Venezia*. Roma: Istituto Poligrafico dello Stato. Aymonino, Carlo, 1977. *Lo studio dei fenomeni urbani*. Roma: Officina. Rossi, A. 1966. *L'architettura della città*. Padova: Marsilio.

[5] See for instance: Panerai, P., Depaule, J., Demorgon, M. and Veyrenche, M. (1980) *Eléments d'analyse urbaine*. Bruxelles: Editions des archives d'architecture moderne.

[6] See Christopher Alexander's concept of "wholeness". Alexander, C., Ishikawa, S. and Silverstein, M. (1977) *A Pattern Language*. New York: Oxford University Press. Alexander, C. (1979) *The Timeless Way of Building*. New York: Oxford University Press.

tion of much larger composite units, which is the source of consistent structural unity.

Having said that, it becomes clear that Urban Morphology is not driven by antiquarian interests: it is not merely about superficial stylistic features of individual buildings (although this may be a secondary aspect of it), but about the three-dimensional articulation of architectural structures and their interconnections within the overall urban context that holds them together. Similarly, the focus is not exclusively on outstanding monuments (that may have special merits and rights of their own), but on the urban fabric as a whole, which puts much emphasis on the conservation and/or incremental renewal of housing areas in historic districts – a matter often neglected in conventional conservation schemes.

Accepting that Urban Morphology seeks to take hold of both the constitutional qualities and of the evolutionary forces that animate the urban fabric and drive its metamorphosis in time and space, it is obvious that its operational procedures must adapt to and conform with the nature of its subject. Indeed, those vital forces cannot be comprehended by dealing with the macro-townscape of historic cities only, nor by artificially dissecting their architectural micro-elements. They can only be understood and put to work by addressing the interdependencies that exist between basic cellular components and their wider urban context, that is the modes of articulation that govern their clustering into larger urban sub-units in ways that eventually result in a unified but highly differentiated urban structure. Grasping these internal differentiation and integration processes – and the structural hierarchies operating within them – is an essential prerequisite for dealing with the city as a lively, 'organic' entity. For it is the interacting character of spatial components at various levels, that generates the spark of life in the urban tissue and makes it so different from 'mechanical' urban artefacts.

Here, we touch upon the fundamental difference between Urban Morphology and conventional modern planning and design methodologies. Large-scale physical planning is far too abstract to come to grips with the small grain of historic urban structures. Its sectorial and functionalist approach cannot cope with the structural integrity of 'organic' entities, since it tends to impose alien interventions from outside, rather than working from within. Meanwhile, the conventional architectural design approach tends to be object-driven and isolate single urban components from their urban matrix. Similarly, classical conservation projects usually deal with detached monuments and neglect the context in which they are embedded.

Urban Morphology, instead, attempts to deal precisely with the lost intermediate dimension, as constituted by the live web of the urban fabric with its internal correspondences, connections and articulations. It is these that inform and control urban growth and transformation processes, define the qualitative sense of place and infuse meaning to complex physical structures. Hence the often observed parallel between Urban Morphology and linguistic syntax: single letters only acquire sense once they are combined into words, and single words acquire meaning once configured and contextualised in sentences. Urban Morphology thus works with a vocabulary of basic urban components that could be described as a 'family' of archetypical forms that are variegated in size, style and sequence according to a given topographic, socio-cultural and physical context. Ideally, its methodology includes the following levels of analysis:

a) **Topographic Analysis**
   With rare exceptions, historic cities have emerged and developed in particularly significant strategic locations, such as hilltops, mountain ridges, harbours, river loops or important regional transportation nodes. The characteristics of such location – whether seen as constraints or opportunities – tend to have a strong impact in terms of orientation, density, terracing patterns, circulation networks, etc., and become a strong component of the local sense of place

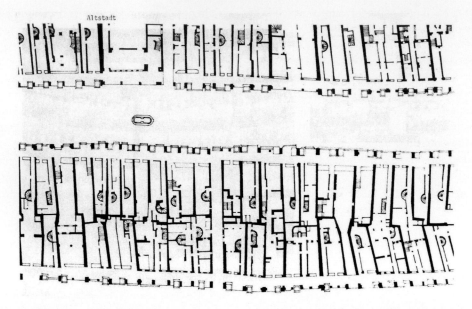

Figure 3.1 Survey of the urban fabric of the historic centre of Bern (Switzerland), showing its homogeneous morphology, as grown and maintained over centuries. (Source: Paul Hofer, Berne Studio, ETH-Zürich 1974/75).

or *genius loci*. In fact, one could say that it is the 'resistances' encountered in a natural site that have often shaped the identity and the distinctive features of a given city. (Figures 3.1, 3.2 and 3.3 a–c).

b) Historic Macro-Analysis

While the topographic features in a way represent a timeless component, the history of cities through the ages and periods of their existence brings in a time factor. This becomes manifest in various phases of growth, in superimposed historic layers, and sometimes in violent disruption of the urban fabric through natural catastrophes, war damage or deliberate urban interventions involving large-scale redevelopment schemes, as have occurred since the seventeenth century in Europe, but particularly in the nineteenth/twentieth century, as well as during the colonial age in non-European regions.[7] Such radical transformations produced clearly visible 'fault lines' in the historic fabric and often result in the co-existence, side by side, of different types of urban structures, each one featuring its own morphology and requiring different treatment. The best way to analyse corresponding historic sequences is through interpretation of cadastral maps of various periods and through seeking archaeological evidence of previous urban strata.

c) Architectural Micro-Analysis

Once the macro-analysis has determined more or less homogeneous sections of the urban fabric (in some fortunate cases the totality of the urban fabric may represent homogeneous features), the 'grain', that is the typological composition of such homogeneous sectors needs to be investigated. This will include plot-by-plot cadastral anal-

---

[7] Here, one may think of the transformation of medieval Rome by Pope Alexander VII during the years 1655–1667, or the transformation of Paris by Baron Haussmann in the 1860s – a model that was followed in many Third World cities during the period of colonialism.

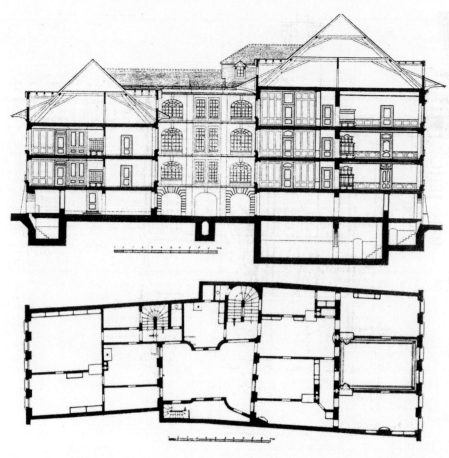

Figure 3.2 Detailed section through a typical residential/commercial unit of the plan shown in Figure 3.1. (Source: idem).

ysis, research on three-dimensional architectural typologies (archetypes) used for housing, public buildings, and industrial or commercial facilities, as well as characteristic street networks and public open spaces. Of particular importance will be to understand how the various components can vary, how they connect, how they combine or cluster into urban sub-units and what is the dialectic relationship between macro-form and micro-units. Such studies may be complemented by 'figure-ground' or 'solid-void'

analyses that establish the qualitative rapport between built form and open space,[8] and by recording stylistic typologies of architectural elevations (or typical period ornamentation) applying to architectural archetypes.

d) **Socio-Cultural Content**

To fully appreciate the formative rationale underlying the recorded architectural typologies, it is important to understand the cultural codes and social patterns that have conditioned the archetypes of the physical

---

[8] See also: Trancik, R. (1986) *Finding Lost Space*. New York, Wiley. He discusses the lost interrelation between buildings and open spaces in modern town planning.

Figure 3.3 (a–c) Typological variations of a characteristic shopping arcade relating to Figures 3.1 and 3.2, the variations being due to the adaption of the basic prototype to different contextual situations and topographic levels. (Source: idem).

shell and provide meaning and identity to corresponding physical structures. Some of these patterns are (or have been) rooted in religious or spiritual principles that are translated into shared modes of social conduct. While the religious background may have weakened in more recent times, such patterns often endure as cultural traditions that continue providing comfort and identity. Other living patterns may be linked to more mundane contemporary trends of civilisation, but may be accommodated in the given historic structures without major conflicts – except for modern modes of individual vehicular transportation that in most cases cannot be fully absorbed by the pedestrian network without destroying the very essence of historic cities. Therefore,

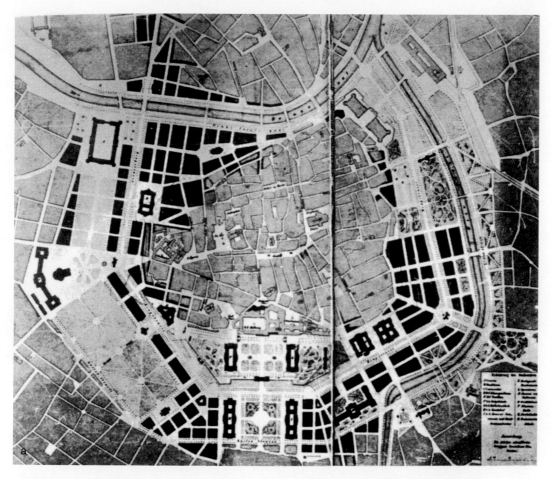

Figure 3.4 (a, b) Plan of the inner city of Vienna, showing juxtaposed urban morphologies of different scale and character, with the preserved medieval nucleus in the centre and the surrounding nineteenth century development of the 'Ringstrasse' (and beyond) that was planned and built partly on the site of the former moats and fortifications. (Source: Carlo Aymonino, 'Lo studio dei fenomeni urbani', Roma 1977).

centrality and a high-quality social and architectural environment will have to be traded off against a deficit in immediate accessibility. (Figure 3.4 (a,b))

e) **Further Opportunities**

Given the fact that morphological and typological analyses are based on plot-by-plot cadastral analysis, the resulting detailed mapping of historic cities can be put into use for complementary analytical and operational information. Eventually, such material will provide a computerised data bank that can be used for continuous monitoring, as well as the steering of consistent urban evolution. Information recorded and entered on plot-by-plot basis should include:

- type of building;
- historic age;
- architectural importance – from major monuments to contextual value;
- physical condition – from well-preserved to ruined;
- previous and current use, such as religious, institutional, commercial, industrial, residential; and
- ownership (including residents' status).

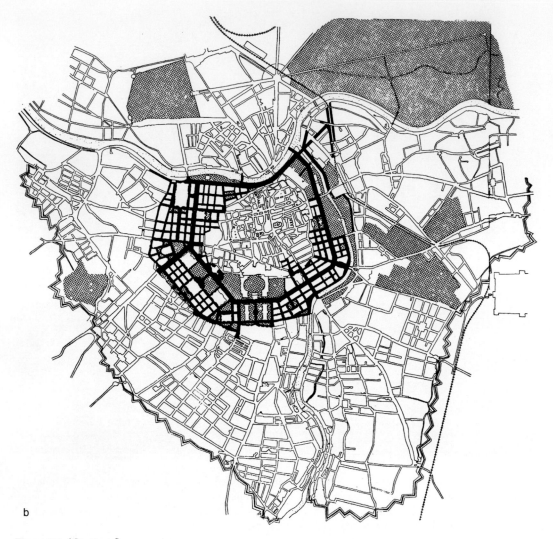

b

Figure 3.4 (*Continued*)

Once fully available, such detailed analytical information must be regularly updated. Properly cross-referenced, information from different categories will inform the mode of intervention best suited for each specific plot or building. Such interventions may range from scientific preservation (e.g. for important historic structures in retrievable physical condition) to typological substitution (e.g. for houses in ruined physical conditions and with contextual value only). Creative reading of the available analytical material will also enable planners to establish pro-active area conservation plans or adapted urban rehabilitation projects that preserve the continuity between old and new (Figure 3.5).

Advantages and Problems of the Urban Morphology Approach

The evident advantage of Urban Morphology, particularly if extended into the described oper-

Figure 3.5 Aerial view of the historic fabric of Aleppo (Syria), before the current war, showing its 'organic' cellular tissue that provides variety of individual architectural forms on the basis of consistent structural affinities between its components.

ational and interventionist dimension, is that it can steer urban evolution from within, acting at the level of smallest individual components which are the micro-generators of incremental change and 'organic' transformation. Moreover, this approach is the only one that enables planners to transcend the limitations of the classical urban conservation approach: firstly, because Urban Morphology looks not just at individual monuments, but at the historic fabric as a contextual continuum that integrates all individual components – including housing, which accounts for substantial but often neglected parts of historic cities; and, secondly, because it

acknowledges that historic cities cannot be frozen as antiquarian artefacts, but that they have to renew themselves day by day.[9]

Through providing consistent guidelines for comprehensive revitalisation, including restoration of monuments, upgrading of housing restoration, adaptive re-use of obsolete facilities and substitution on ruined plots, Urban Morphology resolves the key problem of historic cities by staying alive under changing contemporary conditions, and yet being faithful to their identity and traditional structuring principles – values that account for their character and their enduring attraction. Historic cities are thus enabled to

---

[9] In this context one needs to remember Friedrich Nietzsche's statement at the beginning of his book: Nietzsche, F. (1873) *On the uses and disadvantages of history for life* (Hollingdale Translation): '*We want to serve history only to the extent that history serves life: for it is possible to value the study of history to a degree that life becomes stunted and degenerate – a phenomenon we are now forced to acknowledge in the face of certain striking symptoms of our age.*'

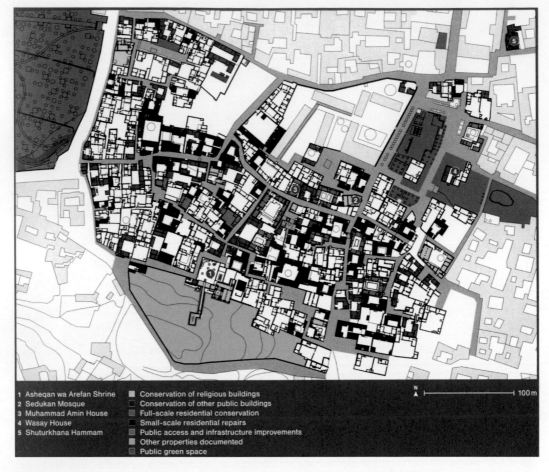

1 Asheqan wa Arefan Shrine   ■ Conservation of religious buildings
2 Sedukan Mosque   □ Conservation of other public buildings
3 Muhammad Amin House   ■ Full-scale residential conservation
4 Wasay House   ■ Small-scale residential repairs
5 Shuturkhana Hammam   ■ Public access and infrastructure improvements
                             ■ Other properties documented
                             □ Public green space

N ▲   ├─────────┤ 100 m

Figure 3.6 Urban rehabilitation project for the Ashekan wa Arefan district in Kabul, Afghanistan, based on morphological and typological surveys and corresponding intervention criteria. (Source: AKTC).

overcome the daunting dilemma between the two extremes of sterile conservation and radical redevelopment, and to ensure their long-term viability as part and parcel of larger urban systems (Figures 3.6 and 3.7 a,b).

Admittedly, certain problems may arise regarding the identity issue, due to the fact that historic cities today (particularly in the non-Western world) often no longer represent fully homogeneous entities in terms of built form and social content. Many of them are inhabited by hybrid communities – hybrid in the sense of adhering to ancient beliefs and traditions as well as modern Western ways of life, or hybrid in the sense of including different ethnic and/or reli-

gious allegiances. In such cases, it will be important to find common denominators and to resist sectarian political and ideological biases that in modern times tend to abuse the sense of identity. Architectural forms are usually open to different cultural interpretation, as long as they appeal to collective human values. In the past, cities with multi-ethnic or multi-religious background (such as Mostar or Aleppo, for instance) have succeeded in developing shared spatial identities. They have demonstrated that Muslims and Christians not only could successfully cohabitate, but that they made use of the same architectural vocabulary – except for their distinct religious buildings.

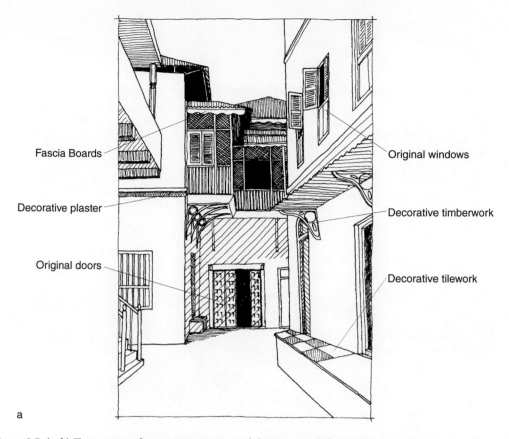

Fascia Boards

Decorative plaster

Original doors

Original windows

Decorative timberwork

Decorative tilework

a

Figure 3.7 (a, b) Two excerpts from a conservation and design manual for the Old Stone Town in Zanzibar, with exemplary guidelines concerning typical elevation elements. (Source: AKTC).

Other problems connected with the implementation of the Urban Morphology approach relate to manpower and institutional capacity. Due to the complexity of the methodology and to its emphasis on the urban micro-scale, the full process, from starting analytical surveys to reaching operational working tools, takes time. It also requires enduring commitment on the part of those carrying out the process, in order to guarantee equal and consistent criteria of evaluation both in the analytical and operational phase. Staff must be field-based, in order to monitor and supervise on-going activities on the ground – an obligation that is not always appreciated in developing countries, where 'white collar' desk-work has a higher social status than fieldwork. Institutional continuity is equally important, since consistent monitoring is a long-

term task. Ideally, a special agency dedicated to historic districts should be in place, to conduct both research and implementation, and it should have sufficient legal authority and be independent from changing political constellations (Figures 3.8 a,b and 3.9).

These are conditions that, unfortunately, do not always prevail in countries blessed with a rich urban heritage. Often the historic city is dealt with by the planning department of the municipality, which is submerged with large-scale urban development and infrastructure problems and does not have the time, nor the manpower, nor the know-how, nor the financial means to deal appropriately with historic districts. Technical support through international organisations, bilateral donors or private funding agencies promoting urban conservation and

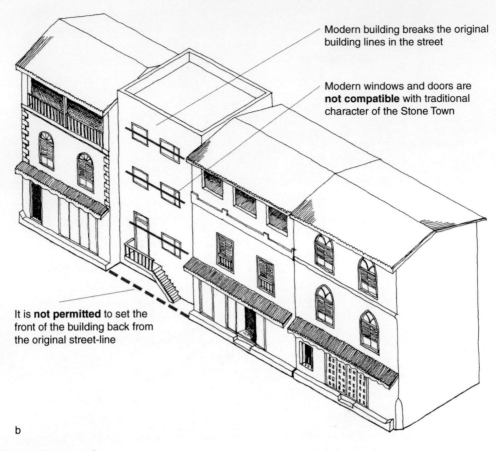

Modern building breaks the original building lines in the street

Modern windows and doors are **not compatible** with traditional character of the Stone Town

It is **not permitted** to set the front of the building back from the original street-line

b

Figure 3.7 (*Continued*)

related socio-economic development projects can help, as shown for instance by various interventions of the Aga Khan Historic Cities Programme in the Islamic world. However, it is essential that such partnerships give high priority to training of local staff (including capacity-building in management of historic areas and revitalised historic buildings), in order to ensure future sustainability of initiated projects. Restored action areas (including public open spaces) and re-used monuments are often capable of generating direct or indirect economic benefits, which would need to be skimmed off and re-injected into the project by appropriate legal and financial mechanisms – be it for ensuring continued maintenance of restored structures or for funding the required local management unit (Figures 3.10 a,b and 3.11).

Relevance within the Historic Urban Landscape Concept

The Historic Urban Landscape approach has been developed in the context of the UNESCO World Heritage List that covers a vast number of historic cities of different cultural origin, of different periods and of different state of physical preservation. Urban Morphology is applicable to most of these cases, since as a 'neutral' and universal tool it carries no ideological bias of its own with it, nor does it impose alien structures and

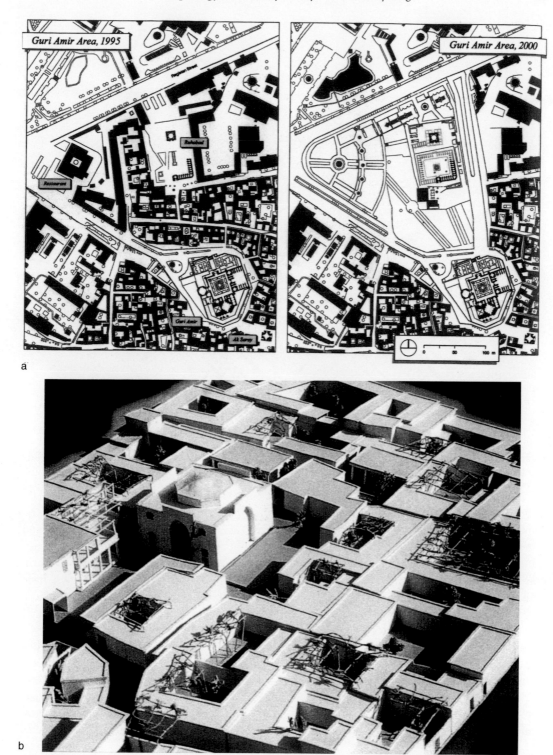

Figure 3.8 (a) Negative example of a large scale planning intervention by the Uzbek government that led to the loss of substantial parts of the historic fabric in favour of anonymous and meaningless public open space (b). (Source: AKTC).

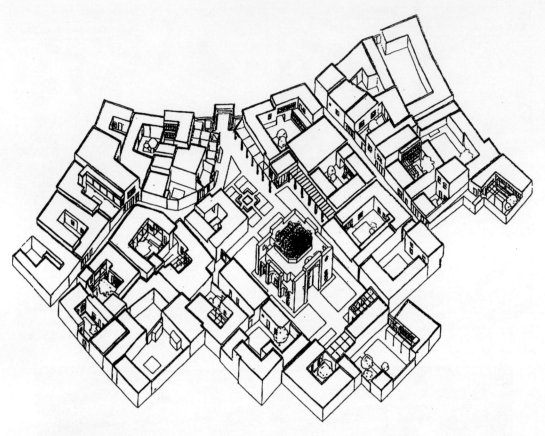

Figure 3.9 Morphologically faithful infill and rehabilitation proposal for a partly destroyed residential area in historic Samarkand. (Source: AKTC).

modes of intervention.[10] Its methodology is purely empirical, geared to the individual conditions, and deriving its conclusions from the specific local context it analyses – whether it deals with European or Asian sites, with medieval cities, or with nineteenth century suburbs. Indeed, its objective is to detect, support and enhance the urban characteristics that are responsible for a city's or a district's identity, without, however, claiming to 'freeze' the urban fabric in an arbitrary stage of its on-going evolution. Through the on-going regeneration of the tissue cell by cell, the underlying urban frame-

work as a whole will be confirmed and strengthened.

Some World Heritage cities, such as Bern, Orvieto, Fez, to quote a few examples, have remained relatively homogeneous, be it because of their topographic site conditions, or be it because of lack of extensive modern development around them. In such cases, the morphologic approach can be extended to the complete urban form. Other cities, such as Vienna, Paris or Cairo, show a composite urban pattern that features historic typologies of different ages, either side-by-side or interspersed, which makes

---

[10] This 'neutrality' seems particularly important with reference to historic cities in the developing world. Today they are subject to heavy globalisation pressures that in fact perpetuate Western colonial ideologies in new disguises.

Figure 3.10 (a) Urban renewal from within, based on existing environmental, architectural and community potentials and (b) view of the historic district of Darb al-Ahmar in Cairo before and after intervention. (Source: AKTC).

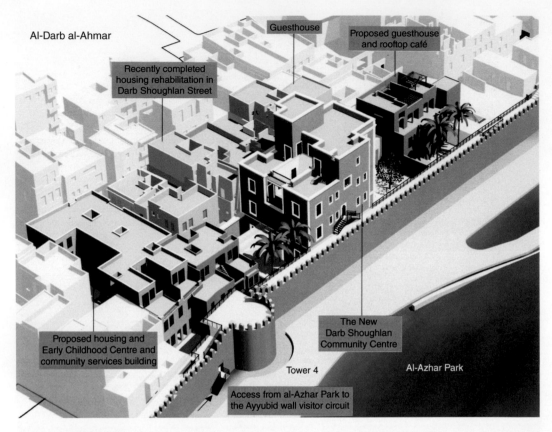

Al-Darb al-Ahmar

Guesthouse

Proposed guesthouse and rooftop café

Recently completed housing rehabilitation in Darb Shoughlan Street

Proposed housing and Early Childhood Centre and community services building

The New Darb Shoughlan Community Centre

Tower 4

Al-Azhar Park

Access from al-Azhar Park to the Ayyubid wall visitor circuit

Figure 3.11 Multifold intervention on a housing cluster in Darb al-Ahmar, combining conservation, upgrading infill and community development projects. (Source: AKTC).

them become 'collage cities'.[11] There, morphologic approaches will have to be applied to parallel historic urban systems, each one in its own right.

Given the fact that most historic cities today happen to be located at the centre of a growing metropolis, they suffer from the clash with adjacent, more recent central business districts or with modern urban extensions that mushroom on their peripheries. Since Urban Morphology can be applied to urban structures of many periods and can even be projected into contemporary design, it provides a good basis for achieving consistent formal order in juxtaposed urban systems and, eventually, for establishing compatibility, if not harmony, between old and new. A necessary precondition, however, will be that the authorities are able to maintain strict development control in areas exposed to heavy real estate speculation – be it by enforcing appropriate legal back-up in terms of building regulations, and/or by creating economic mechanisms, such as differentiated property tax systems, transfers of development rights and cross-subsidising of historic areas from income raised in the Central Business District.

[11] See: Rowe, C. and Koetter, F. (1978) *Collage City*, Cambridge, Massachusetts, MIT Press.

Interview with Wang Shu
Searching for a Chinese Approach to Urban Conservation

Architect and Professor at the China Academy of Art in Hangzhou, China.
20 February 2012, Hangzhou, China

Question: *Foreign observers are always struck by the magnitude and speed of the urban development process that is taking place in China, by the obliteration of traditional city forms, and by the uniformity of modern urban areas. I would like to ask your views on the process that has led to the loss of the largest part of China's historic urban fabric. This was one of the oldest and largest urban civilisations in the world, and yet, very little is left of this important heritage. How has it been possible to destroy in just a few years the result of centuries of urban history?*

Answer: The destruction of the traditional urban fabric is a relatively recent phenomenon: even the Chinese Revolution and the Cultural Revolution had left a great part untouched. The great destruction started in the 1990s, when the fast economic development driven by market liberalisation started. Practically most of the old buildings have been destroyed, including my own work before that time! This fracture with our tradition had already been denounced in the 1930s by a great scholar, architect LIANG Sicheng, the founder of the Tsinghua University School of Architecture and Planning in Beijing and a friend of the American planner Clarence Stein. He wrote a paper predicting that the future would bring great destructions. And they came, with World War Two, the Revolution and the contemporary urban explosion. He saw that modern architecture was more interested in techniques and technology and did not care about tradition, culture and the arts. In the past 30 years, 90% of the traditional buildings in China have been demolished, entire historical cities have been wiped off the map in one go.... With them, culture disappeared, and people have now lost the memory and feeling of traditional urban life. Even if some of the old buildings have remained, they are empty shells....

Question: *And yet Chinese value tradition and are interested in their past. Why was this not reflected in urban design?*

Answer: Chinese value a lot tradition, but we consider mostly our history, while urban design is seen only in relation to modern architecture. It is a bias that perhaps only now people start understanding. There is a new interest in traditional urban environments, but I am afraid that this remains an empty concept, as these are not anymore living environments. In fact, for most people, the urban experience is not linked to culture and tradition; it's just a modern environment, most of the time repetitive and anonymous.

Question: *You have developed a different approach in your design work. How did you form your concepts and your language?*

Answer: In the 1980s, before the big economic changes, there were many discussions among architects on the role of tradition in the development of an architectural language. There were many ideas and experiments, among which I would like to mention the work of WU Liang Yong. He did very important restorations of the Beijing *hutongs*, showing how it was possible to modernise old buildings without losing the traditional design features. Also at Tongji University in Shanghai, FENG Ji Zhong started experimenting with traditional materials and built some very small pavilions: small but important. Unfortunately, this reflection ended with the economic boom. At that time, I was thinking that we needed to find a way not to lose our tradition, while at the same time avoiding to copy Western architecture. We needed to find a new Chinese architectural language. I was very radical at that time, and in 1986 and 1987 during some Congresses in Beijing and Chongqing I stated publicly that in China there were no architects! This did not make me very popular with the Government and the profession.... But nothing really happened after that time. In the 1990s, the profession took another direction, the one of business. People stopped thinking and just went for the money. Only a very small group continued the reflection. Among those, I think the most interesting are the architects JANG Yung He of Beijing and LIA Jia Kung of Chengdu, whom I met in the 1990s. They do small projects, but very significant.

Question: *Do you feel isolated or do you have a network of colleagues you work with?*

Answer: I feel a bit isolated, but not alone! In fact all my work is done with my wife LU Wenyu. I also have a lot of friends around the world.

Question: *So, when the business mentality prevailed in the 1990s, how did you react?*

Answer: At that time I decided to stop working. I only did reading and writing, and I wandered a lot around the West Lake in Hangzhou.[12] I tried to forget everything I knew and taught, and I tried to have a real life experience, finding the real materials, the original building techniques, the relationship between buildings and life. I tried to rediscover the relationships between the arts, the techniques and the builders. I tried to re-establish the relationship with tradition, beyond the teaching of architecture I had received. I made many trips to the villages, where the traditional forms are better preserved, and I was able to learn and develop my own language, a language that tries to connect tradition and modernity.

Question: *What do you see as the greatest challenge for contemporary architecture?*

Answer: The most difficult thing is to create a feeling of real space, a space for life. Modern architecture has lost this relationship, space is abstract and uniform, and there is no connection with people. I try to design a small world, I do not design buildings, I design life. In architecture, every element has a meaning: doors, stairs, corridors, yards, etc. These elements have to be conceived in relation to people. Architecture is about culture, meaning, not just doors and stairs! When I walk around Paris, I see that the city blocks have big doors: each one of these doors represents a community, a group that has formed there, with its relationships, its culture. I insist: architectural language must have a relationship with culture!

Question: *How can this approach be transferred from architecture to urban design?*

Answer: It's a similar process. We cannot just start doing planning and design for a city without knowing it. The first step is to understand the meaning of a place: many things existed before us, we have to retrace and understand them. Design cannot be done without a perfect knowledge of what existed before. Many architects think that architecture comes from the sky! I do not. A good example of this approach is the work I have done for the historic centre of Hangzhou. It has been a very complex work, involving the rehabilitation and design of the main historic axis, the Zhongshang street,[13] a street with many traditional buildings and characterised by a number of 1920s and 1930s commercial buildings. In spite of its historical importance, the street had lost its role as the centre of

---

[12] West Lake or XĪ Hú (Chinese: 西湖; literally 'West Lake') is a famous fresh water lake located in the historic area of Hangzhou, the capital of Zhejiang province in eastern China. The lake is divided by the causeways of *Sū Dī* (苏提 / 蘇堤) , *Bái Dī* (白堤), and *Yánggōng Dī* (杨公堤 / 楊公堤). There are numerous temples, pagodas, gardens, and artificial islands within the lake. West Lake has influenced poets and painters throughout the ages for its natural beauty and historical relics, and it has been among the most important sources of inspiration for Chinese garden designers, as evidenced by the impact it had on various Chinese classical garden. It was made a UNESCO World Heritage Site in 2011, and was described as having 'influenced garden design in the rest of China as well as Japan and Korea over the centuries' and as reflecting 'an idealised fusion between humans and nature' (Source: Wikipedia).

[13] Zhongshan Road in Hangzhou is the major arterial running north-south through the city. Originally called the Imperial Street during the southern Song dynasty, the road was re-named after Sun Yat-sen during the Republican era (Source: Wikipedia).

the city was moved closer to the West Lake. The street declined and several high-rise modern buildings replaced the old ones. In spite of this, it remained an historic area, and because in the past 20 years Hangzhou has demolished 90% of its traditional buildings, the citizens wanted to preserve the street. There was a big debate: some people wanted to reconstitute the street in a kind of Ming-Qing dynasties style, and demolish all the rest! Others wanted to give it a 1920s look and demolish the Chinese style buildings.... Each one of these solutions would have been a fake. So the City Government and particularly the Mayor, WANG Guoping, invited me to make a proposal. My approach was to maintain the layers of time, and preserve the 'mixed' situation that we had received from the past. I saw the intervention not as a 'revolution' but as an 'evolution'. I put down 6 conditions to accept the commission: 1) that before doing any design work, I needed 6 months to study the place, as it is a very complex area; 2) that no people had to be removed, as I wanted to preserve the real life of the area; 3) that I was allowed not just to design for the preservation of the façades of the buildings, but of all the elements of the place, like courtyards, building types, public space, etc.; 4) that I could decide on all buildings, even the illegal ones, and not just on the monuments; 5) that I could design new buildings in the historic area; and 6) that I needed three years of time for the implementation, in order to do a quality work (in fact, we did it in two years).

Question: *This was a clear departure from the established practice. China is full of fakes, like the reconstructed 'traditional' neighbourhood of the Qianmen Street in Beijing just south of Zhēngyáng mén. What was it that pushed the Government to allow you to implement such an innovative approach?*

Answer: The Mayor was also a planner and he understood that this approach met the desires of the people. But it was not a simple job; we had to take many hard decisions. And we could not implement the programme as fast as the Government wanted. I prepared the Master Plan, but not all the projects are mine, there were several architects working. I had the overall responsibility and I had to adjust the plans all the time, to meet the needs of the people. I also designed a number of small buildings along the street; these are modern constructions that are important to correct some mistakes of the previous interventions. The original street, or at least the one that was designed in the 1920s, was 12 metres wide, but in the 1990s many buildings were replaced and the street in many parts had been enlarged to 24 metres, to make more space for the cars! I thought that it was very important to preserve the spirit of the place, so I introduced these small buildings that play the function of re-establishing the proportions of the street. They also function as 'observatories' as people can go up to the second floor and watch the street from there. I also rebuilt the three City Gates that had been demolished. They are very important, as they structure the space and provide the place with a strong identity. They are also part of the memory of this city that was very important in the past, as the capital of the Song Dynasty, between the 10th and the 13th centuries. When the Yuan Dynasty established by Genghis Khan took over, the walls of the city were demolished and commercial streets were created that penetrated the fabric of the city. But the memory of the Gates remained and influenced the design. I also used water as a unifying element in the design. Water is very important in China, traditionally there has always been water in the city, and the river was the main road. Water connects, it is life.

Question: *Integrating modern design and tradition is not at the core of the theory and practice of modern architecture. Which principles you followed in order to make this connection possible and effective?*

Answer: I think that the most difficult thing is to understand the patterns of life; we have to understand that the past is not dead, it's a continuous development. The Zhongshan Street was a very good example of this continuity, as people cared a lot about that space, even beyond my design. We had to find ways to make the reuse of buildings and spaces possible, as there are new functions. The design had to be very flexible, and this is a principle I always apply in my work. For instance, in the campus of the Xiangshan School at the China Academy of Art in Hangzhou, all the functions have been decided after the building was finished! However, in China there is not yet a capacity to make modern architecture compatible with the conservation of historic areas. We need to understand that changes have to gradual, and not brutal and sudden.

Question: *You are very interested in Chinese painting and you also practice painting. How does this influence your work?*

Answer: This is a complex theme. I think painting allows you to understand the meaning of nature, and this is the main goal of Chinese painting. It's a philosophical approach to nature, a way to contemplate. In our tradition, painters did not work for a public or to become famous, they did it for themselves, to be able to understand nature. I practice calligraphy every day. It helps me seeing design not as a static process, but as a dynamic one, as the identity of a place includes the dimensions of time and the plurality of the views of the users, citizens and visitors.

Case Study *Bologna: From Urban Restoration to Urban Rehabilitation*[14]

Patrizia Gabellini

Architect and Professor of Urban Planning, Politecnico of Milan, Italy

From Monuments to Urban Centres

In considering the approach to the conservation of historic areas in Europe, with a particular reference to Italian architecture and planning, we can identify three main historical periods, each with its own design principles: the nineteenth century, the first decades of the twentieth century, and then the period after the 1960s.

In the second half of the nineteenth century, parallel to the major urban transformations linked to industrialisation, the emphasis was put on the unsuitability of the old urban forms for the new urban functions, and the result was the demolition of the old fabric and the preservation only of the isolated 'monuments'. At the end of the first decade of the twentieth century, the emergence of a new historical sensitivity prompted Gustavo Giovannoni's innovative thinking on the environmental value of the 'old town', that he thought could be thinned out while maintaining the urban context necessary to support the meaning of the monuments.

Since the 1960s, the preservation of the historic parts of cities attracted the attention of analysts and planners, in a crescendo of initiatives culminating in the Symposium on Urban Heritage organised by the Council of Europe in 1975. The recognition of the special nature and of the fragility of the 'historic centre' due to extensive processes of decay and abandonment prompted the development of legislative measures and planning tools for the conservation of historical buildings.

Monuments, old towns and historic centres, express different visions which, over a century of urban history, have been matched by different planning approaches: first an extensive urban renewal process that has disrupted the pre-existing city, then a process of micro-changes, and finally the protection and conservation of areas carried out with philological attention.[15]

From the Historic Centre to the Historic City

A new planning concept came about in Italy at the end of the last century, at the time of the making of the Master Plan of Rome, the Italian city with the greatest history.

The studies carried out at that occasion, as well as the design practice accompanied by theoretical reflections allowed replacing the concept of 'historic centre' with the one of 'historic city'. This semantic change reflects an expansion of the concept of 'historicity' to include all parts of the city with historic quality present in a wider territory. Historicity, freed from a predetermined time frame (how many years are needed to define a building or an urban area as historic?) is understood as a 'recognised value' of the past, linked to uniqueness, utility, testimony, beauty.[16]

Historicity becomes an attribute that applies to everything that you do not want to lose because it is part of the memory and identity of an area. It follows that the value is not only limited to the city core, typically the 'historic centre', as the idea of historic city requires extension and articulation.

---

[14] First published in Aa.Vv., 2012. *Dossier: Città storica città metropolitana.* 'I martedì di San Domenico', Bologna, N. 7: 10–12.

[15] The development of the methodology of typological intervention was under the theoretical and practical leadership of Bologna. See for this: Cervellati, P. and Scannavini, R. (1973) *Bologna: politica e metodologia del restauro nei centri storici.*, Bologna, Il Mulino.

[16] See: Gasparrini, C.: 'Strategie, regole e progetti per la Città Storica' and Manieri Elia, M.: 'La Città storica struttura identificante', both in *Urbanistica*, N. 116, 2001. In Evangelisti, F., Orlandi. P. and Piccinini, M. (eds) (2008) *La città storica contemporanea*, Bologna, UrbanCenterBologna.

Historic centres differ in size and in their typological and morphological structures, historic cities also are a mix of different parts, each of which is worthy of attention. Compared to the city core, the 'historic city' is discontinuous, multi-faceted, scattered in the territory and mixed with ordinary parts. The historic city includes the ancient nuclei (even those who have never become 'centres' of modern settlements); the nineteenth century workers neighbourhoods; settlements with special functions; the garden cities; the public housing neighbourhoods that have embodied the ideas of the modern city (rational or organic), as well as the small settlements and buildings, isolated or in clusters that dot the rural landscape. These areas are different as regards layout, land subdivision, articulation of built and open areas, presence of ordinary and special buildings, and layering of uses, which speak to us of their past and require specific projects to ensure their future.

This repositioning marks the end of a planning approach based on districts: preserving the historic city involves a plurality of actions and requires rehabilitation projects that embrace all its different components. In this sense, urban regeneration becomes a logical corollary to urban conservation (Figures 3.12, 3.13 and 3.14).

The Case of Bologna

Bologna has played a leadership role in the conservation movement of historic centres since the 1960s and can be considered a good example of the evolution of the approach to the historic city, as reflected in the planning instruments developed between 2007 and 2009: the Structure Plan and the Planning and Building Regulations, prepared according to the Planning Law of the Emilia Romagna Region N. 20 of 2000.

Figure 3.12 Aerial view of the core of the historic centre of Bologna. Photo courtesy of Comune di Bologna, Assessorato all'Urbanistica e all'Ambiente.

Figure 3.13 The Historic Centre of Bologna in the 1960s Conservation Plan. Photo courtesy of Comune di Bologna, Assessorato all'Urbanistica e all'Ambiente.

The first step of the Structure Plan was the identification of the two 'Cities of the Via Emilia' (West and East) to shift the focus to the urban and metropolitan dimension of an historic city organized along the main historical infrastructure of the Po valley (the Roman Via Emilia), to confirm that its development strategy was connected to a wider spatial system. This structural link of the historic city with the rest of the territory is a basic condition to give a new meaning to the rehabilitation strategy.

The second step of the Structure Plan was the identification of a plurality of 'historic areas': the historic centre, the garden cities of the foothills, other compact urban settlements. Each of these areas requires a different planning framework, in order to value the specific typological and morphological characteristics and identify the possible uses.

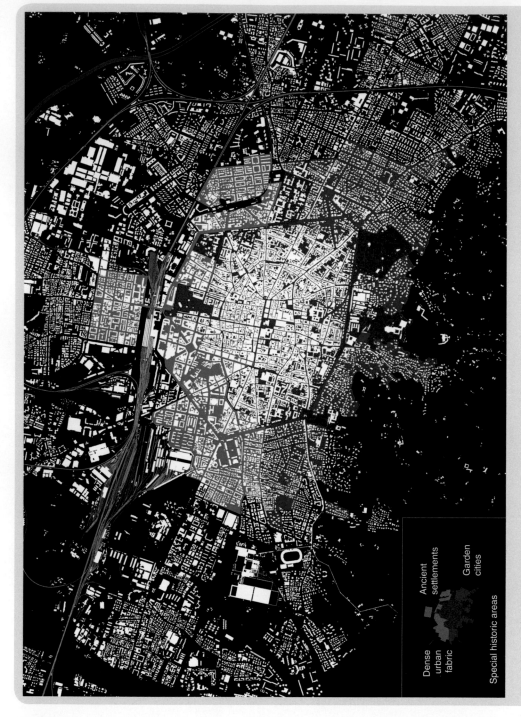

Figure 3.14 The Historic City as seen in the new Bologna Master Plan. Graphic by Patrizia Gabellini.

The third step, implemented by the Planning and Building Regulations, involved a census of the historical heritage in the entire municipality, that led to adding the new zones comprised in the so called 'consolidated areas' to the parts of the city already recognized as historic. These parts are the result of modern city planning, of interventions done in the fifties, sixties and seventies.[17]

Today we have theories, methodologies and techniques that allow us to include the historic city in the rehabilitation process at the regional scale, and to use it as a powerful antidote to homogenizing trends. In spatial planning, in fact, these settlements become important resources to promote diversification and adaptability, as their multiple forms can sustain different reuse models able to support a variety of needs and lifestyles.

---

[17] Bonfantini, B. and Evangelisti. F. (eds) (2009) *Bologna. Leggere il nuovo Piano urbanistico. Psc + Rue + Poc*, Municipality of Bologna, Ferrara, Edisai.

4 Historic Cities and Climate Change[1]

Anthony Gad Bigio

Urban Advisor and Adjunct Professor of Urban Resilience and Low-carbon Cities at the George Washington University Graduate Program on Sustainable Urban Planning in Washington DC, USA

We have embarked globally on a path of unsustainable development. Our lifestyles, the way we produce goods and services, are all part of a system that is completely unsustainable. I see solutions to climate change leading to a much larger philosophical shift in the way human society develops. We need a new matrix to define what human progress is.

Rajendra Pachauri

The Emerging Challenges

It is beyond scientific dispute that global warming is occurring at a faster pace than previously predicted, and that its impacts on the planet are taking their toll on the natural environment as well as on human welfare systems. Cities, where already more than half the population of the planet resides, are contributing to climate change as they currently account for about 70% of greenhouse gas emissions, but they are also being increasingly impacted by the consequences of climate change. Urbanisation, which will see an additional two billion people residing in cities by 2050, will dramatically increase this twofold interaction between cities and climate change.

The rate at which such impacts manifest themselves is part of the uncertainty inherent to climate change. Such uncertainty is also related to the pace at which greenhouse gas emissions will – or won't – be curtailed. The most drastic emission reduction policies, if enforced right now, might be able to contain the increase in atmospheric warming within 2°C, the envelope within which, according to most scientists, humankind can still adapt without major disruptions. In the absence of radical abatement policies, global warming may increase to 4°C or

[1] This chapter draws from: Bigio, A., Ochoa, M. C. and Amirtahmasebi, R. (2014). Climate-resilient, Climate Friendly World Heritage Cities, *Urban Development Knowledge Series Papers*, Washingon: The World Bank. The author is indebted to his colleague Katie McWilliams who prepared the maps for the original report and paper.

Reconnecting the City: The Historic Urban Landscape Approach and the Future of Urban Heritage, First Edition. Edited by Francesco Bandarin and Ron van Oers.

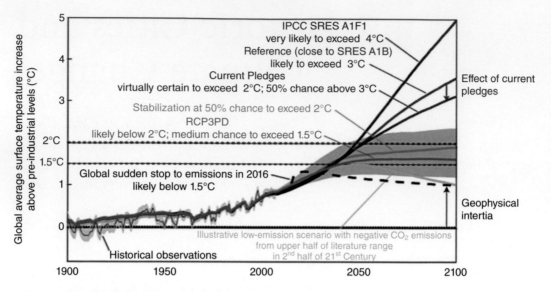

Figure 4.1 Global average surface temperature increase. Source: 'Turn down the heat: why a 4° world must be avoided': a Report for the World Bank by the Potsdam Institute for Climate Impacts Research and Climate Analytics, November 2012. Reproduced by permission of the World Bank.

higher by the end of the century or sooner, with catastrophic consequences for the planet (Figure 4.1).

The impacts on cities directly related to climate change, include sea-level rise, coastal erosion and marine inundation; increase in the frequency and intensity of coastal cyclones and hurricanes; intensification of the urban heat-island effect and of periodic heat waves; more frequent and intense episodes of precipitation and related flooding and mudslides; changes to the water cycle with consequences on fresh water availability; the deterioration of urban eco-systems; and the increase in vector-borne diseases with impacts on urban public health. Such climate hazards are compounded by natural hazards already threatening cities, such as earthquakes and tsunamis. It is sometimes impossible to distinguish the root causes of a catastrophic event, such as a major flood, which can be considered either as a 'natural' or as a 'climatic' event, or both. The results, of course, can be equally devastating.

Urban risk assessment methodologies have been developed in order to identify and measure the likelihood and intensity of risks deriving from natural hazards and climate change impacts on a given city. The urban risk is considered to be the result of the magnitude of the hazards, compounded by the level of exposure of the local population and assets to the hazards, and the degree of vulnerability of population and assets by discreet segments or locations.

Historic cities are often subject to a high level of urban risk. This is due to their often coastal location, a consequence of their foundation as maritime trading hubs, as well as to the special vulnerability of their heritage assets and fragile urban fabrics. In developing countries, historic urban cores are often the location of poor and marginalised groups, who reside and earn their livelihood in often degraded, decaying and congested areas, of particular vulnerability to the impacts of climate change.

The assessment of urban risk is the basis to identify measures to increase the urban resilience of a given city. Urban resilience is defined as the capability of urban systems to prepare for and respond to the risks and impacts of natural hazards, climate variability, and climate change.

World Heritage Cities

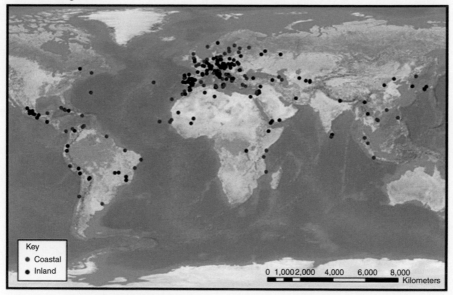

Figure 4.2 World Heritage Cities examined in the study. Source: 'Climate-resilient, Climate-friendly World Heritage Cities', reproduced by permission of the World Bank.

It can be achieved via 'soft' measures, such as land use and urban planning, community awareness and preparedness, monitoring of hazards and risks, early warning systems, emergency and evacuation plans, as well as 'hard' or physical interventions, such as the retrofitting of critical infrastructure, adapting buildings and urban spaces, managing retreats and relocation from at risk areas, and maximising eco-systems services.

Urban risk assessments and urban resilience plans have to be adapted to the very special nature of historic cities and of urban cultural heritage assets. The process of urbanisation has transformed most historic cities into historic cores of much larger agglomerations, ranging from a few hundred thousand inhabitants to many million. The assessment of urban risk and the response in terms of adaptation to climate change and resilience to natural hazards will have therefore to be developed for the entire agglomeration, with a special lens and attention to the valuable historic parts of the city. This is congruent with the Historic Urban Landscape approach, which recommends an integrated vision and management of the entire city, encompassing its historic components and recognising their uniqueness.

Exposure of World Heritage Cities to Multiple Hazards

The 238 World Heritage cities represent a special cluster of urban centres (Figure 4.2). They are identified as part of the UNESCO World Heritage List, where they were progressively inscribed in the 40+ years since that specific Convention was established, at the request of the Governments of the countries they belong to, and with the approval of the World Heritage Committee, which is the custodian of the list. They represent about one-fourth of all World Heritage sites, which include natural sites as well as cultural ones, and the reasons for their inscription are related to the unique character of the urban fabric and of the historic buildings that compose them.

Like all cities in the world, World Heritage cities are vulnerable to the increasing impacts of climate change, which compounds their exposure to natural hazards. Climate change risks vary very much by location. Coastal cities (e.g. Cartagena), which represent one full third of World Heritage cities, are exposed to increasing sea-level rise, storm surges, marine inundation, and coastal erosion risks. Other World Heritage cities may be exposed to the consequences of glacier melt (e.g. Quito), or of increasing ambient temperatures with impacts on human health via heat waves and increased ambient pollution. Urban flooding and landslides are prevalent risks for many World Heritage cities, as climate change in many parts of the world induces increased precipitations and extreme weather events.

The vulnerability of World Heritage cities to natural hazards and climate change is higher than the one of modern cities, on account of the particular fragility of their urban fabric and of the ancient buildings, which they are composed of. The vulnerability of a city is measured via its physical hazards, but also via the ability to respond of its residents and local institutions. In general terms, World Heritage cities in Europe and North America are inhabited by well-off residents, real-estate values are high, and regulations and institutions are in place for their protection and conservation. Conversely, in emerging economies and developing countries, World Heritage cities are mostly inhabited by poorer populations, real-estate values are low, and resources for conservation are insufficient.

A World Bank team conducted a detailed review in 2011 of the risk exposure of World Heritage cities, using the Global Risk Data Platform PREVIEW of the United Nations Environment Program (UNEP), initiated in 1999 by UNEP/GRID-Geneva. The Global Risk Data Platform has now evolved following all standards for Spatial Data Infrastructures (SDI) and providing all the web-services in compliance with the Open Geospatial Consortium (OGC). The data currently present in the platform has benefited from new developments made for the Global Assessment Report on Disaster Risk Reduction version 2009 and updated for the 2011 version.

The outcomes were developed by a large, interdisciplinary group of researchers from around the world, making global disaster risk more visible as a key step towards mobilising the political and economic commitment needed to reduce it. Methodologies on hazards modelling were reviewed by a team of 24 independent experts selected by the World Meteorological Organisation (WMO) and the United Nations Educational, Scientific and Cultural Organisation (UNESCO). The platform[2] is supported by the UN International Strategy for Disaster Reduction (UNISDR). The Global Risk Data Platform provides world maps generated on the basis of a Multi-Hazard Index, compounding risks of cyclones, earthquakes, floods and landslides, and classifying all world locations in five categories of risk: (1) low; (2) moderate; (3) medium; (4) high and (5) extreme.

The World Bank team geo-referenced each of the 238 World Heritage cities based on latitude and longitude coordinates. The points were then overlaid with the global Multi-Hazard Risk grid (Figures 4.3 and 4.4). The risk value at each point was extracted and visualised according to the five categories of the Global Risk Data Platform. The specific exposure of World Heritage cities to flooding and landslides (two risks caused or compounded by climate change) was also calculated following the same methodology (Figure 4.5).

It is important to underline that the Global Risk Data Platform provides information only on the intensity of the hazards to which each location is exposed. A complete assessment of risk, as mentioned before, includes the assessment of exposure of assets and population, and of their relative vulnerability. Therefore neither

---

[2] The Platform is accessible at: http://preview.grid.unep.ch

Multi-Hazard Risk for World Heritage Cities

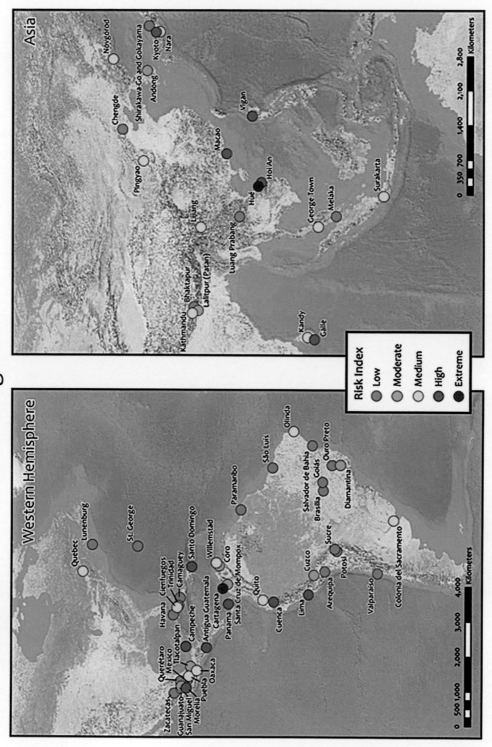

Figures 4.3 and 4.4 Multi-Hazard Risk for World Heritage Cities. Source: 'Climate-resilient, Climate-friendly World Heritage Cities', reproduced by permission of the World Bank.

Multi-Hazard Risk for World Heritage Cities

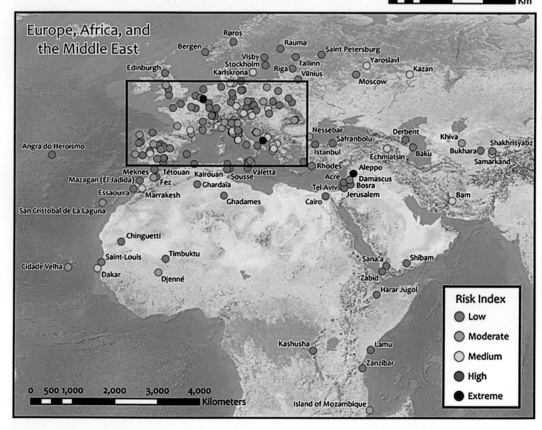

Figures 4.3 and 4.4 (*Continued*)

Flood Risk To World Heritage Cities

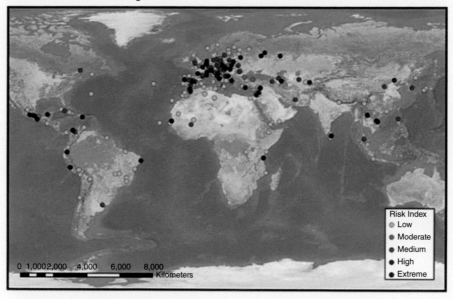

Figure 4.5. Flood Risk for World Heritage Cities. Source: 'Climate-resilient, Climate-friendly World Heritage Cities', reproduced by permission of the World Bank.

Table 4.1 Top ten World Heritage cities by multi-hazard risk.

Ranking	City	Country	Region
1	Hué	Vietnam	Asia and the Pacific
2	Aleppo	Syrian Arab Republic	Arab States
3	Cartagena	Colombia	Latin America and Caribbean
4	Amsterdam	The Netherlands	Europe and North America
5	Mostar	Bosnia and Herzegovina	Europe and North America
6	Morelia	Mexico	Latin America and Caribbean
7	Santo Domingo	Dominican Republic	Latin America and Caribbean
8	Paris	France	Europe and North America
9	Lima	Peru	Latin America and Caribbean
10	Vilnius	Lithuania	Europe and North America

the maps, nor the rankings, should be taken as much more than a first approximation, one which is nevertheless very useful in identifying which World Heritage cities more than others have to be concerned with natural hazards or climate change impacts.

The following rankings are directly derived from the UNEP/GRID described above, as applied to the geo-referenced World Heritage cities. It should also be noted that alternative methodologies may generate different rankings

of World Heritage cities most at risk from the ones shown in Tables 4.1, 4.2 and 4.3.

Historic Cities and Urban Resilience

Ensuring the resilience to natural hazards and adaptation to climate change of historic cities is a complicated challenge. This is particularly true in emerging economies and developing countries, where national institutions and financial

Table 4.2 Top ten World Heritage cities by flood risk.

Ranking	City	Country	Region
1	Hué	Vietnam	Asia and the Pacific
2	Aleppo	Syrian Arab Republic	Arab States
3	Kandy	Sri Lanka	Asia and the Pacific
4	Baku	Azerbaijan	Asia and the Pacific
5	Hoi An	Viet Nam	Asia and the Pacific
6	Lima	Peru	Latin America and Caribbean
7	Cartagena	Colombia	Latin America and Caribbean
8	Saint-Louis	Senegal	Africa
9	Moscow	Russian Federation	Europe and North America
10	LuangPrabang	Lao People's Democratic Republic	Asia and the Pacific

Table 4.3 Top ten World Heritage cities by landslide risk.

Ranking	City	Country	Region
1	Mostar	Bosnia and Herzegovina	Europe and North America
2	Santo Domingo	Dominican Republic	Latin America and Caribbean
3	Dubrovnik	Croatia	Europe and North America
4	Oporto	Portugal	Europe and North America
5	Cuenca	Ecuador	Latin America and Caribbean
6	Antigua	Guatemala	Latin America and Caribbean
7	San Marino	Republic of San Marino	Europe and North America
8	San Cristobal	Spain	Europe and North America
9	Berat	Albania	Europe and North America
10	Macao	China	Asia and the Pacific

resources for urban development and historic city conservation may be limited. Local Governments responsible for historic cities already have the special responsibilities of managing the local cultural heritage assets, protect them from encroachment and decay, and contribute to their conservation as part of the economic, social and cultural life of the city.

These tasks come in addition to the general responsibilities of managing urban growth, delivering urban services, mobilising financial resources, and ensuring urban livability of the entire urban agglomeration. Increasing urban resilience to natural hazards and adaptation to climate change of historic cities are additional tasks for which Municipalities need all the support they can get, from national governments and from international institutions alike.

Adaptation and Resilience Action Plans define responses to climate change and natural hazard risks. They are the result of a combination of technical assessments, consultations among local and national stakeholders, and careful operational and financial planning. Once defined, climate action plans need to be supported politically, institutionally, and financially, in order to get implemented. In the case of historic cities, climate action plans have to:

- **Concern the entire urban agglomeration to which the historic city belongs**: Many of the hazards and risks concern the entire agglomeration and the responses must equally be found at the urban scale, while addressing the specific vulnerabilities and risks of the fabric, buildings and human activities of the historic city. UNESCO's Historic Urban Landscape Recommendation should serve as guidance and provide guidelines in this respect.

- **Integrate resilience and adaptation within urban development and conservation plans**: Rather than planning separately in order to deal with natural hazards and climate change risks, resilience and adaptation responses should be incorporated in the ordinary urban planning instruments and in the conservation plans and regulations.
- **Be based on priority risks and incorporate specific future climate scenarios for their locations**: While some risks are common to clusters of historic cities, such as coastal cities, ultimately each city is a unique case which requires careful forecasting of future climate change impacts for the location, which can only be obtained with down-scaling models and detailed risk assessments.
- **Integrate resilience and adaptation to mitigation of green-house gas emissions**: Many actions that are required to improve the adaptation of a World Heritage city to climate change also contribute to reducing its GHG emissions. For instance, thermal insulation of the building envelopes to deal with increased ambient temperatures will also reduce the energy consumption, thus emissions.

Historic Cities and Climate Change Mitigation

While cities, including historic cities, have to adapt and become overall more resilient to a combination of natural hazards and climatic risks, they equally should play a major role in the reduction of greenhouse gases they emit. Ultimately, no adaptation is manageable if the trajectory of GHG emissions continues unabated and causes an unprecedented warming of the planet beyond what is considered as the acceptable limit of 2°C.

However, not all cities emit the same amount of GHG per capita. That amount will depend on many factors, including local climate and geographic location, average GDP per capita, lifestyle of the local population, patterns of production and consumption, land use, urban form, density, transport systems, and sources of energy, among other factors. Historic cities, or cities which have a historic core, tend to have lower GHG emission per capita. For instance, where transport-related energy-consumption is used as a proxy for GHG emissions, European cities score better, when compared with American or Australian cities, given their higher urban density and historic nature.

While in many cases the historic cores represent only a fraction of the entire urban agglomeration, they have often driven the patterns of urban growth and determined the ways in which the cities have kept on expanding. Given its urban features, even if embedded in a much larger agglomeration, the historic city maintains unique characteristics which preserve it as a naturally energy-efficient and climate-friendly urban area, which combines cultural qualities with urban livability. Such characteristics should be preserved and inspire the design of future urban development, which should be based on low GHG emissions criteria. Among such characteristics are:

- **Dense network of paths**: Historic cities tend to have a more intensive and narrower road network, as well as smaller-size blocks and buildings, all of which favour less use of space and energy per capita. Greater urban density generally translates into a more vibrant city life as well.
- **Buildings with passive temperature-regulating systems**: In the absence of abundant energy sources to heat or cool, historic buildings have generally been designed and built with passive systems and materials providing thermal inertia, and therefore are highly adapted to local climatic conditions while emitting relatively low amounts of carbon.

- **Zoning that fosters accessibility to mixed services with less vehicle travel**: Historic cities favour mixed land-uses, which minimise the distance from residential to productive or commercial areas, which results in compact centres with short commutes. These characteristics make the historic cores attractive and sought after by residents.
- **Mobility by sustainable modes**: The compact structure of historic cities allows for pedestrian and non-motorised mobility and for the use of public transit as the preferential transportation modes, as opposed to the use of private motorcars, in view of the size of the road network and limited parking opportunities.
- **Adequate vehicle travel demand management**: Historic cities precede the automobile era, therefore their streets offer limited space to accommodate on-street parking. Limiting on-street parking is considered to be the second-best travel demand management strategy after congestion pricing, that is charging for individual vehicle access to the historic city.

Converting central areas into pedestrian zones is the 'low-hanging fruit' of GHG mitigation in historic cities. Creating such car-free environments promotes energy efficiency and economic growth, while creating space for citizens to experience the historic core. Many historic cities have made great strides as of the 1960s to transform their historic centres into pedestrian zones in which to different extents and under different regulations private vehicles have restricted circulation. J.H. Crawford, the champion of the 'car-free' concept, has developed a list of noteworthy care-free areas. To be included in the list a sizeable part of the city should be car-free or should have a car-free area that is of an exemplary nature for the national context. World Heritage cities dominate the list, with cities such as Vienna or Graz in Austria, Strasbourg, Lyon, and Mont St. Michel in France, Rhodes in Greece, Budapest in Hungary, Siena, Rome and Florence in Italy, Guanajuato and Mexico City in Mexico, Istanbul in Turkey, and Tel-Aviv in Israel.

The implementation of the Historic Urban Landscape Recommendation can therefore become the conduit for a broader integration of climate change adaptation and mitigation concerns in the management of the existing city and in the planning of future urban development. The special vulnerabilities of historic cities and historic urban cores to climate change impacts and to natural hazards have to be taken into account, as well as the opportunities they offer towards a low-carbon future, in the achievement of which cities in general play a major role.

Historic Cities and Climate Action Plans: The Case of Edinburgh, Scotland

Edinburgh is a congruent juxtaposition of organic medieval and planned neoclassical town urban systems. The combination of Old and New Towns of Edinburgh was listed as World Heritage in 1995. The Old Town represents the Scottish enlightenment while the new town exemplifies its thinking and ideals. The buildings in the Old Town have continuous and unified façades, neatly bordering the streets. Behind the streets, the urban fabric tends to be more fragmented and organically shaped, with random enclosed gardens scattered across the fabric. On the other hand, the buildings of the New Town were built upon a plan with a precise street network, which is laid out in a hierarchical way. The buildings shape a harmonious façade with consistent material (sandstones and slate roofs) and shape.

What makes Edinburgh a unique historic city is that the World Heritage site includes a large portion of the city centre, which functions as a living urban environment (Figures 4.6 and 4.7). The World Heritage status has helped the city with economic development and has highlighted the importance of the city as the political and economic heart of Scotland. A city of nearly 500 000 inhabitants, Edinburgh is the second largest tourist destination in the UK generating

Figure 4.6 A view of the centre of Edinburgh, Scotland, inscribed in the UNESCO World Heritage List.

about £2 billion income a year from tourist receipts.[3] The city constantly renews itself by long-term investments in building conservation and quality new buildings.

Risks

The Edinburgh Management Plan acknowledges that there are threats to the World Heritage site posed by the impacts of climate change. Among these impacts are fires and floods. While fires are not predictable, the flood risk has been studied and identified in a limited part of Edinburgh's World Heritage site around the Dean Village and Stockbridge. It is estimated that Scotland will

experience about a 20–30% increase in peak river flows by 2080.[4] In addition, climate change can impact the architectural quality of the World Heritage city. For example, it can speed up stone decay, or can cause possible damage to the integrity of historic buildings from mitigation actions to reduce emissions.

Actions

In order to address the risk of fire damage to the Old and New Towns of Edinburgh, the City of Edinburgh Council has prepared an information catalogue, which outlines methods to reduce risk of fire, fire safety management, and suppression

---

[3] *The Old and New Towns of Edinburgh World Heritage Site. Management Plan 2011–2016*. Edinburgh: Edinburgh World Heritage, Historic Scotland and the City of Edinburgh Council, 2011.
[4] http://www.scotland.gov.uk/Resource/Doc/195446/0052419.pdf

Figure 4.7 The centre of Edinburgh, Scotland.

and detection systems. In addition, the Council works with the Fire and Rescue Services to develop and manage a database on historic buildings. This database provides fire-fighting crews with information on the importance and value of category 'A' listed buildings.[5]

Flood risk mainly comes from 'the Water of Leith', which is the main river of Edinburgh. This threat became clear when after a period of heavy rainfall in April 2000, the riverbanks broke and about 500 neighbouring historic properties were damaged. In order to prevent similar events, the city developed a flood prevention scheme in 2003 (with a revision in 2007) with regards to

effects of climate change in the next 200 years. The 'water of Leith' management plan acknowledges the role of river, not only as a natural resource but also as an integral part of the World Heritage of Edinburgh. While aiming at maintaining a clean urban river, the plan emphasises promoting awareness about Leith's natural, cultural and historical heritage. The management plan also aims at identifying, safeguarding, and promoting features of archaeological and cultural importance.[6]

The Water of Leith scheme emphasises building flood protection walls, sound reservoir management and maintenance of flood plains, by

---

[5] The Edinburgh World Heritage Site Management Plan has divided the buildings in the Old and New Towns into 3 categories: A, B, and C, based on their architectural and historic significance.
[6] The Water of Leith Management Scheme. http://www.waterofleith.org.uk/management

restricting development where necessary. Water reservoirs were designed originally to supplement low water-flows for mills during dry seasons. Today, they are used to protect the city against downstream flooding. By discharging extra water from the reservoirs, they can play an active role in flood protection. The water scheme also includes a rain retention section, which suggests contour ploughing, tree planting, urban agriculture, and creation of wetlands. The water scheme will also contribute to improved water quality, while establishing limitations against additional development, which may lead to a significant increase in the risk of flooding.

The city of Edinburgh has prepared an extensive management plan for the World Heritage site. This plan specifically addresses climate change issues under the Scottish Government's agenda regarding climate change. The Scottish Climate Change Programme was published in March 2006, with ambitious goals of leading the Northern European Climate Change agenda. The goal of the plan is to reduce the current carbon emissions by 26.7% by 2050 (compared to the 2005 baseline). The World Heritage Site Management Plan also establishes a relationship between the natural and cultural heritage of Edinburgh and acknowledges the role of natural heritage in the universal value of Edinburgh, which consists of architectural spaces and gardens. Therefore, the open spaces within the site and those on its edges, along with Water of Leith contribute to this setting.

The Edinburgh Biodiversity Action Plan is another instrument designed to provide a framework for ensuring that the integrity of Edinburgh as a whole is conserved as a World Heritage city. The Old and New Towns of Edinburgh World Heritage Site Action Plan has further actionable measures. It plans to address the issues of fuel poverty and changes of energy behaviour in the Edinburgh World Heritage city and to develop mechanisms for adapting historic buildings to reduce carbon emissions.[7]

In 2006, to assess energy demand for heat and electricity and develop emissions reductions actions, the City of Edinburgh Council commissioned a report called 'Powering Edinburgh into the twentieth century'. The report assumes that the target for building-related emissions is the same as the target for overall emissions, including those from transport and industry. Due to population increase and demolition of some old housing units, Edinburgh is expected to need 27 200 new dwellings by 2025. Meanwhile, the city plans to improve insulation in existing units to decrease heat demand by 10%, while the average domestic boiler efficiency is assumed to increase to 86%. Electricity demand per dwelling is assumed to remain the same. The new dwellings are expected to have considerably lower heat demand and 20% less electricity demand, while their average domestic boiler efficiency is assumed to be 92%. Overall, Edinburgh is expected to have 25% reduction in annual heat demand and 30% reduction in electricity usage by 2050.[8]

'Powering Edinburgh into the twenty first century' suggests that the most certain way the city can meet its emissions reduction goals is to proceed with the Decentralized Energy (DE) scenario.[9] This scheme includes gas-engine combined heat and power (CHP). CHP has several environmental benefits. The plan suggests that Edinburgh work on a community heating system (CH) to encourage the usage of CHP. The historic sections of the city can be barriers to this initiative. However, the plan suggests that the operation cannot be more invasive than establishing pipe utilities or communication networks. Community Heating networks have been successfully installed in other historic cities in Europe such as Copenhagen, Amsterdam and Paris.

[7] *Edinburgh World Heritage Site Action Plan*. Edinburgh: The City of Edinburgh Council, August 2011.
[8] *Powering Edinburgh into the XXIst Century*. City of Edinburgh Council, WWF Scotland and Greenpeace, November 2006.
[9] DE includes a mix of decentralised energy sources and conventional centralised energy generation.

Interview with Filipe Duarte Santos
Looking at the Challenges of the Urban Century

Professor of Physics and Environmental Sciences at the University of Lisbon and Director of the Research Centre SIM – Systems, Instrumentation and Modelling in Environmental and Space Sciences and Technology

22 April 2013, Lisbon, Portugal

Question: *Your work* Humans on Earth – From Origins to Possible Futures *(Berlin: Springer-Verlag, 2012), in which you take a broad and sweeping look at the evolution of the human species in relation to the history of science and its role in shaping our living environment, can be placed among similarly ground-breaking works where the combination of knowledge from different disciplines results in new insights into established theories and histories, such as James Lovelock's* Gaia Hypothesis *(1972), Ken Wilber's* A Brief History of Everything *(1996), Jared Diamond's* Guns, Germs and Steel *(1999), Brian Fagan's* The Long Summer *(2004) and Marc van der Erve's* A New Dimension of Time *(2007). What motivated you to write the book?*

Answer: The sociologist Max Weber, in his lectures on 'Science as a Vocation' (1918), stated that the knowledge resulting from the application of scientific methodology would lead to a systematic and global disengagement, and ultimately to the loss of any other values or meanings than those which are strictly technical and utilitarian. While we may argue that this actually has happened, there's nevertheless an urgent need to underline the huge success of science and technology for over three hundred years, which has contributed to the enormous improvement of our lives during this period. However, our remarkable civilisation – which is flourishing mainly in the industrialised countries – is achieved through intensive energy use made possible by the availability of relatively cheap fossil fuels. This is now coming under severe pressure, with more countries entering the industrialisation phase and larger population groups enjoying greater wealth, foremost in Asia, but also in practically all other regions of the world outside of the West. The increasing scarcity of fossil fuels for energy, with no viable alternative as yet to replace them, will coincide with continued population growth, projected to reach nine billion in 2050, and increasingly severe weather phenomena related to climate change, such as droughts and flooding, which will put large population groups at extreme risk. All this will create the likelihood of multiple interlocking crises that will put our civilisation to the test.

To be able to meet these future challenges, it is essential to know the long history of our past and the conditioning factors of the present. As I explain in the book, a very significant part of our common future is inevitably constrained by the biological characteristics of our species Homo Sapiens, but probably a larger part will depend on our capacity for cultural evolution. But this cultural evolution is not assisted by the coexistence of contradictory values associated with the currently dominant model of development based on growth, the worldwide dissemination of neoliberal forms of capitalism, and supra-national institutions influencing economic policies, with very limited success. Thus, it is critical to promote information, communication, dialogue, and sharing of experiences between disciplines and the diverse perceptions on the environment and human development, which will be a very important contribution to the creation of a well-founded and diverse global conscience, with adequate strength and resilience to face current and future environmental challenges. In this there's a need for scientists to make a stronger effort to inform, communicate, and explain to the general public what they are doing, and on top of that to highlight and defend the methodology of science to arrive at conclusions, as opposed to pure conjecture and populist speculation.

Question: *What do you suggest is the role of heritage?*

Answer: There are vast areas of human experience that are impenetrable to abstract analysis and mathematical reasoning is incapable of learning or understanding the contradictory nature of human experience, which lies outside the scope of scientific knowledge. But heritage and culture are essential for inspiration and innovation in the search for solutions to environmental and societal crises. After all, as I write in the final pages of the book, artistic creation is one of the defining characteristics of our species.

Question: *Let's go back to the opening of your book, where you quote Karl Popper in saying that 'Optimism is our duty. We are all co-responsible for what is coming'. Thinking about this powerful statement, I cannot help but wonder about the large discrepancy between our magnificent technological progress and our rather poor sociological progress, with continued family, clan and ethnic group thinking and behaviour, which is at the root of most, if not all conflicts. Given the amount of time and money spent, annually and globally, on weapons as compared to education, isn't pessimism actually the norm?*

Answer: Research has indicated that the development of a more interactive and complex social life of Homo Sapiens was based on alliance building and collective responsibilities in the group for activities such as hunting, food sharing, protection, and intergroup fighting, which improved the capacity for survival and opportunities for reproduction. However, the process eventually became irreversible because, by reaching a certain level of social complexity, new survival pressures within the groups would have worked in such a way as to increase complexity even further and the lineage of Homo Sapiens got trapped in an evolutionary ratchet from which there was no going back. Success meant that our main allies and opponents became members of our own communities; in fact, we became our own prisoners, in a self-sustainable and one-dimensional process determined mainly by internal mechanisms of social behaviour rather than by the external natural environment. A recent study by Bowles (2009) supports the idea that violent intergroup competition and lethal conflict may have contributed in significant measure to the proliferation of a genetic and cultural trait to behave altruistically towards the other members of one's own group. Altruism in its origins, according to this point of view, is a kind of response and adaptation to the rigours of warfare. Thus, social adaptation indeed has been a long, complex and arduous process of selection, and is not characterised by the quantum leaps in technology.

Question: *So, it comes back to cultural evolution again – isn't that in fact the key message of your book?*

Answer: One of the key messages, indeed. In the first chapter I posit that the attainment of ethical concepts and practices and progressive forms of altruism, related to a given culture, requires a continuous process of learning and integration into that culture. Altruism for outsiders to the social group of one's belonging is a relatively recent type of behaviour in the history of mankind, which is not innate but results from an educational and cultural process. From a wider point of view, the communication, acceptance, and practice of the basic social principles of modern society, such as tolerance, equality, democracy, and human rights, is only possible through a permanent effort of cultural learning. Education, science, communication and cultural dialogue, all of them UNESCO's fields of competence, are thus essential for social advancement.

Question: *As regards UNESCO's Historic Urban Landscape approach, the critical idea behind it is connection: of different professional disciplines concerned with the management of the (urban) environment, of different communities and their value systems, of the past and the future, of culture and science – or more accurately: re-connection, as this was generally the case in pre-industrial times, but which got severed by Modernism.*

Answer: The Historic Urban Landscape approach, as I see it, is part of applied research, where the objective is to search for processes, products, or instruments that will be useful in specific economic, functional or social contexts. Its emphasis on cultural diversity to seek solutions rooted in the local context is extremely relevant since each culture appropriates and integrates the technologies it uses in its own way, striving to develop them in directions that are more beneficial to the pursuit of its general and specific interests – including policies and actions for heritage conservation and urban development. Historic Urban Landscape is a new paradigm and in this regard it's important to remember Kuhn's observation (1962) that the emergence of a new paradigm, which replaces a previous one, depends on sociological forces also, capable of determining when and how it may be fully accepted. This represents a profound break from the strict concept of science in which the discourse of scientists bases itself on undeniable evidence of an experimental or observational nature. As a matter of fact, the dialogue and controversy between scientists in the social context in which they find themselves at the very least influences the time required for a new theory to be accepted.

Question: *One of the critical steps in implementing the Historic Urban Landscape approach is the establishment of vulnerability of heritage resources, both in terms of socio-economic pressures and climate change. Climate change research and the development of strategies for adaptation and mitigation is part of your current research activities, and the reason for which you are sitting on the United Nation's Intergovernmental Panel on Climate Change, that is now writing its 5th Assessment Report to be published in 2014.*

Answer: Indeed, we have developed scenarios, impacts and adaptation measures for climate change in Portugal (2002), for the town of Sintra, a World Heritage site (2009), and most recently as part of the report Urban adaptation to climate change in Europe, published by the European Environment Agency (EEA) in March 2012.

Question: *The Historic Urban Landscape approach is still poorly understood – the primary reason for us to compile this book, while its six steps for implementation are perhaps so straightforward that they may be seen as too obvious, and thus overlooked. What is your opinion on the Historic Urban Landscape approach?*

Answer: The Historic Urban Landscape approach in my view is a highly relevant attempt to put the management of urban environments on to a more systematic footing. The systematic application of science and technology to seek sustainable solutions for development will demand new conceptual structures and new methodologies, as well as a new willingness by scientists of the natural and social sciences, engineers and planners to integrate the natural and social sciences, which is proposed also in Historic Urban Landscape. The role of science is to handle the functional complexity of so-called socio-ecological systems, a term coined by Jeffrey Sayer and Bruce Campbell in 2003 to indicate systems that unite human communities and ecosystems. These systems are generally characterised by multiple and wide-ranging spatial and temporal dimensions of mutual interaction and response to internal and external stimuli. Responses to such a situation are difficult to model and simulate, because they are more or less delayed, frequently non-linear and generally affected by significant uncertainties – I guess what you have described as the moving target of contemporary urban conservation. But in spite of these complexities and uncertainties, and the time it will take for this approach to be fully integrated into planning strategies, Historic Urban Landscape is probably going to prove a next significant step in urban management.

5 The Intangible Dimension of Urban Heritage

Rohit Jigyasu

UNESCO Chair Professor at the Institute for Disaster Mitigation of Urban Cultural Heritage at Ritsumeikan University, Kyoto, Japan, and Senior Advisor at the Indian Institute for Human Settlements (IIHS) in Bangalore, Karnataka, India

You can't love a city if you have no memories buried there.

Marina Tavares Dias

Introduction

Essentially, people are actors in heritage, as they continuously reshape the physical form and the environment of the city. Therefore, cultural heritage cannot be extricated from the total urban reality, precisely because history cannot be the sole criterion for extricating heritage from its social fabric. The Historic Urban Landscape approach suggests identification, conservation and management of historic areas within the broad urban context, by considering the inter-relationships of their physical form, their spatial organisation and connection, their natural features and settings and their social, cultural and economic values.

Cities have always been important places where intangible heritage values linked to collective identities are found and represented. Social values, communities' identities, and civic pride are present not only at public and private monuments and buildings, but also in specific places and moments of the collective life.

However, intangible heritage is usually not properly taken into account in the urban and regional planning process. There is a lack of understanding of the complexity and pervasiveness of intangible values, and of their direct relationship with the physical structure of the city. Traditional knowledge tends to be seen as separate from the physical fabric, while the intangible values (individual, social and institutional) gradually vanish with the loss of the meaning of urban spaces that forms the basis of the *genius loci*.

The Historic Urban Landscape approach aims to give intangible heritage values their rightful importance in the process of interpretation,

Reconnecting the City: The Historic Urban Landscape Approach and the Future of Urban Heritage, First Edition. Edited by Francesco Bandarin and Ron van Oers.
© 2015 John Wiley & Sons, Ltd. Published 2015 by John Wiley & Sons, Ltd.

planning and conservation of historic cities. This is to ensure that a full understanding of the role that intangible heritage values can play is associated with planning methodologies and decisions related to the management of the urban environment. In the light of this approach the following key questions will be addressed:

- What are the intangible heritage values of the historic urban landscape?
- Which values are usually represented and conserved and which are not?
- How do intangible heritage values affect, or how are they affected by, urban management practices?
- Which tools can be used and developed to identify and assess the impact on intangible heritage values?
- How can we integrate intangible values in urban planning and management?

We will explore these questions based on examples derived from the historic cities of Varanasi (Kashi) in India and Patan in Kathmandu Valley in Nepal, providing operational indications in order to orient innovative planning and management processes.

Defining Intangible Values in Historic Urban Landscapes

The scope of cultural heritage is no longer limited to monuments, archaeological sites and collections of objects. It also includes traditions and living expressions inherited from our ancestors and passed on to our descendants, such as oral traditions, performing arts, social practices, rituals, festive events, knowledge and practices concerning nature and the Universe and the

knowledge and skills to produce traditional crafts, collectively defined as 'intangible heritage values'.[1]

Intangible heritage is inclusive, representative and community based, and while fragile, it is an important factor in maintaining cultural diversity in the face of growing globalisation. An understanding of the intangible cultural heritage of different communities helps with intercultural dialogue and encourages mutual respect for other ways of life.

The importance of intangible cultural heritage is not the cultural manifestation itself, but rather the wealth of knowledge and skills that is transmitted through it from one generation to the next. The social and economic value of this transmission of knowledge is relevant for all social groups, whether minority or mainstream.

Heritage has two fundamental dimensions, for example as in the human soul and body, where the soul is intangible and the body the tangible part. It is interesting that the reality of the tangible depends upon the intangible soul. Conversely the intangible soul will become meaningless and unrealised if it does not work through the tangible physical body and world. Therefore, the tangible and intangible parts are mutually dependent and complementary.[2]

In historic urban landscapes, the intangible heritage is preserved by the local communities in the form of their knowledge and skills and also through rituals, festivals and other social activities. The tangible components of historic urban landscapes in the form of the built and/or natural environment often play an irreplaceable role in the manifestation of intangible cultural heritage as public buildings, housing and spaces for carrying out collective social activities at city or neighbourhood unit level. Often heritage

---

[1] UNESCO (2013) *Intangible Cultural Heritage?* Available at: www.unesco.org/culture/ich/doc/src/20435-EN.pdf
[2] Goswami, S. (2007) The Conservation of the Intangible Heritage, *Proceedings of the 12th International Conference of National Trusts*, 3rd to 5th December 2007, New Delhi, INTACH.

values for the inhabitants are closely related to the collective memories that are associated with these buildings and spaces. The continuity of these memories is part of the social capital in traditional communities. Social capital is defined as 'social networks or norms'[3] to build up reciprocal interrelationships among individuals in a certain district, to establish common goals and harmonious development. It is an important form of capital with potential power similar to economic capital, environmental capital and cultural capital. Sachs[4] argued that the crucial issue for social sustainability is to sustain existing social structures, territories, and identities. Social capital therefore assesses the social quality of living heritage sites.

Public spaces can take the form of public squares or routes, which are tied to people through intangible heritage defined by religious or spiritual associations and belief systems or memories linked to events that are shared at various levels ranging from a larger community to distinct social groups or families and clans, and even to the individual level.

Intangible heritage in historic urban landscapes also shapes the way boundaries are determined at various levels, ranging from neighbourhoods to cities and even regions. The boundaries are implemented and reinforced through processional routes or spaces that are universally identified, often translating complex metaphysical concepts into more mundane boundaries in various cultural and religious traditions.

For example, the spatial manifestation referring to the integration of the cosmos and human being is known as Mandala in the Hindu belief system, and in ancient times this was incorporated to develop an ideal city in accordance with the cosmological arrangement of space,

better known as sacred space. The sacred space includes spaces that can be entered physically, imaginatively and visually. As in many old and sacred cities in the Oriental world, metaphysics based on cosmological principles has been a dominant force in shaping the historic urban landscape of Kashi (Varanasi), one of India's oldest and most sacred cities. It is one of the ideal cities of celestial archetype where material expression paralleling macrocosmos and microcosmos is still in existence. The spatial context and experiential form of the mandala is symbolised by tiers and layers of the circumambulatory routes, associated auxiliary shrines and divinities, attached to a bounded space and fixed route. In Kashi, many sacred routes and territories are defined and followed by pilgrims in different contexts. However, the five routes are given special consideration in terms of cosmogony and sacred geometry. The Panchakroshi route demarcates the territorial limit of the cosmic territory of Varanasi, referred to as 'Kashi Mandala' (Figures 5.1 and 5.2).

The four remaining routes form an irregular shape, while the enclosing external route is identical to a circle. The four inner sacred journeys meet at Jnanavapi, the *axis mundi*. The temple of Madhyameshvara serves as the centre of the outer circle. The whole system becomes complex, because of the above two axis mundi. The five sacred territories are mythologised as the archetypal manifestation of macrocosm (gross elements, e.g. sky, earth, air, water and fire) corresponding to the five sacred routes (mesocosm). The mesocosm at the next level symbolises the microcosm of the cosmic man, Purusha, associated with the transcendental power and sheath, head, legs, face, legs, face, blood and heart.

The spiritual homology of the sacred territories further shows the interlinking relationship

[3] Putnam, R. D. (2000) *Bowling Alone: The Collapse and Revival of American Community*. New York, Simon & Schuster: 19.
[4] Sachs, I. (1996) *What State, What Markets, for What Development?* Social Indicators Research 39. Sachs, I. (2006) *New Opportunities for Community Driven Rural Development*. Sao Paulo: Institute of Advanced Studies of the University of São Paulo.

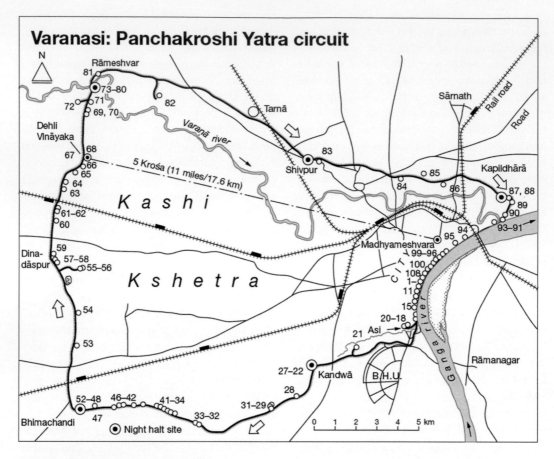

Figure 5.1 Panchakoshi Yatra of Varanasi.

between human beings and the cosmos, occur-ring in a strong state of connection to the sacred, where '*one sees one's own soul*'.

The idea of sacredness demarcated by these routes suggests order (spatial limit), wholeness (cosmological representation) and power (faith system).[5] These three perspectives of sacredness may be compared to the notion of territoriality, which provides a form of classification by bound-ary, and a form of enforcement or control by belief systems that represent intangible heritage in a historic urban landscape. Interestingly, no

noticeable cultural change has occurred in the Panchakroshi pilgrimage. In fact, the cultural aspect of this sacred journey has remained basi-cally the same, but its socio-structural aspect has undergone important changes over time, with the growth of good transport facilities.

These boundaries have often consciously or subconsciously contributed towards protecting local eco-systems, thereby ensuring the sustain-ability of historic urban landscapes. For example, in traditional Newari settlements like Patan, physical boundaries were well defined and rein-

[5] Singh, R.P.B. (1991) *Pancakrosi Yatra*, Varanasi: Sacred Journey, Ecology of Place and Faithscape. Singh R.L. & Singh, Rana.P.B. *Environmental Experience & Value of Place*. The National Geographical Society of India, Banaras: 49–51.

Figure 5.2 Place where the deceased of the Hindu faith are cremated in Varanasi.

forced through particular natural or built components and rituals. Stupas at four cardinal points of Patan demarcated the boundaries of the city, and because most of these settlements were planned on non-irrigable lands, this strategy indirectly preserved agricultural plots, which in turn has not only protected the ecological footprint of the settlement but also reinforced the primary occupation of the local people – agriculture – thereby contributing towards building resilience of communities.

These relationships reinforced by intangible heritage values determine how communities view and use natural and human resources such as land, water, landscape and buildings as assets that go beyond mere economic values. As such, these are considered not just for sale or consumption but for protection and maintenance while deriving multiple benefits from them.

Most often this is achieved by gaining social acceptance through rituals, practices and belief systems that serve as guidelines, far different in their nature and scope than currently practised rigid urban controls that specify regulations, often outlining what not to do rather than what to do.

In fact, the way people perceive nature depends on culturally defined value and belief systems that form an important, often intergenerational, source of information. Some of this valuable information, relating in particular to its spiritual dimensions, may not yet be considered in current ecosystem management. Part of the reason for this may be that such knowledge is inaccessible and difficult to understand for outsiders including conventional conservationists and ecosystem managers. Hence, accounting for the various world views and their corresponding

Figure 5.3 Nizamuddin Dargah in the historic urban district (Basti) of Delhi is an important urban space that has retained its character due to intangible heritage values associated with a famous Sufi Saint.

cultural and spiritual values in the practice of ecosystem management constitutes a challenge for managers, policy-makers and local people alike.[6]

Another important aspect of intangible heritage is its dynamic nature, continuously evolving and adapting to changing needs (Figure 5.3). Therefore it is traditional, contemporary and living at the same time and its manifestation in historic urban landscapes should be seen from the perspective of adaptation and change of local

communities to the social and economic context. In fact, considering that intangible cultural heritage is constantly recreated, the term 'authenticity' as applied to tangible cultural heritage is not relevant when identifying and safeguarding intangible cultural heritage.[7] Verification of authenticity in a historic urban landscape takes place out of a collective, that is, inter-subjective recognition sought by the society based on a set of rules and mutable values over time. Value is conferred on place through the activities of the

[6] Verschuuren, B. (2007) An Overview of Cultural and Spiritual Values in Ecosystem Management and Conservation Strategies. In Haverkort, B. and Rist, S. (eds.). *Endogenous Development and Bio-cultural Diversity, The Interplay of Worldviews, Globalisation and Locality*. Leusden, The Netherlands: Compas/CDE, Series on Worldviews and Sciences, No. 6: 299.

[7] Inaba, N. (2009) Authenticity and Heritage Concepts: Tangible and Intangible – Discussions in Japan. Price N.S. and King, J. (eds.). *Conserving the Authentic. Essays in Honour of Jukka Jokilehto*. Rome: ICCROM.

past and present, of memories of knowledge, and of socio-cultural relationships, which occur in space and time in the historic urban landscape.[8]

Urbanisation Processes and Impacts on Intangible Values

Urbanisation is one of the key factors that is increasing vulnerability and risks to people, properties and economy. The world is passing through a great urban upsurge: the number of people living in cities equalled those in villages in 2007 and has been rising ever since. In fact, it has been projected that 1.29 billion people will be added to the city population from 2007 to 2025. Forty eight cities in the world have already reached a density level of more than 15 000 per sq km. Interestingly, all of them are in developing countries. Since cities concentrate people, properties, infrastructure and capital stock, they are increasingly vulnerable to natural and human induced disasters.

As the 2011 *Recommendation on the Historic Urban Landscape* states:

Rapid and uncontrolled urbanization may frequently result in social and spatial fragmentation and in a drastic deterioration of the quality of the historic urban environment and of surrounding rural areas. Notably, this may be due to excessive building density, standardized and monotonous buildings, loss of public space and amenities, inadequate infrastructure, debilitating poverty, social isolation, and an increasing risk of climate-related disasters.[9]

Urban growth is transforming the very essence of many historic urban landscapes. On the one hand, urbanization provides economic, social and cultural opportunities that can enhance the quality of life and traditional character of urban areas; on the other hand, the unmanaged changes in urban density and growth can undermine the sense of place, the integrity of the urban fabric, and the identity of communities. As a consequence, historic urban landscapes are losing their functionality, traditional role and populations, thereby negatively impacting the intangible heritage that have characterized them for generations.[10]

In fact, intangible heritage has contributed towards preserving traditional urban morphology even under tremendous pressure of urbanisation. Take the case of the Kathamandu valley in Nepal (Figure 5.4). Due to rapid urbanisation in the valley, the contemporary city of Patan has now expanded beyond the traditional boundaries defined by the four cardinal stupas, spilling into the valley and across fertile agricultural land (Figure 5.5 a,b).

New residential and commercial developments spring up even before the infrastructure is in place. Many traditional urban systems and local ecological processes, such as the water system of *Hitis*, the traditional step wells, are being damaged beyond repair or have fallen into disuse.[11]

Fortunately the open spaces of the historic city remain surprisingly intact, as do the temples and monasteries, community rest houses, water structures and *Chaityas* (small shrines). In fact cultural heritage in Patan survives, in large part, because it still serves the collective social and religious purposes of communities that determine intangible heritage values. As mentioned in the previous section, processional routes – used for various rituals celebrating the life cycle – have also helped to ensure that the historic urban fabric endures. These routes play an important role in preserving the boundaries of urban form and, consciously or not, protect the agricultural lands.

[8] Jamal,T.B. and Hill, S. (2004) Developing a Framework for Indicators of Authenticity: the Place and Space of Cultural and Heritage Tourism, *Asia Pacific Journal of Tourism Research* 9/4: 353–371.

[9] UNESCO (2011) *Historic Urban Landscape Recommendation: para. 2.*

[10] UNESCO (2011) *Historic Urban Landscape Recommendation: para. 17.*

[11] Jigyasu, R. (2010) *Urban Cultural Heritage for Sustainable Resilience: the Case of Patan in Kathmandu Valley, Nepal.* Urban Risk Management in South Asia: Launch of Global Campaign on Making Cities Resilient. New Delhi: SAARC Disaster Management Centre.

Figure 5.4 Darbar Square: the public open space in the historic town of Patan in Kathmandu Valley.

Recognition of Intangible Values in Existing Urban Management Systems

Unfortunately, current urban planning systems in most countries fail to recognise and protect intangible aspects of heritage. The Master Plans tend to consider only the built heritage represented predominantly by historic buildings. In some instances, historic urban areas are identified for protection as 'special areas'. However, even in such cases, only physical attributes are considered for protection and conservation measures give undue weight to preserving the material aspects of heritage without adequate attention to the associated meaning of traditions, rituals and practices for the community that represent intangible heritage.

This has direct implications for the conservation planning approaches for historic urban landscapes. On one hand several physical attributes that appear to be humble but may be of great significance to the local community because of their associated intangible values are not considered for protection. On the other hand, conservation becomes limited to physical interventions and does not consider the social and economic measures/incentives necessary for retaining the traditional social fabric and preventing gentrification of historic urban areas, and also for reinforcing the links of the local community to tangible attributes of the city by maintaining traditional socio-cultural activities associated with them. Moreover, the mixed activities that are characteristic of historic urban

Figure 5.5 (a) Due to social and religious functions, semi-public open spaces in Patan have remained intact while surrounding buildings have undergone massive transformations. (b) Semi-public open spaces characterise traditional Newari settlements in Kathmandu valley as in the case of Bungamati.

areas seldom get recognised in the Master Plan that promotes rigidly determined single land use zoning. In fact, the concept of 'multifunctionality' of cultural landscapes can help to envision landscapes that cross urban/rural divides in a sustainable and integrated way – characterised by wholeness and 'ecospirituality' that developed in the cultural history of landscapes. That is how the idea of 'wholeness' ('cosmality') is transformed into 'holiness' ('sacrality') – evolved and represented with sacred ecology and visualised through the cosmic frames of 'sacredscapes' in the Asia-Pacific region.[12]

Traditional urban planning principles that take into consideration the local organisation of communities when replaced with alien principles result in problems of management. In Patan, *toles* are traditional neighbourhood units organised around spaces that are used collectively by the surrounding communities. Replacing them with a contemporary ward system, which demarcates boundaries by roads without consideration of traditional boundaries, has led to untold problems in maintenance and management for the municipality as the participation of traditional social groups organised around traditional neighbourhoods is difficult to galvanise for collective action.

Documentation and Impact Assessment of Intangible Heritage Values

There is little or no documentation of intangible heritage in cities, since the focus is primarily on the recording of physical attributes of the urban fabric. Even the documentation of physical attributes seldom goes beyond the physical recording of built form and architectural and aesthetic values associated with them, to include intangible heritage values linked with cultural practices and belief systems of the local community.

Heritage impact assessment is increasingly being recognised as a tool aimed at evaluating effectively the impact of potential development on the heritage values of a site. Thus the heritage impact assessment process considers the impact of any proposed project or change on the attributes that carry these values, both individually and collectively, by assessing the adverse impact on the appearance, skyline, key views and other attributes that contribute to the heritage values of historic urban landscapes.[13] Besides heritage values, economic, social and environmental impacts of the proposed project related to the overall context of the site should also be considered for holistic assessment of impacts.

It is found that intangible heritage associated with communities living in historic urban landscapes, which may extend beyond narrowly defined historic areas, are generally not substantially considered as various tools and methodologies for impact assessment, which tend to heavily focus on the impact on the physical attributes of the designated historic urban area. True consideration of intangible heritage would necessitate inclusion of all social, cultural and economic parameters that are associated with the intangible heritage of historic urban landscape either on their own merit or in relation to the tangible attributes that are needed as containers for intangible values.

It is important to note however, that in some instances these parameters may in fact pose a negative impact on one or more attributes at the cost of others, thereby necessitating balanced decision-making that takes into consideration impacts on all associated intangible values and other benefits that proposed developments may

---

[12] Singh, R.P.B. (2013) *Sacred Landscapes*, Cosmos and Shared Wisdom: the Asian Vision. Keynote address at the 50th World Congress of IFLA (International Federation of Landscape Architects), 9–12 April, Auckland, New Zealand.
[13] ICOMOS (2011) *Guidance on Heritage Impact Assessment of World Heritage Properties*.

Figure 5.6 Ritual paths define the boundaries of traditional settlements such as Bungamati in Kathmandu Valley.

bring to dynamic heritage values that are constantly evolving (Figures 5.6, 5.7 and 5.8).

'Heritage' – Elitist or Inclusive?

There needs to be a larger recognition of heritage among urban managers – not just as a relic from the past to be cherished and preserved for posterity, but as an asset that could play a more proactive role in ensuring the continuity of skills and knowledge, positive change that ensures fulfilment of new needs and aspirations as well as long term sustainability.

Concern for continuity is one of the key elements missing in the conventional conservation approach. Indeed, it is believed that the traditional continuity between the past and the present is always broken, and that heritage can be frozen in time and place. This is based on

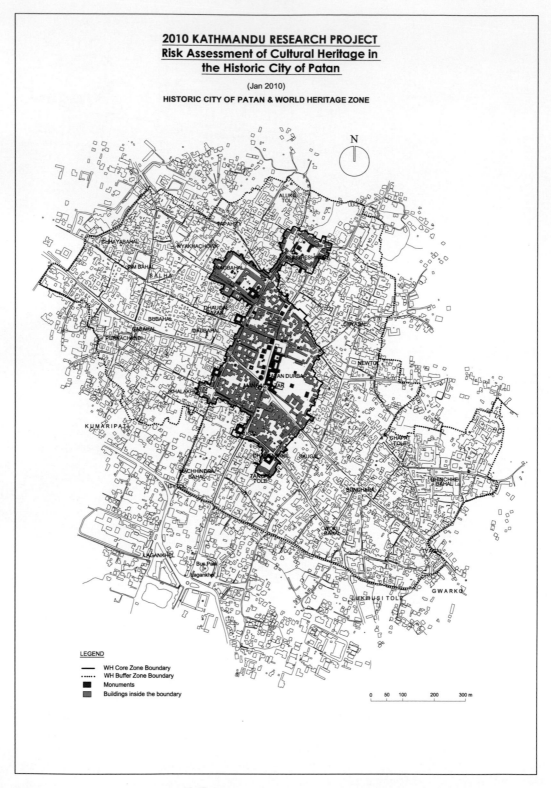

Figure 5.7 The historic city of Patan in Kathmandu Valley with a distinct morphology characterised by interconnected open spaces.

Figure 5.8 Uncontrolled urbanisation in Kathmandu Valley is destroying the ecological relationship of the settlements with fertile agricultural land and water resources.

many assumptions and also on the basis of time which is considered as a linear function (arrow), as in western thought, though this is not universal. Therefore continuity of traditions and traditional management systems will be an important aspect we need to recognise in our effort to better conserve and manage heritage.[14]

Neither fossilised traditional practices nor radical contemporary ideas with no link to local context or realities are the answer for sustaining living urban heritage, which is not just a static compilation of architectural typologies from a bygone era but rather the repository of a rich vocabulary of design that needs to be reinterpreted, readjusted and integrated as a dynamic system. Such an approach would address both the increasing vulnerability of the traditional built fabric that results from the loss and degradation of traditional knowledge, and the need to search for solutions creatively, through continuous trial and error. It is this evolutionary process that defines the true essence of traditional knowledge that should be integrated in larger urban planning and management systems. Moreover these should also recognise and protect attributes with 'experiential dimension'

---

[14] Wijesuriya, G. (2013) *Tapping Stone and Brick. In Search of Traditional Skills and Their Continuity* (Redefining Traditions of Heritage), *Proceedings of the International Conference, Human Resource Development for Transmission of Traditional Skills – National Approaches and Their Application to Stone and Brick, 6–8 December 2011.* Shanghai, Cultural Heritage Protection Office of Asia-Pacific Cultural Centre for UNESCO (ACCU): 3–16.

that connects beliefs and practices to the tangible aspects of space.

This would require redefining and repackaging intangible heritage concerns in various urban development sectors not merely through physical planning but also through measures aimed at regeneration of traditional livelihoods, ecological planning and sustainable development and disaster risk reduction.

Role of Intangible Heritage in Building Disaster Resilience of Cities

In the face of disasters, traditional communities in historic urban landscapes often develop a vocabulary of resilient features in the urban environment that intentionally or unintentionally contribute towards prevention and mitigation, emergency response and recovery.

For example, in case of Patan in Kathmandu Valley, traditional settlements have well established networks of rest places (*Patis/Sattals*) and water sources, wells, stone water spouts (*hitis*), water tanks and ponds, that are strategically located at open squares and at street junctions and village entrances. These serve as places for settlers and visitors to carry out daily activities. In the event of a disaster, these resting places can also be used for sheltering the injured, while water sources used for drinking can double as a local fire hydrant.

These public places typically used for community gatherings, playing traditional music or just congregating can also help maintain a rapport among the local people facing catastrophe. In this way the tangible attributes carrying intangible/social values have the potential to enhance cooperation among residents during a crisis and may well serve as sites for disaster preparedness training.

Intangible heritage manifested in traditional management systems also has tremendous potential in securing collective action for post disaster recovery. In Kathmandu Valley communities use or own the land around the town collectively through a religious trust known as Guthi, which provides resources for the maintenance and repair of temples, public buildings and infrastructure after disasters.

The rich expression of heritage is also a powerful means to help victims recover from the psychological impact of the disaster. In such situations, people search desperately for identity and self-esteem. Traditional social networks that provide mutual support and access to collective assets are an extremely effective coping mechanism for community members.

Integrating Intangible Heritage Values in Urban Planning and Management

The pressures on historic urban areas will surely continue to grow as land values continue to rise and land use controls are modified. If these areas are to survive, they must continue to play important roles in the urban development process by improving the quality of life of local inhabitants through increasing their livelihood opportunities and addressing their basic needs. At the same time viable roles should be found for these areas within the overall economic development while ensuring that these areas continue to play a crucial role in maintaining historical continuity. This will require support to demonstrate their capacity to positively contribute to their future development. Their heritage values, indigenous knowledge systems, economic potential and social structure can only be identified, elaborated and disseminated if the structural integrity of these historic areas is improved in a holistic manner.

Planning tools should help protect the integrity of the tangible and intangible attributes of historic urban landscapes. They should also allow for the recognition of cultural significance and diversity, and provide for the monitoring and management of change to improve the quality of life and of urban space.

One of the essential pre-requisites for this to happen is mapping and documentation of cultural and natural characteristics of historic urban landscapes by the agencies responsible for planning and management. Although a lot of work has been done on recording the physical attributes, recording intangible heritage values leaves much to be desired.

The planning approach should shift away from master planning with rigid land use zoning that has dominated planning practice for the last few decades to a much more holistic territorial approach that seeks to recognise multiple relationships that tie residents to their environment both in materialistic as well as non-materialistic terms.

The principle of territorialism as a counterpoint to functionalism addresses:

- different forms of social integration, which implies understanding of the past and present informal and formal institutional structure;
- resource, technology and skill links contrasting functional and territorial perspectives and traditions;
- land, land-use and relationships within society on the way they organise land tenure and ownership;
- access to, quality of and public nature of open space and the built environment; and
- reinforcing mutual interdependence of cities to their rural hinterland through livelihoods, use of land and other natural resources, and through the cultural and religious associations.

This interdependency can be strengthened through innovative solutions that link urban and rural planning strategies in a regional context.

Territorial relations in historic urban landscapes tell us about their resource base and their daily cooperation and struggle. It is this very resource base which is the community strength and their main potential for livelihood improve-

ments. On this base, strategic principles and actions are drafted where efforts to strengthen cultural integrity shall contribute to reducing vulnerabilities and ensuring sustainability.[15] While tangible relationships are beginning to get recognised through the ecological planning approach, intangible/spiritual relationships have not received the attention that they deserve.

In addition to these, an increasing challenge is the application of territorial approach to changing people–place or identity relations in multicultural societies in rapidly transforming historic urban landscapes in a globalising world.[16]

Mainstreaming Intangible Heritage Through Sustainable Livelihoods and Cultural Tourism

It is paramount to find ways and means to regenerate traditional livelihoods in order to protect intangible heritage in historic urban landscapes. The local economy that is often informal in nature and substantially based on traditional livelihoods should be the basis for rehabilitation of the area and financial, technical and social tools and systems should be created to support local economic development in the area. One of the ways in which this can be achieved is by supporting small cottage and tourism based local industries and encouraging them to cater to new needs.

Cultural tourism presently accounts for 40% of world tourism revenues. Intangible cultural heritage, which sustains living cultural expressions and traditional 'know-how', as well as performing arts, generates substantial revenues and employment from tourism. Therefore identifying tourism as a subsector for investment would

[15] Bjonness, H.C. (2008) *Changing Territorial Values in Urban Conservation. From Patan, Nepal to Gyantse, Tibet Autonomous Region*, Proceedings of the International Conference of ICOMOS, ICCROM and Fondazione Romualdo Del Bianco, 2–4 March 2–7, Florence.

[16] Bourdieu, P. and Hillier, J. (eds) (2002) *Habitus: A Sense of Place*, Ashgate: Aldershot: 43–52. Maalouf, A. (2000) *On Identity*, London: Harvill Panther.

encourage investment in infrastructure and stimulate local development.

Sustainable tourism and the culture and creative industries are strategic outlets for income generation and poverty reduction. Cultural industries require limited capital investment and have low entry barriers. Effective promotion of the cultural industries is likely to have direct impact on vulnerable populations, owing to the significant reliance of the culture economy on the informal sector, where poor and marginalised populations, including women, often find employment, and can thereby stimulate social inclusion while maximising jobs and trade opportunities.

However it is important to recognise that tourism is not the only solution to historic urban landscapes' viability or utility to future development. Indeed, it is to look at the local skills sets as a basis to create opportunities for existing crafts communities to apply their technologies that have evolved over time and evolve new products and designs for the future.

Redefining the Role of Professionals

Integrating intangible heritage into urban planning and management would necessitate increased collaboration of heritage professionals with city managers, planners and development practitioners working in various sectors, who need to pay attention to the traditional systems and socio-cultural practices embedded in the urban fabric that have contributed to resilient communities by re-imagining and re-aligning them as integral to the contemporary life. This is a creative challenge that we must accept as professionals to ensure continuity and positive change in historic urban landscapes,

which would protect and sustain intangible heritage.

An important step in this direction would be a shared understanding of the vocabulary used by heritage professionals and those from other sectors associated with urban planning and development so that an informed dialogue can be initiated. At an institutional level, this would require greater collaboration between public organisations or departments dealing with cultural heritage protection and management and those responsible for various urban development sectors such as housing, infrastructure, livelihoods, health, etc., as well as community based organisations and non-governmental organisations.

The local community is the true bearer of intangible heritage. Integrated actions aiming at safeguarding simultaneously elements of the intangible cultural heritage and associated cultural spaces need to be rooted in the values and wishes of the communities or groups concerned. Therefore professionals should closely engage with all sections of the local community from various social, economic, age and gender groups and at appropriate scale to understand and mainstream intangible heritage values. This would necessitate a dialogue with communities in language that is easily comprehended by them.

It should be clear from the outset that we are dealing with living cultural resources. It is not really desirable to 'freeze' these places in time and pretend that they are still 'alive', rather what is more important is that historic urban landscapes are seen as evolving. As they change, adaptation is to be encouraged within a set of performance guidelines that will recognise various intangible heritage components as assets and resources for the sustainable future of local inhabitants.

Interview with Lisa Prosper
Interpreting Cultural Landscapes as Expressions of Local Identity

Director of the Centre for Cultural Landscape at Willowbank, Queenston, ON, Canada
16 July 2013, Queenston, Canada

Question: *Canada has a rich tradition and international reputation in heritage conservation, arguably stemming from a cross-over between European and American experiences and approaches, and in particular its exploration and application of Cultural Landscape theory is having a significant impact on the national practice of heritage conservation. What is the role of the Centre for Cultural Landscape at Willowbank, under your direction, in this?*

Answer: At the Willowbank Centre for Cultural Landscape we explore the notion of cultural landscapes, in particular the relationship between people, places and the construction of identity. Traditionally, monuments, sites and artefacts have been seen as signifiers of collective identities, acting as the vocabulary for constructing narratives of national and civic belonging. In effect, a semiotic understanding of cultural property has generally informed the tools, mechanisms, and concepts of heritage conservation. This approach generally overlooks the ways in which identity is produced through the day to day inhabitation or practice of place. In fact, the intersection between lived experience, identity and heritage is only beginning to be explored and genuinely valued within the heritage field. In this context, the notion of the cultural landscape is important because it demands an understanding of heritage as grounded practice and not just as 'thing'. It is an important concept if we wish to tap into the full spectrum of heritage – in its multiplicity of forms, expressions and functions – and develop a set of tools that conserve what matters to communities.

Our Centre aims to provide a platform for an expansion of thought and practice that incorporates new heritage realities. To facilitate this, we invite heritage practitioners and thinkers from around the world to give lectures and engage in discussion with Willowbank students as well as the broader community of heritage professionals and policy makers. Our goal is to broaden the scope of heritage conservation so that it becomes more inclusive and responsive to community needs and wishes. The fact that these activities happen within the framework of our school means that our students and their activities are part of this new thinking in heritage. This is critical, as they will soon occupy positions in which they will be able to influence heritage practice.

Question: *As regards the role of the World Heritage Convention, under which in 1992 Cultural Landscapes were adopted as a distinct category, this has certainly resulted in a significant increase in the popularity of this heritage type, if judged by the rate of inscription. In your view, what has evolved or changed over the past 20 years in our understanding of Cultural Landscapes?*

Answer: The adoption of cultural landscapes as a distinct category of heritage has led to the designation of many cultural landscapes and focused attention on the term. However, many of the cultural landscapes designated by UNESCO share the same characteristics as a medieval cathedral or roman ruin – that is to say, they have a decidedly monumental or artefact quality that stems from the physical modification of the natural environment. I think this runs somewhat counter to the original intention behind the adoption of cultural landscapes as a type of heritage, which was to broaden the scope and acknowledge a whole realm of heritage outside the traditional categories of monuments and sites in vogue since the adoption in 1964 of the Venice Charter. In fact, it could be argued that the inclusion of cultural landscapes as a category of World Heritage has actually meant that the incredible potential they offer to recognize otherwise unrecognised heritage and inform the evolution of heritage thinking has been reduced through incorporation into traditional perspectives and preoccupations.

Going forward, I think that the ideas behind two of the sub-categories of cultural landscapes in the UNESCO context – 'continuing and associative cultural landscapes' – have considerable potential that could be creatively expanded upon. These sub-categories lend themselves to exploring compelling examples of the ways in which the grounded experience of place over time results in cultural identities that may be expressed in a wide range of both material and immaterial forms, such as narratives, memories, events, traditions, representations and day to day rhythms. Continuing and associative cultural landscapes can be found in different cultural contexts and environments that are variously natural, agricultural and urban. In other words, the word 'landscape' can be broadly interpreted. Ultimately, expanding the register of continuing and associative cultural landscapes means understanding the ways in which communities attribute value and meaning to place over time that in turn structures who they are. At Willowbank, we think this perspective is important because it can inform vibrant and sustainable communities with a strong sense of inheritance.

Question: *Through this pertinent statement and position, your descendancy from and identity as part of Canada's First Nations, the East Coast community of the Mi'kmaq to be precise, comes clearly to the fore, for instance also in the way you refer to heritage practice.*

Answer: Intuitively, I see heritage as a practice: for me it's a verb, not a noun. Further to this, heritage has little to do with history or the past, rather it has more to do with the present and the continuing practice of cultural traditions. For example, Aboriginal cultural landscapes rely on the continuity of the cultural practices that created them in the first place – harvesting of plants and animals, seasonal travel and gatherings, reliance on traditional knowledge and the perpetuation of oral narratives and histories. In such a context it becomes clear that the cultural inheritance of traditional communities is not held together by things that are protected, but by traditions that are constantly practised, a fundamentally different point of view. This is something that the western heritage tradition could stand to learn from the Aboriginal perspective.

People tend to argue that between tangible and intangible heritage, the latter is more vulnerable – which is not the case. Tangible heritage may appear robust and literally set-in-stone, but built structures degrade over time under natural processes and require constant investment and effort to be maintained and adapted to contemporary uses. On the other hand, although intangible heritage such as storytelling may appear inherently evanescent, there is in fact remarkable resilience in the traditions and practices that structure identity, especially if we accept that they are a constant work in progress, adapting to changing circumstances.

Question: *In his remarkable book* A New Dimension of Time *(2007), Marc van der Erve uses scientific reasoning to state that time comes in two dimensions, namely time in the spatial domain, as interval, which is derived from the relationship of distances travelled, and time in the non-spatial domain, as non-interval, which refers to simultaneity, that is events happening at the same time as interval-less instants. The first dimension of time (time as interval), he argues, manifests itself primarily at the local scale, while the second dimension (time as non-interval) at a global scale. He uses this position to open up a new realm of interpretation, which might have significant consequences for heritage perception and identification also?*

Answer: Interestingly, when one closely examines Aboriginal conceptions of heritage, the past and the present do not come across as distinct spheres. Instead, Aboriginal heritage resides in the lines of continuity between past and present, which is to say in the relevance of traditional practices and perspectives to modern identities and livelihoods. For example, whereas the western perspective on heritage tends to value physical artefacts, Aboriginal perspectives tend to value the skills, knowledge and processes through which objects are made. Heritage is thus a living thing, which bridges past and present.

Question: *Given the close relationship between Cultural Landscapes and Historic Urban Landscapes, in conception and practice, as regards the need to ensure inclusiveness of diverse cultural traditions,* *values, and community livelihoods in the search for empowerment, resilience and sustainability, what would you consider essential for Historic Urban Landscape in terms of theoretical underpinning and practical application?*

Answer: As an approach, Historic Urban Landscape represents a significant shift in perspective. It strives to understand and perpetuate the broader context in which meaning, value and identity are produced and reproduced by multiple communities over time in urban spaces. Historic Urban Landscape essentially adopts a cultural landscape approach by looking beyond discrete historical buildings, towards the multiple forms of material and immaterial heritage that shape urban space. It recognizes both the role of built heritage and lived experience in the production of value and meaning in place. Ultimately, I think the Historic Urban Landscape goes a long way towards acknowledging diversity and complexity with respect to urban heritage, rather than favouring singular narratives that highlight narrowly defined groups. It expands the way in which we conceive of heritage and therefore informs the way we approach heritage.

Case Study *The Traditional Chinese View of Nature and Challenges of Urban Development*

Feng Han

Professor at the Department of Landscape Architecture and Assistant Dean of the College of Architecture and Urban Planning, Tongji University (CAUP), Shanghai, China.

Introduction

Language is magic but sometimes opaque, especially differences between languages, or more accurately, between different cultural contexts. The English terms 'cultural landscape' and 'historic urban landscape' encounter great difficulties in being understood in China. A landscape, from a Chinese view point, is *a priori* a cultural construct, which prompts the question why 'cultural landscape' is expressed in double terms. As for 'historic urban landscape', this is even more difficult for Chinese to comprehend, since it requires taking a landscape beyond trees, mountains and water, and to understand urban dynamics as a 'landscape'. From the traditional Chinese perspective, Augustin Berque's concept of landscape may give the closest explanation to what the Chinese mean:

> Landscape is not the environment. The environment is the factual aspect of a milieu: that is, of the relationship that links a society with space and with nature. Landscape is the sensible aspect of that relationship. It thus relies on a collective from of subjectivity ... to suppose that every society possesses an awareness of landscape is simply to ascribe to other cultures our own sensibility.

There are three key points in the above concept related to Chinese understanding. Firstly, landscape is not all about outside physical surroundings. Secondly, landscape is a nature-related concept. Thirdly, landscape is a spectrum, from intangible subjective nature-related cognition to tangible human-altered nature. Therefore, the essence of landscape is the result of human-nature interaction and a cultural and social constructive process. Based on this understanding and after long debates, China finally accepted the term of 'cultural landscape', which is defined as the 'combined works of nature and of man' in the *World Heritage Convention*, and in 2010 China nominated its first World Heritage Cultural Landscape: the West Lake in Hangzhou, which was inscribed in 2011, almost 20 years after the World Heritage Cultural Landscape initiative. This late arrival of Chinese cultural landscapes is a notable observation in World Heritage practice.

UNESCO's *Recommendation on the Historic Urban Landscape* recognises an urban area as a cultural landscape and encourages 'landscape' as a holistic approach to deal with the historical urban context dynamically. Although it is hard for the Chinese to accept these urban areas or centres without marked natural features as 'landscapes', the idea of historic urban landscape with historical context and relationship focus still interests China. From a broad environment-philosophical perspective, human-nature relationship is essential for sustainable urban development and such holistic human-nature relationship focus is strongly related to Chinese landscape ideas throughout Chinese urban history. From this point of view, the Chinese landscape idea shares a similar perspective and methodology with Western concepts of cultural landscape and historic urban landscape. To explore the Chinese landscape idea may inspire and contribute to those human-nature relationship oriented and historical and contextual-based processes, such as urban development and heritage conservation.

Landscape: The Traditional Chinese View of the Human-Nature Relationship

In a Chinese dictionary, it is hard to find a single word that corresponds to the English word 'landscape' with associated Germanic root *'landschaft'*. Instead, we have to find several words in order to cope with the layering of meanings of landscape. But interestingly, all words are connected with nature. *jiang shan* (江山, river and mountain) might be synonymous to *'landschaft'*, which symbolises territory, region, tract of land defined by geographical natural features such as rivers and mountains.

The landscape idea associated with pleasant natural scenery is referred to as *feng jing* (风景) in Chinese. In ancient Chinese, *feng* (风) means atmosphere or air and *jing* (景) means light. Both *feng* (风) and *jing* (景) are sensory and intangible natural features. From the fifth century these two words were put together as a term referring to a view of pleasant natural scenery. There is another important word referring to landscape: *shan shui* (山水, mountain and water), which is also the renaming of nature. *Shan shui* (山水) became *feng jing* (风景) and symbolised ideal nature, and could be dated back to Wei Jin, Southern and Northern Dynasties (魏晋南北朝, 220–589 CE). This is a milestone that marked that 'nature' became an independent aesthetic objective, which is about one thousand years earlier than in the West. *Shan shui* culture (山水文化), *shan shui* painting (山水画), *shan shui* poetry (山水诗), *shan shui* gardens (山水园) and *shan shui* travel (山水游) later developed into an independent cultural branch and became the most characteristic of Chinese culture.

Compared to these Chinese terms, *jing guan* (景观), today's common Chinese translation of 'landscape', is trying to cover all the above historical areas and even extending to broader areas, including all physical surroundings. Such development of understanding is relatively recent and strongly influenced by Western human geography and landscape architecture, but the meaning of *jing guan* (景观) itself is far different from its original meaning 'view of scenery', which was firstly used in the eighteenth century's Qing Dynasty. Therefore, it is understandable that for landscape as an urban approach, the concept of *jing guan* (景观) is problematic in China.

Despite semantic difficulties, from the above we can see that nature had been a notable important object in Chinese culture throughout its thousands of years of history. As the result of interacted human-nature relationship, the Chinese landscape represents Chinese perspectives of nature, a cultural filtered or perceived nature, which is strongly underpinned by Chinese philosophy centred on a human-nature relationship. Chinese philosophy has greatly shaped Chinese landscape ideas and landscape activities. Instead of abstract metaphysics, Chinese philosophy is considered to be practical concerning human life and playing an important and unique role in Chinese daily life. Philosophy was an understandable life guide for everyone to follow and had permeated all political, social and cultural aspects. It would be helpful to obtain some knowledge of Chinese philosophy in order to understand Chinese landscape.

In general, there are two main schools of philosophy in Chinese history, orthodox Confucianism (儒家) and supplemental Daoism (道家), which originated about 2500 years ago. Confucianism (儒家) took responsibility for politics and ethics because of its social involvement, with positive and morally cultivated attitudes. In Confucianism, 'the wise man delights in water, the good man delights in mountains' (仁者乐山, 智者乐水), where nature is greatly valued for humanised ethical and moral qualities. Nature is the place for moral cultivation. Daoism (儒家) prevailed due to its negative outlook on human society and its romantic retreat in nature. Daoism attests that within nature lies the essential ontological values and great beauty. Daoism associated with the recluse, retirement to the mountain, the worship of rural life, the pursuit of spiritual freedom and the romantic personality, and the banishment of all worldly cares and worries, derives the most characteristic charm of Chinese culture, which is the natural and rural ideal of life, art and literature. Nature was an independent aesthetic object in Daoism.

From the above we see different value orientations, but they did not prevent Confucianism and Daoism from reaching common ground. Both of them posit that 'oneness with nature' is the eternal objective of human life. Humans have been part of nature and human-nature has always been a whole. Such humanistic, ethical, aesthetic and romantic beliefs of nature underpinned by Chinese philosophies had been the foundation of Chinese landscape practice for the latter two and half millennia. 'Harmony' became a golden principle to deal with the nature-human relationship, including the urban-nature relationship.

The Power of Landscape in Chinese Historic Cities

In ancient China, how to respond to nature and build harmonious relationships with nature, physically and spiritually, had been a key topic in urban construction throughout history. To achieve this, natural,

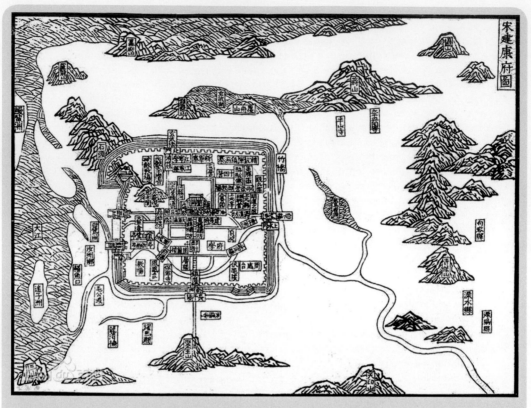

Figure 5.9 Map of the City of Jiankang (建康, Ancient Nanjing in Southern Song Dynasty), 1129 CE.

cultural and social contexts always needed to be examined. Contrary to the West, the most characteristic element of the layout of Chinese ancient cities is that it is open and stretches into nature from different directions horizontally instead of vertically, in order to embrace nature as much as possible (Figures 5.9 and 5.10). Among those Chinese principles dealing with the daily nature-human relationship, *feng shui* (风水) was the most well-known to the West, where natural fabric and settings were carefully examined for human settlement. Such a nature integrated tradition created great potential for iconic landscape construction in cities. Consequently, many noted ancient cities, such as Nanjing (南京), Yangzhou (扬州), and Hangzhou (杭州), are determined by their historic *shan shui* (山水) landscapes, as for example the Xuanwu Lake (玄武湖) for Nanjing, the Slender West Lake (瘦西湖) for Yangzhou, and the West Lake (西湖) for Hangzhou.

Examining ancient Chinese maps, it can be established that ancient cities were never isolated from natural surroundings or located at dominating positions; instead, their locations were carefully selected within natural settings so that urban features could be inter-weaved with nature to form an organic whole. *Shan shui* (山水), mountains and water, were always two highlighted symbolic elements of nature integrated into cities. Water and mountains not only convey natural and geomorphological information, and as geographical coordinators so define urban territories, but also function as indispensable cultural-attached icons stretching into human settlement to demonstrate the active philosophical pursuit of 'oneness with nature' in worldly life. Notably, the most characteristic cultural images of these cities were idealised and represented in many famous Chinese landscape paintings (Figures 5.11, 5.12 and 5.13) and landscape poems. At this point, Chinese landscape culture is unique and a great treasure for historical recordings. In ancient China, landscape paintings were a blueprint for constructing an ideal worldly wonderland.

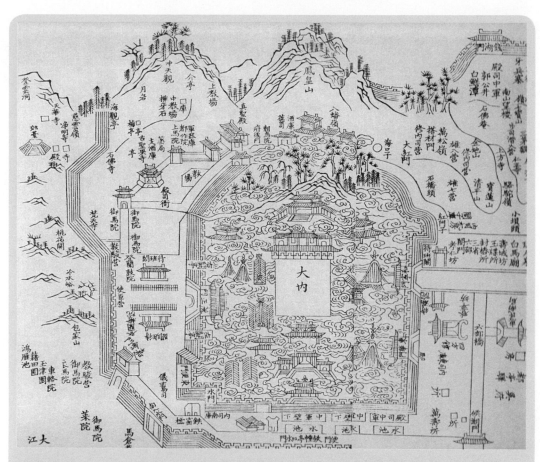

Figure 5.10 Map of the Core City of Lin An (临安, Ancient Hangzhou in Southern Song Dynasty), Southern Song Dynasty, 1268 CE.

The ancient city of Hangzhou, where the World Heritage Cultural Landscape of West Lake is located, sits at the mouth of Hangzhou Bay (杭州湾) and is connected with the Yellow Ocean in the middle of China's eastern seaboard. Qiantang River (钱塘江) runs by the city and into the bay. With an urban history of 2200 years, and once being the capital city of the Southern Song Dynasty, Hangzhou perfectly manifests the wisdom of utilising natural resources and the cultural input process of humanising nature. West Lake was constructed from a lagoon near a seashore bay into a fresh water reservoir to meet social imperatives and then turned into a symbolic landscape. With the continuous input of Chinese culture, the lake and its mountain surroundings have been perfectly integrated into the city and constructed as a classical masterpiece of an urban design landscape (Figure 5.14). The layout of the city, 'three-side lake and mountain, one-side city' (Figures 5.15 and 5.16), not only maintained and enhanced the natural character but also endowed it with a natural-ethical and aesthetic power underpinned by Chinese philosophy. It achieved the most romantic and aesthetic objective of Chinese culture: creating a worldly heaven for life, for harmony and enjoyment, being with nature forever in a glorious city. With its exquisite *shan shui* (山水) landscape, Hangzhou is labelled as the 'Heaven on Earth' and is considered an icon of Chinese culture.

Such a humanistic perspective on nature and its practice is manifest in many Chinese historic cities. The famous 'spring water city' Jinan (济南), the capital city of Shandong Province (山东省), is another

Figure 5.11 Ci Yun Xiang Xue (慈云香雪), one of the 24 Scenic Views of the Slender Lake of Yangzhou, Qing Dynasty.

Figure 5.12 The West lake in the Four Seasons (Spring), Liu Songnian (刘松年) Southern Song Dynasty.

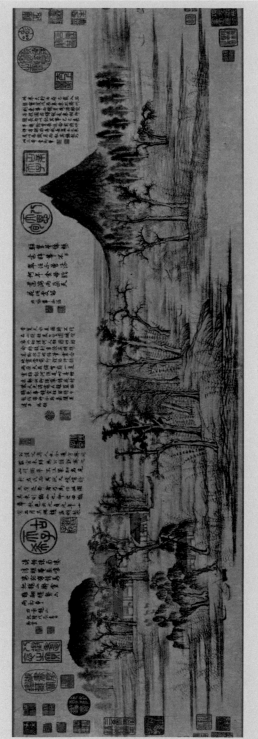

Figure 5.13 Que Hua Qiu Se Tu (Autumn view of Mount Que and Hua 鹊华秋色图) Zhao Mengfu (赵孟頫) , Yuan Dynasty. Image of the ancient city of Jinan (济南).

Figure 5.14 The poetic shan shui (山水) landscape of the City of Hangzhou.

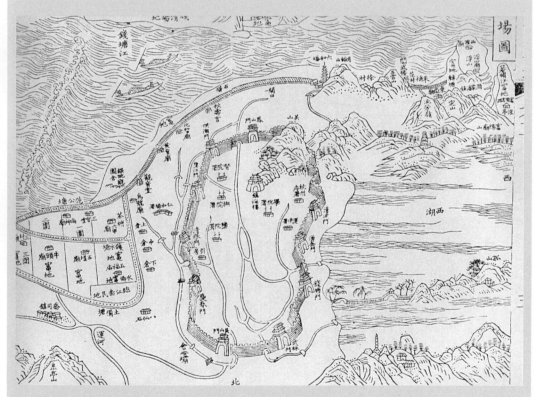

Figure 5.15 Map of ancient Hangzhou, Qing Dynasty, 1724CE.

Figure 5.16 Image of the city of Hangzhou in the 1930s.

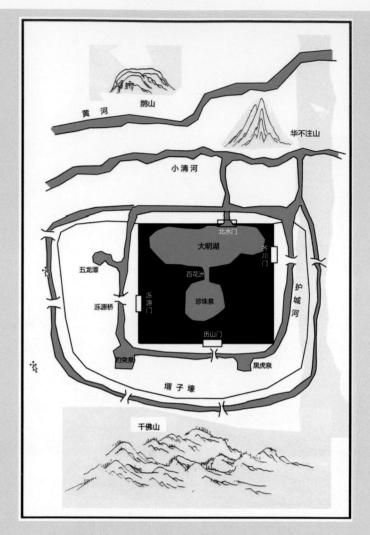

Figure 5.17 Mountain, spring, lake and city, the four components of the city of Jinan (济南).

outstanding example. According to legend, ancestors inhabiting the waterside discovered this treasure land with abundant underground spring water 5000 years ago. In the following thousands of years of urban history, water remained the soul of this city. The spectrum of water, from physical to spiritual, had been systematically integrated into the urban development. Mountain, city, lake, river and spring water had been an organic whole and formed fabulous waterscapes. The underground spring water is nurtured by mountain forests in the north, bubbles out in the central city area through underground geological pressure and flows through the city, runs into the lake in the north and finally back to the river (Figure 5.17). Mountains are sacred and thus temples were constructed for the worship of water and ancestors. The core urban area has been surrounded by a water moat and was beset with famous spring water gardens. Springs are bubbling and running along streets and wells sit in residential yards for daily use. There is even a spring water pool for public recreation in the centre of the city. However, physical and functional use is not enough. The spring water lake, almost half of the city, was beautifully constructed into a similar poetic water body as West Lake. Mountains lying to the north were

Figure 5.18 Local people get bubbling spring water from the moat as daily drinking water.

culturally elevated into an iconic landscape of the city (Figure 5.18). Once again it demonstrated the everlasting and eternal pursuit of Chinese traditional culture: living with nature artistically. Such nature-rooted respect and high level cultural pursuit was a powerful engine for urban development in the past and it shaped clear identities for cities.

Challenges of Landscape Conservation in Chinese Historic Cities

Historic Chinese landscapes embody great traditional wisdom, especially regarding an essential sustainable human-nature relationship and elevated quality of life. But this precious heritage is under huge pressure and threat from rapid urbanisation in China. As the world's biggest developing economy, China's economic success is admired around the world. Great progress has been achieved in the reduction of poverty for 600 million people within 30 years. 2011 is noted as a remarkable year in which the Chinese urbanisation rate increased to 51.27%, over 50% for the first time in history. Meanwhile, however, frequently uncontrolled development is accompanying rapid urbanisation that next to benefits also causes great environmental, social and cultural crises. The most challenging issues regarding the historic urban landscape are threefold.

The first challenge concerns the continuing deterioration of the environment and the human-nature relationship. The traditional environmental ethics to respect nature have been lost. Water, soil and air pollution caused by an arrogant and devastating exploitation of nature is resulting in a notable social outrage and is seriously threatening public health and safety. The movement of 'developing new towns and new districts' prevails. Mountains have been removed and rivers filled in order to create more land. Water and soil have been polluted by illegal industrial waste disposals in order to lower costs and increase profits, directly threatening food safety. In the last decades, the speed and quantity of urban development was much more important than the quality of life, and now the time has come to pay the price.

The second challenge is the loss of identity of cities. Historic cities were deeply rooted in the local natural and cultural context and consequently gained distinct identities differing from place to place. Today's urban development, underpinned by the power of capital, is keen to copy those 'successful' and 'famous' cities without considering the local context. Southeast water towns were even copied in the arid west in order to have a 'new fresh face' and to attract tourism. This results in 'thousands of cities with the same face', from urban structure to architectural forms. Such development not only caused severe ecological impacts on the local natural system, but also weakened local cultural confidence and erased local history. Ultimately, we will fail to trace back where we are from and who we are.

The third challenge is the lack of capacity to implement measures for sustainable urban development in planning, management and administration. The prevailing mode of urban conservation and development planning in China is 'to protect the old and construct a new'. Now almost every historic city has a new town or a new district. Unfortunately, there is hardly any linkage between the old and the new context. Conservation is only applied within the boundary of designated old areas. 'New Town' or 'New District' means possibilities for all kinds of development, which are restrained in an old historic centre. In Jinan, spring water is still running underground, but it is disappointing to see that the historical wisdom of spring water has disappeared in Jinan's urban planning for the new parts of town and the new districts. The same thing also happened to Hangzhou, the historical urban landscape structure of 'three-side lake and mountain, one-side city', which was maintained for 1000 years, is now being changed into a 'one-side lake and mountain, three-side city'. In the new town we can hardly recognise that we are at the 'Heaven on Earth', as nothing is different anymore from any other urbanised city (Figure 5.19). Landscape, once playing an important role in coordinating and elevating the human-nature relationship and representing an aesthetic quality of the city, has now been marginalised to be only the non-cultural 'green' in urban planning. There is an urgent need for an appropriate and multi-context, specific values-based urban planning approach and processes to deal with the layering of historical natural, social and cultural values.

The Opportunity of Implementing the Historic Urban Landscape approach in China

Fortunately, the above challenges and relevant crises have been noticed in China. The central government is promoting the 'ecological civilisation and building a beautiful China' with the outlook of Chinese historical-environmental ethics and high quality of life. It is noted that China's GDP growth has slowed down and that it is transforming to a 'new urbanisation' away from the current low efficiency and low quality mode. Profound changes are taking place in all areas, including in urban development. It is without a doubt that urbanisation is the most potential mode of development for China in the next few decades and there are great opportunities for exploration. At this point of view, the historic urban landscape arrives just at the right time for China. The provided holistic approach to link the old and the new, conservation and development, nature, culture and society, is what sustainable urban development needs. Its context-based dynamic perspective is especially attractive for developing countries.

The Historic Urban Landscape also provides a compromise for the conflicts between urban heritage conservation and development. Heritage conservation maximises the opportunity to learn and benefit from past wisdom for future use. The Chinese traditional perspective of nature provides a holistic approach to deal with the fundamental human-nature relationship. Chinese nature-related historical urban areas present the ideal of high quality of urban life with an accumulation of natural wisdom and cultural input. Methodologically, it could work closely with the historic urban landscape.

Recently, Historic Urban Landscape has already attracted some pioneer cities in China. The China Roadmap for implementation of the Historic Urban Landscape is being developed by Tongji University and WHITRAP in Shanghai. Pilot projects are being undertaken in several historic cities. It is expected that the Historic Urban Landscape approach will contribute to an influential Chinese transformation towards 'new urbanisation', because the success of Chinese sustainable urban development will not only benefit the Chinese, but the whole world.

Figure 5.19 New high-density residential development along the Qianjiang River in Qianjiang new town, Hangzhou.

6 Planning and Managing Historic Urban Landscapes[1]

Francesco Siravo
Architect, Historic Cities Programme of the Aga Khan Trust for Culture

> *We are wiping out (the legacy of the past) piece by piece instead of reinforcing and reweaving to accommodate change.*
>
> Roberta Brandes Gratz, The Living City (1989)

Integrated Planning

Surveys, strategic assessments and proposals are strictly related and as a whole constitute the fundamental tenets of an integrated conservation plan. This integrated approach differs in important ways from traditional ways of understanding and planning historic cities. Traditionally, planners viewed historic areas as a collection of monuments and buildings to be preserved as relics of the past, whose value was considered to be totally separate from their day-to-day use and city context.[2] This strict preservation approach – likening the historic area to a museum – cannot work as old structures, divorced from their everyday context, lose their social and economic life and eventually become obsolete. This approach went hand in hand with considering other aspects of the urban context – infrastructure, transport systems, land use, tenure issues – as separate and completely unrelated components. Often carried to an extreme, with different

---

[1] This article is based on notes prepared over the past several years for lectures given at the University of Rome 'La Sapienza' and the University of Pennsylvania, as well as field work carried out for the Historic Cities Programme of the Aga Khan Trust for Culture. Further, I wish to express my appreciation to the Getty Conservation Institute of the J. Paul Getty Trust for the opportunity to share these experiences at a workshop on urban conservation planning held in 2012. This occasion was the catalyst in organising much of the material presented here. I would like to thank the workshop director, Jeff Cody, for his valuable comments and suggestions.

[2] This is essentially a nineteenth century concept of restoration applied to urban ensembles. An example is the restoration of Carcassonne carried out by Viollet-le-Duc in 1853, or the late nineteenth and early twentieth century restoration of historic Split, built within the core of Diocletian's imperial palace. More recent examples of museum-like approaches to urban conservation include the reconstruction and restoration work carried out in Saint Malo, Glen and Warsaw after World War II. This approach was in many respects reinforced by some of the pioneers of the Modern Movement, such as Le Corbusier, who, in his *Plan Voisin* of 1925 for Paris' First Arrondissement, advocated total preservation of the *Île de la Cité* as a place for contemplation and repose, and total demolition of two square miles of traditional buildings north of the Seine.

Reconnecting the City: The Historic Urban Landscape Approach and the Future of Urban Heritage, First Edition. Edited by Francesco Bandarin and Ron van Oers.

departments planning and managing each component independently, this *per force* piecemeal approach can be blamed for the often poor results. To the contrary, the basic premise of successful conservation planning is that the infrastructure, services, monuments, buildings, open spaces, streets and other elements that make up the historic urban area must not be treated separately or out of context. They must instead be re-integrated with the other components of the social fabric (including the so-called 'immaterial' or 'intangible' heritage) in order to understand their interdependent relationships and identify the best way to ensure their continued well-being and long-term physical preservation.

The first step in this planning process is a thorough reconnaissance of the area and its different components, which will serve as the foundation for all further work.[3] The way different urban spaces and elements are organised, used, maintained and changed provides useful clues as to the physical problems as well as the socio-economic phenomena at work. A relatively rapid, yet well-organised physical survey provides the tools needed to assess the prevailing conditions of an historic area or neighbourhood, which in turn is the basis for the planning proposals, both at the city level as well as for specific action areas and demonstration projects.[4] While conducting the investigations, it is important to keep in mind that the survey is not an end in and of itself, and that preparation of the planning proposals is the final objective. In this respect, the survey criteria and topics to be investigated should always have a direct bearing on the future development of planning policies and specific interventions. This initial physical survey must be carried out in parallel with the widest possible survey and consultations with the different interest groups and constituencies occupying a given area (cultural mapping). An understanding of the views and expectations of residents and users is indispensable in appreciating the underlying causes of what does and does not work. It is also essential in charting any future course of action for the historic area.

A general survey of the physical elements together with an understanding of the social and economic dynamics, are the premise for the formulation of a plan. The plan is a necessity: without one it is impossible to formulate a programme of action, and the subsequent implementation of concrete conservation and development measures cannot go forward. In essence, a plan for an historic area should establish priority physical, legal, economic and social objectives, supported by the broadest possible public consensus, and draw up a programme of actions to achieve these objectives. Some of the principal questions a conservation plan will have to answer are:

- What are the problems and the main deficiencies in the present organisation of the historic area?
- What sustains its economy, and what depresses it?
- Which trends to be discouraged and which encouraged in order to improve living and working conditions?
- Which economic activities are compatible with the traditional building fabric and character of the historic area? What actions to attract investment?
- Which services and public facilities lacking and to be provided in future? Where and what kind of new development is acceptable?
- Where to effect conservation measures and restrict new construction? What are the opportunities for immediate action?

---

[3] The importance of identifying and understanding the physical and documentary and community aspects that make a place significant is strongly emphasised as a prerequisite for planning by the Burra Charter (The Australia ICOMOS Charter for Places of Cultural Significance, 1999): '*Study before work is a shorthand way of expressing the logical steps in good conservation practice: first understand significance, then develop policy for managing a place, and then implement that policy and care for the place*' (Australia ICOMOS Inc., (2006) *Introducing the Burra Charter*).
[4] Ideally complemented by a cultural mapping exercise.

- What kinds of financial resources are available and how are these to be mobilised?

In providing answers to these kinds of questions, the plan shapes a coherent vision for the future, translating the social, economic and physical objectives that have been identified into a series of specific actions.[5] Some will be physical interventions – a different traffic network, the rehabilitation of an old building, the re-landscaping of a public open area, or the improvement of surface water drainage. Others will result in the enactment of fiscal or legal measures. Others again may involve defining alternative administrative or management procedures, establishing a public subsidy initiative, or creating a mixed public and private investment programme. A plan cannot be reduced to a single document – a physical plan, an act of legislation, a management programme, and it cannot be defined once and for all. Even though, at times, it may be each of the above, a plan is essentially a framework for different actions, a large container for the complementary tools needed to plan and manage the historic urban landscape. Moreover, it is not frozen in time, but instead should be considered as an ongoing process in which programs and actions are woven together to achieve the specified objectives. Over time, it must be monitored and subjected to periodic review and fine-tuning.

Although definition of the plan may be achieved in a relatively short period, its implementation is always a lengthy process spread over many years. Flexible, adaptable plans are easier to manage over time. A plan worked out to the smallest detail is difficult to manage because real situations change over the years and the opportunities for implementing specific actions or programmes are often very different from those envisaged during the planning phase.

In the end, a rigid plan may be more of an impediment than a positive step forward. For this reason, once the conservation plan's objectives and general policies have been specified, it is best to leave the actual identification of resources and implementation modalities to subsequent detailed planning work, when the most appropriate and expedient measures can be developed.

It is, however, at the stage of implementation that the effectiveness of a plan's overall policies and detailed programmes can be assessed. The sooner the planning process is taken to its implementation phase, the sooner its results can be evaluated and, if necessary, adjusted. In the end, a conservation plan attains its objectives if it leads to improving the historic urban landscape for the people who live, work and have businesses there. These improvements can be measured in physical, concrete terms: better housing, well-cared for monuments and historic buildings, enhanced infrastructure and security, more services, improved public facilities, open areas and recreational spaces. Ultimately, planning a historic urban landscape begins and ends with the town's physical fabric: from an assessment of physical conditions to effecting an improved physical environment throughout the historic area (see, as examples, Box 6.1 on the Plan of Assisi, Italy and Box 6.2 on the Plan of Chester, UK).

Key Aspects of Analysing and Planning Historic Urban Landscapes

The key issues to be addressed when analysing and planning historic urban landscapes are essentially the same as those prevailing in any new urban context.[6] They are universal and dictate the functioning of every society: housing, health, income, education, transportation,

---

[5] As examples of strategy formulation see: Siravo, F. and Pulver, A. (1986) *Planning Lamu, Conservation of an East African Seaport*, Nairobi: National Museums of Kenya. Siravo, F. (1996) *Zanzibar. A Plan for the Historic Stone Town*, Geneva: The Aga Khan Trust for Culture, pp. 117–121.

[6] On the concept of Historic Urban Landscapes see: Bandarin, F. and Van Oers, R. (2012) *The Historic Urban Landscape. Managing Heritage in an Urban Century*, Chichester: Wiley-Blackwell.

Box 6.1 The Plan of Assisi

The Plan for Assisi, an Italian hill town located in Umbria in central Italy, remains exemplary from a methodological standpoint and for having established a number of the fundamental tenets of preservation planning. In particular, the Assisi experience proves the importance of an in-depth study and understanding of the historic fabric, together with the concept of continuous planning and sustained management of the urban heritage. Prepared by Giovanni Astengo between 1955 and 1956, the plan departed from conventional planning approaches of the time, which were aimed at piecemeal conservation of monuments and questionable attempts at transforming historic areas to meet 'modern' requirements, such as the introduction of motorised traffic and contemporary building standards. Contrary to this approach, the Assisi Plan was the first in Italy, and one of the first internationally, to look at the historic city in its entirety, as a total ensemble endowed with specific characteristics and qualities. These were recognised as being quite different from those of the contemporary city, as well as requiring different methods of study and intervention. Some of these ideas had already been introduced by Gustavo Giovannoni during the 1910s, particularly the idea that modern city expansions are incompatible with the nature of historic areas, but they had never been tested through the elaboration of a comprehensive plan and the identification of specific methods of intervention. Assisi provided a first opportunity to work with a local administration, which, at least initially, was open and supportive of Astengo's work. His innovative approach is apparent in the following aspects of the plan:

- The importance given to the process of analysis and evaluation of the historic area. Astengo is convinced that a solid understanding of reality is the essential prerequisite for the formulation of a plan. *To know, understand, evaluate and intervene* are for Astengo the four interconnected phases of a single process that leads to the fundamental choices a plan has to make.
- The consideration given to the historic area's social and economic conditions, in addition to those related to its physical configuration and aesthetic values. This is an aspect that today is taken for granted, but which at the time was not normally considered important. In particular, an understanding of the correlations between the physical condition of the city fabric and the socio-economic phenomena at work were aimed at identifying general trends to be curbed, mitigated or encouraged through the policies and actions established by the plan.
- The inclusion in the plan of areas that are on the edge or beyond the strictly historic sections in an effort to protect the visual relationship of the old city and its context and to resolve the critical problems often associated with transitional zones of city expansion, such as provisions for car parking and attention to the scale and materials of new construction.
- The direct involvement of the local institutions and the permanent establishment of a public planning office, as opposed to the temporary recruitment of specialised consultants. Astengo was one of the first to understand that without such a permanent structure, it is impossible to establish the conditions for the long-term planning and management of an historic area.
- The introduction of the concept of 'continuous planning' where the choices related to the formulation of the plan are verified and adjusted over time to respond to medium and long-term objectives, and where the objectives themselves are periodically re-considered or modified to respond to changing circumstances. In many respects, the current emphasis on sustained management as a key element in the preservation and appropriate development of an historic site, were anticipated by the Assisi Plan.

Sources:

Astengo, G. (1958) *Assisi: piano generale e piani particolareggiati di primo intervento*, Torino: Edizioni di Urbanistica. Calabi, D. (2004) *Storia dell'urbanistica europea. Questioni, strumenti, casi esemplari*, Milano: Bruno Mondadori Editore.
Di Biagi, P. (2002) *I classici dell'urbanistica moderna*, Roma: Donzelli Editore.

security, recreation. But the ways in which these issues present themselves in historic areas, as well as the opportunities and problems found here, are different, as are in most cases the solutions. These issues with comments on how they might be approached in formulating a plan and defining specific implementation programmes in an historic context are now discussed.

Box 6.2 The UK experience and the Chester plan

The late 1960s saw the development in the United Kingdom of a sustained experience in the conservation of historic urban areas. The *Civic Amenities Act* introduced the country to the idea of conservation areas for the first time in 1967. Initially based on an extension of the law protecting individual historic structures (the *Town and Country Planning Act 1944*), the protection of building ensembles gradually grew in scope, and was extended to the protection of entire historic townscapes. The preparation of four pilot studies for the historic cities of Bath, Chester, Chichester and York provided an opportunity to test and fine-tune methods and forms of intervention applied to some of the most significant historic towns in the country. The lessons learned helped in the subsequent extension of the conservation area legislation to over 9000 towns and villages in the UK.

The study for Chester was the most comprehensive and innovative of the four studies. Prepared by Donald W. Insall and Associates in 1968, it was explicitly aimed at providing a comprehensive assessment of the City of Chester and making very detailed recommendations for a general conservation programme, as well as specific interventions within individual sections of the city. The document (*Chester. A Study in Conservation*, HMSO 1968) remains a methodological model for a comprehensive analysis plus proposals extended to an entire conservation area. The following aspects deserve special mention:

- A general assessment of the Greater Chester area, including an evaluation of land uses in and around Chester, considered necessary to set the stage for a closer integration of the historic area and its larger context. Special attention is given to parking and circulation of automobiles, but in ways that, contrary to prevailing approaches at the time, aimed at identifying non-disruptive circulation arrangements. These were not based on road expansions, but on existing street networks and available spaces that were to be re-organised for minimum traffic impact, peripheral car parking, greater pedestrian use and more efficient public transportation. These measures had correctly been identified by the Chester Plan as the appropriate response to the effects of expanding motorised traffic in historic areas.
- A critique of new buildings and prevailing development patterns both within the centre and its proximity. Even though this critique may be considered 'soft' by current conservation standards, the Chester Plan outlines an important and still valid methodological approach. It is based on a good understanding of the nature of the city's historic fabric and the recognition of its incompatibility with the bulk and scale of modern construction. It also pays close attention to the urban landscape as a whole, and to the buildings and environmental elements that contribute or detract from it, in order to make recommendations for its general improvement. The detailed townscape appraisals extended to entire city sectors point out sore points or unsatisfactory conditions, identify significant views and architectural values, make recommendations for new listings, and indicate ways to address or rectify unwanted developments and unsatisfactory trends.
- A programme for the re-conditioning of historic buildings, particularly houses, including the identification of the properties in need of urgent attention, the nature of the works, the costs involved and the ways and means by which the rehabilitation process could be started and carried out over time through continued repairs and regular maintenance (improvements grants, tax concessions, establishment of a building corporation). Paramount in the Chester plan is the recognition of the key-role played in the conservation of historic areas by policies aimed at sustaining the full occupancy and rehabilitation of existing residential and commercial structures. The proposal to establish a corporation tasked with the purchase of listed structures, either under-used or in poor condition, which would then be rehabilitated and leased at a profit, was one of the first instances in which private funds were seen as instrumental in the development of a long-term conservation programme. In this respect, the Chester Plan envisioned the important role private funds could have in the rehabilitation of the historic heritage, until then considered exclusively a burden on public resources.

As a result of the Chester study, the local administration initially established a comprehensive conservation area extended to a total area of eighty hectares, including areas located outside the old city walls and 140 listed structures. In 1971 a multi-year conservation fund was set up to provide support for public conservation work and assist private owners to repair their buildings under the supervision of a conservation officer. Throughout the 1970s conservation-related activities focused on measures to mitigate the impact of motorised traffic, restore major landmarks, control new construction and launch a concerted effort for the rehabilitation of the Bridgegate area,

(Continued)

one of the areas of the city in particularly poor condition, with many under-used and derelict structures. The rescue programme for Bridgegate did not foresee any special publicly-funded initiative, such as public acquisition of properties or the allocation of extra-budgetary funds, but was largely based on the implementation of ordinary rehabilitation measures. These consisted in the gradual upgrading of individual structures and public open areas financed by private owners and the yearly funds made available by the local administration for public works. What made the difference was a conservation-oriented vision of the objectives to be achieved and proper coordination of activities between public institutions and services, and private owners and investors. Two additional conservation action areas were added after 1976. The Chester experience, and the positive results obtained over a period of more than forty years, proves that the conservation of historic areas can be achieved through well designed conservation programmes, good coordination between public and private actors, and the implementation of consistent policies over an extended period of time.

The Chester and related experiences developed at the national level led to the general updating and consolidation of the previous legislation into a single act, the *Planning (Listed Buildings and Conservation Areas) Act* of 1990, and, during the 1990s, to the creation of new instruments, such as the 'town schemes' and 'conservation areas partnerships'. These new planning tools gave greater responsibility to local administrations in the management of conservation areas and a stronger role to the residents in the decision-making process, as well as opening the way for increased private investment in the rehabilitation of the historic heritage.

Source: Insall, D. W. (1968) *Chester. A Study in Conservation*, London: HMSO.

Understanding Land Ownership, Tenure and Land Use

Land issues are a crucial and sensitive aspect of planning and managing historic urban landscapes. Recording and analysing the locations and sizes of existing uses, as well as public facilities – particularly for health, education and recreation – can help planners determine the present and future requirements of a central district. Also an understanding of ownership patterns and forms of tenure is important as these are often complex and affect the kinds of policies and initiatives that can be used to implement planning measures. The patterns of ownership, tenure and use in historic areas are often quite complex and interwoven, resulting from a varied and long-established stratification of uses. Also, public facilities, such as green spaces, schools and health centres, are often underrepresented or non-existent. These conditions require on the part of administrators and planners a special effort to address unresolved, longstanding issues, such as the presence of poorly maintained public properties, the lack of security of tenure for low-income residents, and the identification of underused open spaces and/or buildings suitable to be converted into public facilities.

Integrating Socio-economic Realities

The effectiveness of a plan depends to a large extent on how well its policies and measures respond to the needs and expectations of residents and other users. A reliable social and economic profile of the households residing in the historical area, including information on levels and sources of income, will help determine people's attitudes, abilities and willingness to contribute to upgrading the buildings and improving their surroundings. Other useful information may be solicited from the business community, particularly regarding the potential for commercial growth and increasing employment. This information is important as it will help in designing incentives to attract more business and generally create a policy environment conducive to private investment.

Upgrading Infrastructure

Public utilities are the same in old and new urban contexts: water, electricity, sewage, drainage, paving of streets and roads, street lighting and telephone lines. But in historical areas intervention is markedly different. Normally, the need for capital financing of new installations in old city centres is limited because the infrastructure in most cases is already there. There is however considerable need for funds to maintain and upgrade old systems that are often in poor condition and possibly obsolete. An investigation of these will help determine the capacity and adequacy of existing facilities and ascertain what needs to be done to update and improve the infrastructure. Any changes to existing networks need to be considered carefully because of the possible impact on the surrounding historical structures. New installations may require specially designed elements, whether for aesthetic or structural reasons, as well as environmentally sensitive solutions.

Improving Public Services

Improving the delivery of public services is often indicated by residents and other users of historic neighbourhoods as their highest priority for public action. Problems vary from place to place, but poor services due to lack of equipment, personnel and funds tend to be of concern everywhere. Innovative, more effective solutions can be devised with closer cooperation with community groups to improve delivery of services and facilitate operations and prompt payment of charges. Properly functioning infrastructure networks and consistent delivery of services may be the most important factors in attracting and sustaining a positive economic climate in an historic area.[7]

Re-thinking Transportation Policies

Private cars and modern transport systems are certainly not compatible with historical areas. Laid out long before automobiles existed, the streets of old urban areas cannot sustain a lot of motorised traffic. Accommodating the increasing levels of traffic in town and city centres has required progressively larger investments that have added to the problem rather than solving it. New arteries have been created – often demolishing valuable historic property – to allow yet more cars to be introduced without a real understanding of the consequences. Complicated circuits are invented that become new bottlenecks choked with traffic, leading to higher pollution and vibration levels that undermine public health and loosen building structures. Historically, old towns worked well because spaces and activities were highly integrated. These special characteristics of historical areas are often studied as models of successful planning, and solutions to today's problems must be devised along these same lines, with a careful distribution and integration of different activities. Movement to and from the historical area, and thus the relationship between it and the greater urban context cannot be considered in isolation. This relationship needs to be evaluated vis–à–vis the general planning options for the historical area, to harmonise and re-integrate the transportation system into the planned distribution of land uses and activities. In addition, the use of private cars will need to be discouraged with a series of carefully worked out incentives and controls, including peripheral parking, more public transportation, coordinating private and

---

[7] On the delivery of infrastructure and public services to disadvantaged communities, see: Davidson, F. and Payne, G. (eds) (1983) *Urban Projects Manual*, Liverpool: Liverpool University Press. See also Tayler, K. and Cotton, A. (1993) *Urban Upgrading. Options and Procedures for Pakistan*, London: Published by WEDC for ODA.

public transit systems, as well as exploring pedestrian and other non-motorised alternatives.

Working on Old Buildings

How to work on traditional buildings is perhaps the most difficult problem public administrations have in historic areas. Their response ranges from attempting – and failing – to enforce strict and sometimes obscure regulations, all the way to an irresponsible *laissez-faire* attitude. This is not surprising if one considers how much building methods have changed since the early part of the last century, and the fact that in many countries craftsmen expert in traditional building crafts have now disappeared. This impasse can be overcome first of all by deepening our understanding of the historic fabric. Assessing the types and condition of buildings in historical areas helps determine the adequacy of the existing building stock and evaluate the extent and cost of future rehabilitation work. In identifying and classifying the buildings in an historical area, typical, recurrent features should be noted. The more this analysis is based on real physical features, the easier it will be to establish the criteria and guidelines needed to protect and rehabilitate the old structures. Additional clues are provided by an analysis of the types of transformations carried out over time by residents. This can help in understanding people's motivations for making changes and identifying the most appropriate strategies to counteract unwanted transformations of historical buildings, while at the same time accommodating the needs of the inhabitants. An accurate analysis will also facilitate identification of the best rehabilitation options and strategies, such as incentives for owners or building rescue programmes designed to redress major widespread deterioration. But above all, there is a need to re-introduce traditional construction methods, educate residents and builders with expert advice and demonstration programmes, and train a new generation of craftsmen to carry out maintenance and rehabilitation work. Only a public policy firmly oriented in this direction can reverse the present unsatisfactory state of affairs and rescue the increasingly threatened and dilapidated historic fabric.

Financing Conservation and Development Programmes

No conservation plan can be implemented without funds. Resources need to be identified and mobilised prior to initiating any concrete programme. Financing and financial management is a complex issue, and here we have room only for a few essential observations on the subject (see Chapter 12 for further discussion). The first is that proper use and good administration of public funds on the part of the empowered agency is the key to establishing credibility and attracting funds from external sources and support from the community. Second, in an age of diminishing public funding, there is a need to generate additional resources from the non-governmental sector. Increasing taxes and levying additional fees is one way of doing it, but communities will often resist, especially when the public institutions have little to show for the increased financial burden. A better approach is to explore ways of raising external or private funds, and tapping people's capabilities to contribute directly. Policies to this effect might include provisions to attract more commercial investment in an effort to generate employment and increase the tax base, or, alternatively, delegating responsibility for certain services and amenities to community groups, commercial enterprises or non-governmental organisations.

Governance: The Case for Public Management in Historic Urban Areas

Although traditionally urban planning has been the responsibility of national or local governments, the poor results obtained in dealing with the problems of old city centres has led many to question the effectiveness of public management of historic areas, calling for more private involvement and for the relaxation or total aboli-tion of government controls. Nevertheless, the case for keeping historic areas under the juris-diction of public entities remains strong for at least three good reasons:

- A public institution is – or should be – *impartial*. In a complex environment where different groups co-exist and compete for space and resources, it is important that the institution in charge be perceived as a non-partisan organisation capable of acting in the general interest and giving fair consideration to the needs and expectations of all the dif-ferent parties concerned. A public institution is in the best position to subordinate short-term or individual interests to superior con-siderations, such as the preservation of open areas – a scarce commodity in urban centres – or structures of architectural significance.
- A public institution is also in the best posi-tion to ensure *proper coordination* of plan-ning and management activities carried on by the various government agencies and private groups active in the historic urban area. Coordination is all the more essential as these areas tend to have very high building and population densities, where every small decision will inevitably set in motion a series of physical and political repercussions. The various activities therefore need to be care-fully phased to ensure that the results are appropriate to the historical context and compatible with each other.
- The coordination and implementation of complex planning activities normally takes many years. As in the case of Assisi, only a public body with a permanent mandate and a well-defined brief can ensure *continuity of effort and the sustained commitment* needed to plan and manage a historic area over the long term.

What Kind of Public Institution?

Depending on the importance of the town or city and on the political system prevailing in dif-ferent countries, public planning of an historic area may be the prerogative of the central gov-ernment, or of a provincial or regional institu-tion, a municipality, another form of local government, an independent *ad hoc* agency created especially for the purpose, or any com-bination of the above. The most difficult institu-tional arrangements to manage are those where there are confused or overlapping areas of responsibility between different levels of govern-ment. This situation usually produces conflict and can eventually paralyse planning and man-agement efforts. Other difficulties may arise when the level at which decisions are made is too far removed from the local reality. This is often the case with central or regional government institutions.

The most appropriate level of governance for historic areas is the local or municipal gov-ernment level. Here there is close contact between administrators and their constituencies, the users and residents of the area; it is also where essential community services are dis-charged, and where – particularly in small to medium-sized towns – the administration's ability to respond to new situations or effect nec-essary changes rapidly is greatest. In many coun-tries, however, the local level of governance is also the level where the administration's legal and enforcement powers may be fairly limited, working facilities inadequate, and financial resources lacking. In addition, personnel at this local level are in most cases insufficient, poorly paid, only partially trained, and often subject to considerable political and partisan pressures. For these reasons, the approach that has produced the best results has been that of rein-forcing the local government with the establish-ment of a new specialised conservation and development agency that has full jurisdiction over the historic area as well as special powers, increased resources and highly trained profes-sional staff.

The main advantages of such an *ad hoc* agency – as opposed to a local planning authority with general responsibility for a large geographical area comprising old and new urban sectors – is that the specialised agency will focus entirely on the historic area and treat it as a single, integrated entity requiring a specialised yet comprehensive planning approach. As we have seen, this grows out of the fact that historic urban areas call for policies and planning methods that are different from those applied to new development areas. Equally important is the increasing awareness that, in order to operate effectively, the new agency must be given special implementation powers – without which it will not be in a position to act effectively. The full empowerment of an *ad hoc* institution normally requires strong support from the central government to provide funds, facilities and equipment normally not available at the local level, in addition to special permission to hire external consultants and local professionals at competitive salaries. More importantly, it will require some form of enabling legislation to allow the new institution to assume sensitive legal and financial responsibilities including the ability to buy land, manage public property, collect revenue, and allocate financial resources.

Organisational Framework of the Conservation Agency

In order to fulfil its expanded role, the *ad hoc* agency requires an efficient and diversified organisational structure capable of covering all the different areas for which it is to be responsible. It may be unrealistic to concentrate responsibility for all the activities and individual tasks involved in overseeing the planning, management and daily operation of a complex historic area into the same institution. But it is possible to imagine a smaller agency empowered to represent the historic area's principal interests and constituencies, and having the technical, analytical and advisory expertise in key planning and management sectors.

Although it is difficult to indicate outright the ideal framework for such an institution without reference to a particular situation, one can foresee, in very general terms, a two-tiered organisational structure:

- A coordinating committee responsible for setting the general policies and ensuring broad public consensus for the programmes envisaged by the plan. The committee would be made up of prominent members of the community, representatives of the various constituencies, and officials from ministries and other central or local authorities that provide services to the area. This committee consults on the objectives and reviews the plan during its preparation; endorses it before passing it on for official approval; and, subsequently, during implementation, sets priorities and targets, decides on the general policies to be pursued and negotiates arrangements within the community as well as external parties, donors, and government agencies.

- A strong technical planning office responsible for translating the committee's broad targets and decisions into conceptual and increasingly more detailed plans of action – the office is staffed by a highly integrated group of professionals providing inputs in key sectors. They are not policy makers, rather their role is analytical and advisory, both during formulation of the conservation and development programme and the subsequent implementation phase. These officers will act as the link, the go-between among the different players, and help develop and promote the solutions needed to solve the special problems of the historic area. This corps might include engineers, architects, planners, sociologists, economists as well as traditional construction and conservation experts. Where local expertise is not available, this can be provided by central government employees on temporary secondment, or through external consultants. In the long term, however, local staff must be able to perform most of the necessary tasks, and training and capacity-building should be part of the institution's mandate from the beginning.

Participatory Planning and Implementation Strategies

Centralised planning and attempts to mould the future organisation of towns and their historic areas following pre-set 'ideal' schemes have generally failed for lack of support and because they do not reflect the real problems and needs of the people. Successful plans are those that consider central historic urban areas as a common good and mobilise resources and people to improve them for all concerned. This planning approach requires from the outset the involvement of private and public interest groups operating within the historic area.

To begin with, planning institutions do not have sufficient resources to act alone. In order to mobilise the necessary means and achieve visible results, community groups will need to be involved – both in shaping the plan's objectives and policies and in implementing practical initiatives. Even in cases where the public institution responsible has an inclusive mandate and all the resources needed – a very rare occurrence – it is unlikely that significant progress will be made without the active participation of those directly affected by the plan. This is why it is crucial to create lasting links and partnership arrangements with the various players who have a direct interest in the historic area. Without such arrangements, ideas and plans are not likely to move beyond the planning phase. Broadly speaking, the different groups operating in an historic area can be identified as:

- Other government agencies and institutions, municipal departments or semi-private agencies that provide services or that have interests or properties in the historic area;
- Local constituencies, religious groups, residents, and community organisations sharing similar problems and expectations for improved public services and housing;

- Shopkeepers, artisans, small manufacturers, café and restaurant owners, as well as vendors and people in the informal sector who are concerned about better working conditions;
- Property owners, developers, hoteliers and larger businesses or institutions – banks, insurance companies – which are often eager for investment opportunities and an improved business climate in the historic area;
- International aid organisations and bilateral donors active in different fields such as infrastructure development, housing, health, small business development, conservation and education. All these sectors can be linked directly or indirectly to improving living and working conditions in historic areas;
- Local and international non-governmental organisations (NGO's) operating in various development sectors. These organisations have a particular interest in promoting activities at the grass-roots level and have often established direct links with local communities. They can be instrumental in stimulating public awareness and participation in the planning process.

One way of ensuring public participation in the plan's conservation and development programmes has already been mentioned. It consists of involving at institutional level selected representatives of the various constituencies and other public agencies. Such a high-level steering committee can be instrumental in identifying priorities, seeking the necessary support, and helping mobilise the wide range of resources needed during a plan's implementation.

Other initiatives that can facilitate participation include informal meetings where local elders and leaders, long-time residents and experienced administrators can help with gathering information and identifying specific objectives and implementation modalities. Public meetings open to one and all require careful preparation, but are indispensable in evaluating first hand people's expectations and reactions to different ideas and proposals. Open meetings can also help identify opportunities for direct involvement and partnership.

Partnership arrangements can be developed in many different sectors. Their individual merit will be determined by evaluating their contribution toward resolving issues of public interest – which remains the administration's first priority – and by assessing their impact on improving living and working conditions, safeguarding and enhancing the physical environment and in promoting the social and economic development of the historic area. Below are a few examples of how public agencies and institutions and non-governmental players may cooperate to mutual advantage in historic urban areas:

- A public service agency which has trouble providing its services efficiently or collecting user charges may benefit from collaboration with a constituents' organisation which can facilitate cooperation with the community, whether by assisting with the operation of a given service or by promoting quicker payment by users. For instance residents can be organised to deposit garbage at more accessible collection points, helping the municipality save on staff. The savings can then be invested in increasing collection and improving delivery of the service. In most cases, deteriorating public services are part of a vicious circle that can only be overcome with dialogue and better cooperation between users and providers.
- A group of shopkeepers with shops along the same street can be persuaded to contribute to the cost, provided the work is carried out speedily, of improving the paving or lighting of their street to enhance business in the central urban area.
- Tenants of government-owned buildings may very well have an interest in forming cooperatives and auto-financing emergency repairs or basic service improvements in their homes in return for security of tenancy,

government-funded technical advice and supervision of home improvement works.
- Buildings of historical importance can be rehabilitated and adapted as useful spaces with funds provided by public or external organisations in need of space for their activities. A local dispensary, a small training centre, or a nursery. If done with sensitivity, if proper restoration techniques are used, and if there is an understanding and respect for the original structure, these new occupants will give many more years of useful life to buildings otherwise likely to be neglected, underused, left empty or, worse, demolished.

Beyond the particular arrangements made to mobilise resources and people for different initiatives, establishing a direct and sustained dialogue between the planning institution and the historic area's community can bring about a climate of confidence and trust, and, eventually, the consensus needed to act. [8] As with any political process, public participation is a complex and time-consuming exercise, which requires compromises and a balancing act between the various groups and sometimes powerful interests. Without this dialogue, however, all players are bound to lose, as little or no progress can be made in ameliorating unsatisfactory conditions and translating ideas into concrete actions (a good example of the application of these principles is presented in Box 6.3 , that deals with the case of Amsterdam in the Netherlands).

Conclusion

Attention is often focused on the difficulties of planning in historic areas. Planners and administrators have been disoriented both by the complexities of the social and economic issues, as well as the many unresolved physical problems in older urban areas. The traditional fabric is often incompatible with modern building prac-

---

[8] On the issue of public participation, see: Brandes Gratz, R. and Mintz, N. (1998) *Cities Back from the Edge, New Life for Downtown*, New York: Wiley.

Box 6.3 Public participation and the Amsterdam plan

Amsterdam is without any doubt the city that over time more than any other has maintained a continuity of approach and relied strongly on public advocacy and participation in the development and implementation of a comprehensive plan for its large historic area. It has also been the principal testing ground for some of the most successful policies in preserving the built environment and adding new architecture within a historic context. This extraordinary and fairly unique planning experience is retold by Anthony Tung in his book on the preservation of world cities. It is a story that explains why Amsterdam remains one of the most successful and visually homogeneous cities in Europe, and why it has lessons to offer to any city that wants to develop a coherent planning system that is firmly rooted in its historic past.

Since its inception, the Amsterdam plan reflected a conscious effort to plan and control the form of the city through and by the public administration. Amsterdam's distinct structure came about in the early seventeenth century with the formation of a series of canals at the edge of the original town – a medieval settlement on the River Amstel. The waterways surrounding the town were retraced as a series of encircling canals, giving Amsterdam its distinct horse-shoe shape. Toward the end of the seventeenth century, the city, now including four successive rings of canals, was encircled along its southern edge by defensive bastions. After the middle of the nineteenth century, the city grew concentrically beyond these city walls, first according to the 1876 *Plan-Kalff*, and, subsequently, between 1920 and 1940, according to the *Plan Zuid* for the area to the south and following the Plan West for the area to the west of the historic city and its nineteenth century expansions.

The historic canals were and continue to be lined with docks, streets and blocks of houses, which, because of the concentric disposition of the waterways, are trapezoidal in form. The current size of the historic centre, recognised by UNESCO in 2010 as a World Heritage site, is 680 hectares, including its buffer zone, with 82 000 inhabitants, about 10% of the total population registered within the Amsterdam city limits.

From the outset, modernising pressures in Amsterdam came into direct and dramatic conflict with the intrinsic character of the old city centre. The poor state of the urban fabric and the general call to modernise, including the introduction of cars and the conversion of the canals into streets, were all too correctly perceived as a threat to the old city. Already in 1900, the closing of a canal had led to strong opposition and inspired the birth of the first conservation organisation, the *Amstelodamum Society*, whose participative approach would become a distinctive characteristic of planning in Amsterdam. This was followed in 1918 by the establishment of a private group of concerned citizens, the *Hendrick de Keyser Society*, which began restoring a number of buildings both in Amsterdam and other parts of the country.

In 1921, the public sector became directly involved with provisions in the Amsterdam building code requiring that new developments in the old city be approved by a Committee of Urban Beauty in order to avoid inappropriate transformations in the city centre. By 1933, a public inventory of monuments and significant buildings had been established at the national level, with 4200 structures identified in Amsterdam alone. The Second World War brought all urban recovery programmes to a halt. During and immediately after the war, parts of the historic area were abandoned and monuments lost due to the protracted lack of maintenance and poor condition of many buildings. In fact, more and more people migrated to areas outside the city centre, a pattern repeated in cities all over western Europe.

Post-war government efforts at modernisation within the old city limits, however, met with renewed opposition. Transformations directed at traffic improvements (the closing of the canals) and privately-funded building substitutions were perceived by the remaining residents as ways to upset their traditional environment and disrupt their way of life. They worried about the risks of gentrification, especially expulsion from the centre. In 1952, an open letter sent to municipal officials by residents and preservation groups stated these concerns in terms that anticipated the current attention given to the totality of the urban landscape: '(…) the approach to historical building conservation based on protecting individual properties is unsatisfactory. Amsterdam's townscape (…) does not derive its beauty from the relatively small number of buildings important to the history of art. (This) beauty springs from the totality of her facades, their colouring, their proportions and their interplay with the street-canal environment – including its cobbling and greenery. Preserving remarkable buildings alone will inevitably result in the erosion of the townscape as a whole.' Eventually, the government became engaged in a long-term effort to preserve the core of the city through the establishment of a Municipal Department for the Conservation

(Continued)

and Restoration of Historic Buildings and Properties (1953). This was followed by building regulations attuned to the character of the traditional buildings, with controls on total and floor-to-floor heights, façade proportions and building materials (1957).

The issues of cost and the permanence of the disadvantaged social groups residing in the centre, however, remained unresolved. Both issues were addressed in 1957 with the establishment of *Stadsherstel*, (The City Restoration Company), with the specific aim of rehabilitating historic houses threatened by negligence and demolition. *Stadsherstel* is a limited liability company that works as a not-for-profit organisation, reinvesting its profits in the acquisition and renovation of more houses. These activities are financed with the company's funds as well as investments from various public and private institutions, including the City of Amsterdam, large banks and insurers that are willing to accept lower than usual profits. It is involved not only in the renovation of the buildings, but also in making sure that the original residents remain in the houses after renovation. Finally, the Company acts as a landlord to ensure that repairs and regular maintenance take place.

As reported by A. M. Tung in 2001, the Company started in 1957 with a capital value of approximately $500,000. By the mid 1990s its capital value had grown to over $500 million with approximately 350 buildings (later expanded to 478), including nine churches, a shipyard, two pumping stations, 941 dwellings, 175 shops and commercial spaces and 32 pubs and restaurants (2002). But beyond these numbers, *Stadsherstel's* most significant contribution to the renovation of Amsterdam was its strategy of investment in key locations at the right time in ways that stimulated and supported parallel private investment. The crucial role played by *Stadsherstel* raised the hopes of residents in marginalised communities and initiated a process of gradual but steady rehabilitation through private investment, while at the same time counterbalancing the negative social impact of substituting buildings and gentrification. Its significance as one of the most successful Public/Private Partnerships in urban conservation cannot be overstated. After 1999, the Company merged with the Amsterdam Monuments Fund (*Amsterdam Monumenten Fonds*), thus acquiring expertise in the restoration of monuments and playing an even greater role in advancing its fundamental economic and social aims.

In order to fully appreciate the city of Amsterdam's significance in preservation planning, let us mention another important project: the Reconstruction Plan for *Nieuwmarkt* implemented in the 1960s and 1970s. Also in this case, the sympathetic reconstruction of a dilapidated area, which had originally been the Jewish quarter, located just east of the city centre, began with strong opposition from residents to government plans to demolish parts of the area, including its original canal system, and create a new highway. In spite of this opposition, part of the government plan went ahead, and eventually a compromise had to be struck between the remaining historic structures and the new insertions. This experience demonstrated the disruptive impact of large modern structures in historic settings. However, the *Nieuwmarkt* experience also raised awareness about the need to adopt a different approach and explore ways to keep the underlying historic patterns of the city fabric while integrating it, where necessary, with the sympathetic design of new infill structures.

By 1971, when the *Jordaan* district west of the centre was to be rehabilitated, both new legislation for the protection of the historic areas and a greater awareness of the consequences of unsympathetic modern design were already in place. In fact, the need for careful integration of new designs within a traditional context was now fully acknowledged. Moreover, rehabilitation plans and works were not left only to 'specialists', but were subject to public design reviews and the participation of concerned residents and organizations. In this respect, Amsterdam could count on a long tradition of participation in planning and aesthetic decision-making going back to the last decades of the nineteenth century. This participation stemmed from an early awareness of the city's distinct urban qualities and historic traditions and the need to reconcile these with modern developments, a concept strongly advocated by Hendrik Petrus Berlage, the influential architect who was responsible for the development of Plan Zuid, the new housing expansion south of the historic core (1904). He, in turn, was strongly influenced by the work of the Austrian architect Camillo Sitte on city planning (*City Building According to its Artistic Principles, 1889*). Berlage was acutely aware of the fact that a city is more than the sum of its individual buildings. Convinced that the harmonious development of an overall plan was a necessary pre-condition to achieve quality results in urban settings, he had a great impact on an entire generation of architects who were being trained in Amsterdam's newly established architecture academy. (These same architects would later be involved in the design of the early twentieth-century social housing projects, still well known today as the distinctive product of the Amsterdam School.)

Amsterdam's rehabilitation efforts profited from another key element, the creation of a new permanent position within the City Council's Public Works Department, that of aesthetic advisor, to work with all the players – architects, craftsmen, builders, developers and residents.

Amsterdam's strong tradition of participatory planning and public design reviews is at the core of its long experience in adaptive re-use and contemporary infill design within the city's historic area. Amsterdam has developed an approach that is far from the radical urban renewal and the out-of-scale, wilfully contrasting modern buildings that are typical in so many other cities. Municipal restrictions, far from being perceived as an undue burden imposed on individual creativity, are seen here as a challenge to develop innovative and sophisticated designs that are in harmony with the adjacent historic structures, bringing forward a new vocabulary of compatible forms and maintaining a sense of continuity with the city's past.

Sources:

Tung, A. M. (2001) *Preserving the World's Great Cities. The Destruction and Renewal of the Historic Metropolis*, New York: Clarkson Potter Publishers.

Eggenkamp, W. (2004) *Stadsherstel Amsterdam. A Unique Example of Public/Private Partnership in Restoring a City*. Deben, L., Stalet, W. and van Thoor, M.-T. (eds) *Cultural Heritage and the Future of the Historic Inner City of Amsterdam*. Amsterdam: Aksant Academic Publishers.

tices, not to mention the automobile and modern traffic systems. Too often, the solution that is advocated, particularly in economically emerging countries, is wholesale demolition. Creating a *tabula rasa* and starting from scratch is mistakenly considered the simplest and most efficient solution, and only when it is too late does it become apparent that the situation is even more complex and difficult to manage than before.

The international experience of the last decades, however, has brought a renewed appreciation of the value of city centres.[9] This appreciation has recently been extended to entire urban landscapes, independently of their monumental and architectural merit. People, administrators and planners have begun to realise the inherent value and potential for development concentrated in old towns. It is enough to consider the irreplaceable investment of past generations in creating buildings and infrastructures, the relative ease of movement in these closely connected, compact areas, the strong productive base represented by the variety of people and array of skills that tend to be concentrated in any traditional town. We must also consider the large concentration of financial and communication services, the variety of businesses and opportunities for employment at all levels, the concentration of religious and public structures, and the inherent social stability of these settlements. It is difficult to imagine replicating similar urban entities today, and indeed doubtful that we would be able to create *ex novo* such socially vibrant urban areas or invent a similar quality of architectural expression.

As we have seen, the issues when planning historic areas are universal, the same as in new cities. The difference lies in the approach, the techniques and methods that are applied, and – more often than not – in the highly satisfactory results that can be achieved by planners working in traditional contexts. Planning for historic areas brings into sharper focus what does not work in our modern settlements and forces us to consider the cities created by past generations in a different light. We begin to realise that traditional urban spaces have something more to offer, and that conservation planning offers lessons that can be applied across the board, beyond the confines of historic areas. Town planning in recent years has seen a remarkable convergence of ideas and objectives: planners working on new towns are striving to replicate the same human scale and harmonious environment that conservation planners are trying to preserve in historic urban areas, which are indeed the heart and soul of so many of our towns and cities.

---

[9] For a worldwide review of conservation initiatives, including city centres, see: Stubbs, J. (2009) *Time Honored. A Global View of Architectural Conservation*, Hoboken, New Jersey: John Wiley & Sons, Inc.

Interview with Mohsen Mostafavi
The Challenge of Urban Transformation
Alexander and Victoria Wiley Professor of Design and Dean of the
Harvard University Graduate School of Design, Cambridge, Massachusetts, USA.
5 March 2013, Cambridge, Massachusetts, USA.

Question: *In your work, you have explored new ways of dealing with the future of cities. How have you been able to redefine the nexus between urban preservation and development? This is an issue that concerns urban conservationists, mostly because contemporary structural social changes have created new challenges.*

Answer: I agree that as far as urban areas are concerned, we are facing many new challenges. Many parts of our central cities have suffered urban degradation, as is visible in the United States, perhaps more so than in Europe, where the notion of the historic core is still strong. In the US, the relationship between the central district and the sprawl into suburban areas has dominated the past century, and the result has been the emptying of the heart of the city. There is a clear imbalance of density and intensity of uses between core cities and suburban areas, although the situation is not the same everywhere. In the US, the main variable is the quality of urban governance; we cannot separate the issue of urban preservation from the issue of governance. In American cities there are different types of governance structures, and the professionals and administrators dealing with suburban areas are not those who understand the problems of the core areas, where preservation strategies are based on a mix of conservation and adaptive reuse. This is the key issue today, as it is clearly not possible to 'mummify' the city: we have to learn to deal with it as a living organism that is in constant trans-formation. This was in a way the important concept of the 'propelling monument', introduced by Aldo Rossi. With this term, he defines the concept of monuments that change with time, that are transformed at various stages of their lives as a condition for their very survival.

The conservation discipline has established a different approach, one that I see as problematic: the preservation of the monument to an 'ideal' moment of authenticity. This 'fetish' of authenticity has brought many people to become dogmatic and even fundamentalist in the ways that they inter-pret architectural and urban projects in historic buildings and areas. I think we need to change the mind-set of professionals on this issue. Another aspect is the question of the methodological approaches and sensitivities toward historic preservation in different cultural contexts. I refer, for instance, to the differences existing in approaches to urban preservation between the Anglo-Saxon world and Italy, where the architectural design principles are more sophisticated as compared to the thinking in the UK.

I believe that research work on the different sensitivities to preservation should be at the forefront, but I do not see the literature addressing it. This discussion would also open the way to an analysis of the way projects are received by the public. If you accept the idea of looking at a building to understand its life and evolution, you also are willing to participate in the multiple histories of an architectural project. I find problematic the systematic attempt to erase the multiple histories of a building or a place and reduce them to a single moment of architecture. There must be room for a more systematic discussion on the approaches, from both philosophical and aesthetic viewpoints, and this is completely missing from research work and discussion of projects, where a certain dogma related to authenticity and age value is dominant. We need to reconsider these issues when dealing with urban preservation, so that we are cognizant of the different types and values of historic areas.

Question: *There is a fracture between approaches to the preservation of the historic city and modern contemporary development; today these are two separate camps. How do you see the issue and what can be done?*

Answer: We do have this cleavage today between preservation of historic districts and modern urban planning, and this is not a positive situation. I think the historic city should be seen as a palimpsest, an important device for understanding the functioning of the modern city. On the one hand, we cannot really recreate the historic city as it was in the past, as this would be a nostalgic project and is not even possible. On another side, through analysis of the historic city we can understand the roles played by its different parts, and we can develop a parallel with modern life. This analysis of the city and the way it functions, of the way life was lived, has a lot to offer to modern city planning.

I think that it is a mistake to look at the historic city as a fixed image and try to replicate it as a form of salvation of the contemporary city. This is what New Urbanism tries to do, as it believes in the idea that the city is its image.

We should of course not downplay the image, but this approach has been problematic as it is based on a nostalgic attitude. Its creations are places that are not really cities but suburbs, with the appearance of historic places. What I find positive in the Historic Urban Landscape approach is that it deals with the image, but also with the functioning of the city – what I would call the 'performance' of the city. In the historic city there was a clear relationship between residential functions and public functions, and we need to better understand the ways in which these elements were interrelated. The development of the modern city is fundamentally based on a clear separation of functions, and we need to create alternative approaches in which the concept of hybridisation, of mixing of functions, becomes part and parcel of the organisation of the city and of urban planning.

The current models, at least in America, are either suburban development with its mono-functional residential district or the New Urbanism concept of small-scale communities. I believe that the challenge today is to reconsider the city and think of new forms of urbanisation. Some of the work I have done in connection with landscape and ecology, under the concept of ecological urbanism, is opening new dimensions, discussing the performance of the city in terms of the combination of ethical and aesthetic problems.

When you design something, you represent in some way the form of the building. The problem is that these outlines then become reality – an abstraction becomes real very quickly. These compositions become imitations, and this is why I use the term scenography in a negative way. In the theatre, scenography is a tool to create an environment, so for me ecological urbanism is a way to distance ourselves from the reliance on composition by introducing other factors, such as mobility, resources, the life of the city, the systems of production and distribution. Many of these factors produce their own systems of relationships. This is what I mean by performance and functionality. It's the description of a certain set of attributes that concerns connection. Modernism had the idea of the separation and autonomy of housing from work. In the historic city there is a greater mix, a relationship between different functions. These proximities produce their own meaning, and all of this creates the theatre of the city.

Question: *So how can design learn from the historic city and shape new approaches to the design of the future city?*

Answer: Architects clearly need to work on an image, a representation of the physical world, but I think that we have to suspend the link with the image, even if temporarily – we have to postpone the moment in which you work on the image.

I did a book with Homa Farjadi called *Delayed Space*. The concept refers to the design of mines, like the coalmines of France or Wales. Mines are internal spaces – they do not have a visual dimension – but to mine you have to create an infrastructure that enables the act of mining. Well, our cities have many of these types of functions, such as bus lines, metros, electrical and water infrastructure, sewage, etc. These functions are supported by infrastructure, which is a technical component, not something that has a visual dimension.

The phenomenon of infrastructure urbanisation is a key component of the future city. In the historic city there is a relationship between the infrastructure and the concept of pleasure, as in public spaces such as market squares, for instance. In the context of ecological urbanism, I support the idea that infrastructure should not only be a piece of engineering but also a site of pleasure.

Let's consider, for example, the engineering of the embankments of Paris, which are at the same time a functional infrastructure to control the river and a promenade, where you can see different horizons, some close and some distant, and where cars can drive through. The engineering of the embankment is also a landscape project. This is very important, as infrastructure today does not have aesthetic value, and in a very inexpensive way we can build cites where infrastructure plays a 'liberator' role as it enables people to get to places. These are ways in which we can learn from the historic city and produce new conditions.

Question: *A key issue today is the relationship between sustainability and urbanism. How do you view the connection between historic areas and sustainable planning? How can we use the knowledge of the ancient city?*

Answer: This point is critical. In many historic cities, because of the way they were developed in the past, time spent on travel is much less than in other urban structures. In terms of population density and physical form, the historic city is much more sustainable than the horizontal, low-density city, which as a result of its expansion has created unsustainable ways to living and moving. In America this is the dominant culture because the automobile has been the key determinant of urban growth. We need to learn from the historic city. In San Francisco you see at the moment a return from the suburbs, especially of an older, well-to-do population that is now revitalising the urban core and enjoying the benefits of the historic city.

We have increasing elderly populations in many countries, and in the future the needs of the elderly – linked to health, free time, and their relationships with the young – must be part of the way we think of cities. Cities are also sites of activity, and the design of spaces must help to create a more dynamic and integrated social environment. Sustainability is not only related to physical quantity and emissions but is also a social project, a project of life.

My latest book, *In the Life of Cities* looks at how the city as a physical artefact enables different ways of life, and how the possibility of imagining different kinds of life helps the designer. It's a two-way process: historic cities offer us different theatres of experiences, on one side, and on the other, we can use contemporary design to produce modern spaces with equivalent function.

Question: *Every city is subject to processes of change, yet at the same time we want to preserve the past. This creates a tension that can lead to varied outcomes. If in the future we can expect even greater pressures from population growth, tourism, and land development, how do you think it will be possible to sustain preservation?*

Answer: Obviously a city should have the capacity to transform itself, and designers must be open to this transformation. The Victorian house, a structure that by design has a great flexibility of uses, gives a good example of this flexibility: we have to think of the city in a similar way. Traditional master planning and zoning is essentially a process of phasing of development, where we define at the beginning a complete project and then define the phases of development.

Today we deal with projects in sites that are 'in waiting' – so-called anticipatory planning. We have many examples of sites where the project is not finished – it has a loose fit. Let me make a parallel with agricultural territory. An agricultural landscape is never complete – there are different seasons and different kinds of uses – and perhaps we should think of cities where we have different endings, meaning that we could make design schemes where you can add and transform when and if the need appears. In reality, this is exactly what happened with the historic city, which obviously did not appear instantaneously but rather grew through a process of accretion and change. I see a lot of value in experimenting with multiple endings, with things that are not really complete. In this way we allow for uncertainty, for change to occur.

7 Cities as Cultural Landscapes

Ken Taylor

*Adjunct Professor in the Research School of Humanities,
The Australian National University, Canberra, Australia, and Visiting
Professor at Silpakorn University, Bangkok, Thailand*

Most of us, I suspect, without giving much thought to the matter, would say that a sense of place, a sense of being at home in a town or city, grows as we become accustomed to it and learn to know its peculiarities. It is my belief that a sense of place is something that we ourselves create in the course of time. It is the result of habit or custom.

John Brinckerhoff Jackson

Reflections

The late 1980s and 1990s were particularly fruitful for the heritage conservation discipline in terms of the opening of a critical debate and understanding of the concept of cultural heritage. This was also the time in which a comprehensive definition of, and operational framework for, cultural landscapes was elaborated, building on the innovative thinking and work particularly of geographers.[1]

The inception of the three categories of cultural landscapes adopted by UNESCO in 1992 for World Heritage purposes extended concepts and international cultural heritage conservation thinking and practice to embrace associative values rather than a focus on tangible, physical fabric. It was an initiative that proved to be of great significance also as a driver to re-think other heritage categories and their conservation principles established in earlier periods. Underscoring these changes was the understanding of the significance of intangible values as a driver of heritage, which was aided substantially by increasing understanding of the cultural landscape concept.

---

[1] Taylor, K. (2012) Landscape and Meaning: Context for a Global Discourse on Cultural Landscape Values. In Taylor, K. and Lennon, J. (eds) *Managing Cultural Landscapes*, Routledge Key Issues in Cultural Heritage Series London & New York: Routledge: 21–44.

As my interest in the cultural landscape concept developed from the early 1980s onwards with its potential for throwing open the door of cultural heritage thinking, it seemed paradoxical that the focus of cultural landscape work zeroed in so heavily on the non-urban sphere. Perhaps this was a result of my UK university education in cultural geography, followed by town planning (with a good dose of emphasis on townscape and people), and then landscape architecture. Within geography the class was given the option of studying a final year course in urban geography. It was not simply a study of urban morphology, but also the relationships between people, their experiences and urban form not least through the medium of cognitive mapping. This learning experience, coupled with an earlier introduction to the fascination of reading the landscape, led to my view that understanding urban environments, apart from the incontrovertible fact that this is where most people lived, offered a rich field of study best understood under the cultural landscape umbrella.

In the light of these introductory comments and the imperative to improve management processes, several challenges offer fertile ground for discussion and resolve, including:

- How the notion of landscape embraces, in particular, its importance as a repository of social history and community values;
- How the cultural landscape concept relates to the historic urban environment: what are the similarities and possibly differences that exist between the two;
- How the identity of a city consists of a plurality of identities and traditional value and belief systems, as expressed and maintained by resident communities;
- How to sustain and enhance this as a way to brand the city; and,

- Which practical tools can be developed and integrated into urban landscape planning and conservation practice.

This chapter explores these challenging issues. It also aims to provide operational indicators in order to orient innovative planning and management processes. Underlying the chapter's rationale is the premise that cities do not exist in a cultural vacuum. They are places with a social history and cultural context.

A Paradigm Shift

The introductory commentary is not intended to intimate that built area conservation did not exist. It did, but its focus was architectural conservation or planning ensembles with a concentration on buildings – and all too often famous buildings or monuments – as opposed to looking at built urban heritage as the places where people lived their everyday lives, where social values and a sense of place inhered. A foundation for the latter approach was presaged in the townscape studies of Cullen[2] and Lynch[3] inquiring into how people experience urban spaces. The declaration, therefore, at a conference in Vienna[4] in May 2005 of the UNESCO *Vienna Memorandum on World Heritage and Contemporary Architecture. Managing the Historic Urban Landscape*, was timely. It followed concern by the World Heritage Committee about impacts of modern developments on historic urban areas and compatibility with the protection of their heritage values. This was particularly so with its proposition of the Historic Urban Landscape notion as a tool to reinterpret the values of urban heritage, and indication of the need to identify new approaches and new tools for urban conservation. Of seminal importance was the primal shift in thinking on the

[2] Cullen, G. (1971) *The Concise Townscape*, London: The Architectural Press.
[3] Lynch, K. (1972) *What Time Is This Place?* Cambridge, Massachusetts: MIT Press.
[4] UNESCO (2005) *International Conference on World Heritage and Contemporary Architecture – Managing the Historic Urban Landscape*. UNESCO World Heritage Centre in cooperation with ICOMOS and the City of Vienna.

urban environment away from purely physical architectural fabric to that of one fitting the cultural landscape model as seen in the *Memorandum* which refers to the historic urban landscape as:

> … *ensembles of any group of buildings, structures and open spaces, in their natural and ecological context, including archaeological and paleontological sites, constituting human settlements in an urban environment over a relevant period of time, the cohesion and value of which are recognised from the archaeological, architectural, prehistoric, historic, scientific, aesthetic, socio-cultural or ecological point of view. This landscape has shaped modern society and has great value for our understanding of how we live today.*[5]

Discussions on, and scrutiny of, the *Vienna Memorandum* at a Round Table in 2006 at the University of Montreal[6] reflected '*that it was best seen as a transitional document that supports the gradual shift away from the preoccupation with the historic city as visual object to an interest in the historic environment as a space for ritual and human experience.*'[7]

Building on the *Vienna Memorandum* a major intellectual endeavour of rethinking and broadening of ideas has evolved in urban conservation. Specifically the Historic Urban Landscape approach, through its recognition of the layering of significance and values in historic cities – deposited over time by different communities under different contexts – relates intellectually to the cultural landscape concept. Van Oers lucidly summarises this thinking in the following definition:[8]

> *Historic Urban Landscape is a mind-set, an understanding of the city, or parts of the city, as an outcome of natural, cultural and socio-economic processes that construct it spatially, temporally, and experientially. It is as much about buildings and spaces, as about rituals and values that people bring into the city. This concept encompasses layers of symbolic significance, intangible heritage, perception of values, and interconnections between the composite elements of the historic urban landscape, as well as local knowledge including building practices and management of natural resources. Its usefulness resides in the notion that it incorporates a capacity for change.*[9]

As background to the growing interest in a shifting approach to urban conservation is the global phenomenon of the ever growing concentration of people living in urban areas. UNESCO notes that there is '*a veritable explosion of urban populations, increasing each day [and] populations living in urban areas increase by 1.25 million every single week [and that] the current total is projected to double over the course of the next generation … [with] more than half of the world's most populous cities and urban regions … found in Asia.*'[10]

Whilst universal, the explosion, therefore, has particular relevance for the study and practice of urban heritage conservation in Asia where the impact on Asian cities is manifestly palpable.

The growth of Asian cities is reflective of what is occurring throughout the developing world overall. Zetter and Watson[11] note that globalisation has dramatically impacted city design with two particular negative outcomes. One is the accelerating destruction of the patrimony of

---

[5] UNESCO (2005) Vienna Memorandum on World Heritage and Contemporary Architecture. Managing the Historic Urban Landscape. para. 7.

[6] See Proceedings of Round Table on 'Heritage and the Conservation of Historic Urban Landscapes' organised by the Canada Research Chair on Built Heritage, Montreal, 9 March 2006.

[7] Bandarin, F. and Van Oers, R. (2012) *The Historic Urban Landscape. Managing Heritage in an Urban Century*, Chichester: Wiley-Blackwell: 196.

[8] Discussed at the Expert Planning Meeting on Historic Urban Landscapes, UNESCO, Paris, November 2008.

[9] Quoted in: Van Oers, R. (2010) Managing Cities and the Historic Urban Landscape Initiative – an Introduction. In Van Oers, R. and Haraguchi, S. (eds) *Managing Historic Cities*. World Heritage Papers N. 27. Paris: UNESCO World Heritage Centre: 14.

[10] UNESCO (2008) *Historic Districts for All: A Social and Human Approach for Sustainable Revitalization*, Paris: UNESCO Division of Social Sciences: 4.

[11] Zetter, R. and Watson, G. B. (2006) *Designing Sustainable Cities*. In Zetter, R. and Watson G.B. *Designing Sustainable Cities in the Developing World*. Aldershot, UK and Burlington,VT: Ashgate: 3–18.

indigenously designed and developed urban places and spaces, with culturally-rooted built environments eroding. The other is that the pressures are commodifying the place-identity of historic urban spaces, detaching them from their local, spatial, and temporal continuity, whilst still representing them as preserved authentic artefacts for global cultural consumption.

Asia will continue the trend of the 1990s as one of the fastest urbanising regions in the world. Rapid economic growth meant that 111 of the 140 new large or big cities emerging after 1990 were in the region.[12] Winter and Daly note, *'China has built more housing in the last twenty-five years than any nation in history.'*[13] They further note that, in addition to on-going rural/urban migration, 1.25 billion people will be added to Asia's population by 2025, more than half of which will live in cities.

The shift to a holistic, contextual view of urban heritage to include the idea of landscape as setting for people's lives – and within this the idea of sense of place – is further seen in the initiative of the *Seoul Declaration on Heritage and Metropolis in Asia and the Pacific*. Notably the Declaration, in relation to a wider understanding of heritage, proposes that:

> *These heritage sites contribute to the life and memory of the metropolitan areas by the diversity of their uses....Along with geographical features and the living social ecosystem, cultural heritage contributes strongly to the personality and character of the metropolis. It is a source of a truly sustainable development of the metropolitan areas in Asia and the Pacific in achieving their strategic and economic roles.*[14]

Whilst the *Seoul Declaration* relates specifically to an Asian context, it is worth noting that its five major recommendations are highly relevant to consideration of sustainable urban conservation needs globally within a cultural landscape way of thinking:

1. Cultural heritage should be recognised as a diverse and non-renewable asset, essential to the sustainable and human development of metropolitan areas in Asia and the Pacific.

2. Conservation of cultural heritage should be integral to the development of the city, including policies, programs and projects, from their planning to their approval, implementation and updating.

3. Conservation is comprised of the on-going identification, evaluation, protection and management of cultural heritage supported by the necessary human, scientific and financial resources.

4. Conservation of cultural heritage requires the development and implementation of adapted tools founded on recognised best practice and local conditions and traditions.

5. Conservation in metropolitan areas requires information, involvement and cooperation among the public, private, academic, and non-government sectors as well as citizens and international organisations.[14]

Documents such as the *Vienna Memorandum* and *Seoul Declaration* can be seen as a move *'towards an awareness of the broader social and political histories of an urban environment ... a shift in emphasis towards understanding of urban places as lived spaces and sites of collective identity [within] the broader socio-cultural and political contexts within which heritage sits.'*[15] Such notions link to the concept of the setting of heritage sites and places. In stating the contribution of setting to the significance of heritage monuments, sites

---

[12] UN-Habitat (2008) *The State of the World's Cities 2008/9: Harmonious Cities*. London: Earthscan: 6.

[13] Winter, T. and Daly, P. (2012) Heritage in Asia: Converging Forces, Conflicting Values. In Daly, P. and Winter, T. (eds) *Routledge Handbook of Heritage in Asia*. New York: Routledge: 6.

[14] ICOMOS (2007) *Seoul Declaration on Heritage and in Asia and the Pacific*. ICOMOS Asia and the Pacific Regional Meeting, Seoul, May 29–June 1, 2007. Seoul: ICOMOS Korea, 2007: 6. *ICOMOS News* Vol. 17, June 2008: 5–7.

[15] Panjabi, S. and Winter, T. (2009) Understanding the Tensions in Place: Conflict and Conservation in Kashmir. *Historic Environment*, 22 (1): 20.

and areas the ICOMOS *Xi'an Declaration on the Conservation of the Setting of Heritage Structures, Sites and Areas* further underscores a changing language and paradigm of urban heritage, not least in its reference to urban landscapes:

> The setting of a heritage structure, site or area is defined as the immediate and extended environment that is part of, or contributes to, its significance and distinctive character.
>
> Beyond the physical and visual aspects, the setting includes interaction with the natural environment; past or present social or spiritual practices, customs, traditional knowledge, use or activities and other forms of intangible cultural heritage aspects that created and form the space as well as the current and dynamic cultural, social and economic context.
>
> Heritage structures, sites or areas of various scales, including individual buildings or designed spaces, historic cities or urban landscapes, landscapes, seascapes, cultural routes and archaeological sites, derive their significance and distinctive character from their perceived social and spiritual, historic, artistic, aesthetic, natural, scientific, or other cultural values. They also derive their significance and distinctive character from their meaningful relationships with their physical, visual, spiritual and other cultural context and settings.[16]

These relationships can be the result of a conscious and planned creative act, spiritual belief, historical events, use or a cumulative and organic process over time through cultural traditions.

In the discussion on changing thinking on urban conservation we may ask whether it is coincidental that the word 'landscape' and association with the word 'culture' has taken place, and hence the 'cultural landscape' linked to the urban environment. Here we think of how the conjunction of the word 'cultural' with landscape

infers an inhabited, active/being. Additionally 'culture' like the German *kultur* (and therefore 'cultural') is about development of human intellectual achievement, care (Oxford English Dictionary): hence the German term *kulturlandschaft* from the scholarly work of late nineteenth/early twentieth century German geographers and from which our term 'cultural landscape' originates.[17] *Kulturlandschaft* emphasises human activity – culture – in shaping landscape patterns, including urban landscape patterns, or streetscapes, or cityscapes, whatever we choose to call them.

The Cultural Landscape Model: Landscape as History and Expression of Human Values and Identity

The places where we live are marked by distinctive characteristics. These are tangible, as in the physical patterns and components of our surrounds, and intangible as in the symbolic meanings and values we attach to places, and also to objects and to traditional ways of expression as in language, art, song, dance and so on. In this way physical space, sites and objects become places in the wider cultural landscape setting. They offer a past, are part of the present and suggest future continuity. It is these places with their association of meanings, which give rise to local identity and sense of place of communities.

A common denominator in sense of place and identity 'is human attachment to landscape and place.'[18] Donald Meinig's aphorism that '[l]andscape is an attractive, important, and ambiguous term … [that] encompasses an ensemble of

---

[16] ICOMOS (2005) *Xi'an Declaration on the Conservation of the Setting of Heritage Structures, Sites and Areas.* 2, paras. 1–2. http://www.icomos.org/xian2005/xian-declaration.htm

[17] Taylor, K. (2012) Landscape and Meaning: Context for a Global Discourse on Cultural Landscape Values. In Taylor, K. and Lennon, J. (eds) *Managing Cultural Landscapes.* Routledge, Key Issues in Cultural Heritage Series. London & New York: Routledge: 21–44.

[18] Taylor, K. and Lennon, J. (2012) Introduction. Leaping the fence. In Taylor, K and Lennon, J. *Managing Cultural Landscapes.* Routledge, Key Issues in Cultural Heritage Series. London & New York: Routledge: 1.

ordinary features which constitute an extraordinarily rich exhibit of the course and character of any society' still holds true, as does his rider that '[L]andscape is defined by our vision but interpreted by our minds. It is a panorama which continuously changes as we move along any route.' It can be seen to offer a trajectory of thinking relevant to the historic urban environment, not least because 'we are dealing primarily with vernacular culture' where landscape study is a form of social history. *We regard all landscape as symbolic, as expressions of cultural values, social behavior (sic), and individual actions worked upon particular localities over a span of time.*[19]

Such discourse in turn supports the notion that views landscape as a cultural construct reflecting human values: essentially the cultural landscape concept. Its significance in the urban sphere is that it allows us to see and understand the approach to urban conservation concentrating on individual buildings as 'devoid of the socio-spatial context' and can lead to a deterioration of the wider urban physical fabric.[20] This is in contrast to the cultural landscape approach because of the human values the landscape approach embodies. Greffe posits this urban landscape way of thinking as contrary to seeing the city as a closed view of architectural wonders of historic cities, but rather seeing the '… postmodern city where we are looking for feelings and emotions. The landscape then becomes an experience.'[21]

Central to the comprehensive cultural landscape approach developed since the late 1970s

through the work of cultural geographers and related disciplines such as anthropology is the manner in which it 'has progressively delved into landscape not simply or predominantly as history or a physical cultural product, but also – and more significantly – as cultural process reflecting human action and over time and associated pluralistic meanings and human values.'[22] Here there are connections with W.J.T. Mitchell and the proposition that 'we think of landscape not as object to be seen or text to be read, but as a process by which identities are formed.'[23] The comprehensive cultural landscape model of landscape as cultural construct is illustrated in Figure 7.1.

In this model inherent landscape values and ideologies result in the multifaceted cultural landscape manifested as a spatial and political phenomenon and a way of seeing that is replete with meanings. The urban cultural landscape is therefore also 'the inhabited landscape, the physical world that people participate in directly, modifying it as they are able to according to their needs, aspirations and means.'[24]

A major theme underpinning the cultural landscape paradigm is the relationship or interaction between culture and nature and the association between people and natural elements. Here natural elements are not seen as merely physical entities, but entities and landmarks that reflect deep associations in the landscape and have meaning for people as seen in Figure 7.2.

The inception of the three categories of cultural landscapes for World Heritage purposes in

---

[19] Meinig, D.W. (1979) Introduction. In Meinig, D.W. (ed) *The Interpretation of Ordinary Landscape. Geographical Essays.* New York: Oxford University Press: 1–3 and 6.

[20] Punekar, A. (2006) *Value-led Heritage and Sustainable Development: The Case of Bijapur, India.* In Zetter, R. and Watson G.B. *Designing Sustainable Cities in the Developing World.* Aldershot, UK and Burlington, VT: Ashgate: 110.

[21] Greffe, X. (2008) *Urban Cultural Landscapes.* Brisbane: Griffith University, Faculty of Arts: 1.

[22] Taylor, K. (2012) Landscape and Meaning: Context for a Global Discourse on Cultural Landscape Values. In Taylor, K. and Lennon, J. (eds) *Managing Cultural Landscapes.* Routledge, Key Issues in Cultural Heritage Series. London & New York: Routledge: 21–44.

[23] Mitchel, W.J.T. (1994) *Landscape and Power.* Chicago: University of Chicago Press: 1.

[24] Punekar, A. (2006) *Value-led Heritage and Sustainable Development: The Case of Bijapur, India.* In Zetter, R. and Watson G.B. *Designing Sustainable Cities in the Developing World.* Aldershot, UK and Burlington, VT: Ashgate: 110.

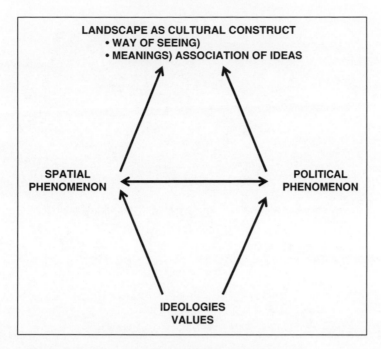

Figure 7.1 Interactive phenomenon of landscape. Copyright of Ken Taylor.

Figure 7.2 View South from Mount Ainslie along the Central Axis of Canberra. Source: Image supplied courtesy of the National Capital Authority. Copyright of Commonwealth. All rights reserved.

Figure 7.3 Royal Barge procession, Chao Praya River, Bangkok.

1992 recognised the culture-nature interaction as summarised by Mechtild Rössler:

> *Cultural landscapes are at the interface between nature and culture, tangible and intangible heritage, biological and cultural diversity – they represent a closely woven net of relationships, the essence of culture and people's identity.*[25]

Bangkok and the Chao Phraya River

The culture-nature phenomenon is just as convincing in the urban setting in relation to urban form and identity. Natural elements contribute to the urban spirit of place, as well as shaping the city's morphology, and have legibility. They are often historically the *raison d'être* for the physical setting of the city, or strong contributor to it. Bangkok established in 1782 on the Chao Phraya

River delta is a case in point. Bangkok without the vibrancy of the Chao Phraya River, where the life and historic pageantry of the city are displayed and experienced, as well as reflecting the historic importance of water transport, is unimaginable. Figure 7.3 highlights this aspect of the city in the Royal Barge Procession celebrating the King's birthday in 2007 at Rattanakosin Island, site of the Grand Palace and many important Buddhist temples such as Wat Phou. At a Royal Barge Ceremony in 1982 marking Bicentennial celebrations of the city Askew notes that, as the Royal Family paid homage to their ancestral spirits at the main palace gate, '*the ceremony of homage was familiar enough to the majority of ordinary Thais who customarily pay respects to the* Chaw Thi *(land spirit) of their home. Homage and respect, key features of Thai social relations (both between people and the*

---

[25] Rössler, M. (2006) World Heritage Cultural Landscapes. *Landscape Research*, 31(4): 334.

supernatural beings that inhabit the world, unseen but nonetheless real) were enacted in the royal ritual, tapping a common cultural source.[26] Equally the modern busy business districts along the River are vibrant scenes reflecting the indelible identity of the city.

Canberra

An example *par excellence* of the interface between culture and nature is the city of Canberra, the federal capital of Australia, a remarkable example of twentieth century city planning. From its inception in the nineteenth century, and before the Walter Burley Griffin entry won the 1911 international competition for the city's design, the concept and ideal of an Australian federal capital envisaged a city in the landscape. This set in train the foundation for Canberra as a remarkable city. In the true sense of the word it is a unique city, for there is no other city like it in the world. Walter Burley Griffin declared in 1912 that he had planned a city not like any other city. These were prophetic words, for its development over the years has maintained its status of being unlike any other. Why is this? There are roads, houses, offices, schools, shops, parks – all the components we associate with urban development – as in any other city.

The underlying reason lies in the way landscape defines and articulates the city morphology starting with the Griffin plan. Changes over the years to the form of the city and hence to the Griffin ideal have taken place. Nevertheless the landscape basis, which binds form and content, remains vividly coherent in the city plan. The form of the physical landscape – natural and created – is a palpable, tangible presence defining the city; but equally so is its content or intangible, symbolic meaning. Underlying the city's spatial structure is the fundamental premise of Canberra as a city in the landscape. Its spatial structure has been progressively and incrementally planned from the beginning to maintain continuity with existing design elements, in particular the hills, ridges, and valleys (see Box 7.1).

From the symbolic heart of the city and the nation in the National Triangle (Figure 7.2) with its serene symmetrical beauty, out through the tree-lined streets, neighbourhood and district parks and open spaces to the hills, ridges, and valleys – the National Capital Open Space System (NCOSS) – it is the landscape nature of the city that predominates physically. In turn this tangible physical presence has inextricable, intangible meanings and values, confirming that landscape is not just what we see, but a way of seeing.[27]

When you look out over the magnificent prospect from Mount Ainslie towards Parliament House (Figure 7.2) across the city to the surrounding hills that form the embracing backdrop for the city, or enjoy the tree-lined streets, gardens, and parks of the suburbs, the landscape itself is more than physical elements. It has a meaning and significance that inform what Canberra is (see also Box 7.1).[28]

Cultural Landscape Characteristics

As interest in cultural landscapes developed, particularly from the 1980s, it became clear that cultural landscapes are a complex phenomenon. They need research information beyond simple historical documentation to embrace detailed

[26] Askew, M. (1994) Bangkok: Transformation of the Thai City. In Askew, M. and Logan, W. *Cultural Identity and Urban Change in Southeast Asia. Interpretative Essays.* Geelong, Victoria: Deakin University Press: 93.

[27] Cosgrove, D. E. (1984) *Social Formation and Symbolic Landscape.* London: Croom Helm.

[28] The above is synthesised from: Taylor, K. (2006) *Canberra: City in the Landscape.* Ultimo, Sydney: Halstead Press.

investigation, including field surveys, using multi-disciplinary input to document landscape characteristics. Documenting landscape characteristics and associated features is, therefore, a fundamental requirement for understanding landscapes, rural or urban. In effect this is taking up the geographer's idea of reading the landscape where '*the basic principle is this: that all human landscape has cultural meaning, no matter how ordinary that landscape may be.*' In this sense '*Our landscape is our unwitting biography, reflecting our tastes, our values, our aspirations, and even our fears in tangible visible form*'.[29]

Landscape characteristics or the forces that shape a cultural landscape can be summarised as follows:

- PATTERNS in the landscape resulting from the PROCESS of landscape making (Spatial Organisation and Land-use).
- NATURAL ELEMENTS.
- CULTURAL TRADITIONS.
- INDIVIDUAL COMPONENTS.[30]

Fundamental to understanding the significance of the cultural landscape paradigm in urban conservation is that landscapes are not static entities. They change through time resulting in us being able to recognise layers in the landscape. They tell the story of people, events and places through time, offering a sense of continuity: it is what Lynch nicely sees as a sense of the stream of time.[31] The concept of 'landscape' therefore infers cultural context, human action and activity and also change over time. From a cultural landscape perspective, the concept of layers as essential to a contextualised understanding of urban heritage in contrast to a fabric

oriented discourse is neatly expressed in the following commentary by Panjabi and Winter:

> As densely populated, historically layered environments, today's cities draw upon their material and social fabric to express a multitude of values – including social equity, multi-culturalism, cosmopolitanism or nation building. Invariably, it is these very values that define the city as place.[32]

Consideration of these factors suggests a number of key issues to help guide urban cultural landscape study:

- Landscapes are a clue to culture,[33] they tell a story which can be interpreted and read.
- Landscapes are inhabited, the world in which we participate, and are therefore subject to change.
- Existence of continuity in the landscape: they represent a composite image (layers) rather than separate dots on a map – everything is interconnected.
- They represent inter-relationships between people, places and events through time.
- They are significant reminders of the past and the present and a guide to the future.
- They reveal social history arousing associative values and interpretative values.

In the light of the above discussion we may appreciate, therefore, that the cultural landscape paradigm presents a firm conceptual basis for urban conservation and operation of the Historic Urban Landscape model. It is a platform on which to build a sustainable approach to urban heritage planning focused on protection of not merely physical built fabric, but equally social values and intangible cultural heritage. The criti-

---

[29] Lewis, P. (1979) Axioms for Reading the Landscape. Some Guides to the American Scene. In Meinig, D.W. (ed) *The Interpretation of Ordinary Landscape. Geographical Essays*. New York: Oxford University Press: 12.

[30] See Page, R., Gilbert, C. and Dolan, S. (1998) *A Guide to Cultural Landscape Reports: Contents, Process, and Techniques*. Washington DC: US Department of the Interior, National Park Service.

[31] Lynch, K. (1972) *What Time Is This Place?* Cambridge, Massachusetts: MIT Press.

[32] Panjabi, S. and Winter, T. (2009) Understanding the Tensions in Place: Conflict and Conservation in Kashmir. *Historic Environment*, 22.1: 20.

[33] Lewis, P. (1979) Axioms for Reading the Landscape. Some Guides to the American Scene. In Meinig, D.W. (ed) *The Interpretation of Ordinary Landscape. Geographical Essays*. New York: Oxford University Press: 15.

cal word under a cultural landscape approach is that of 'values' leading to a value-based approach to heritage conservation, including urban conservation. It is an approach that is, for example, central to Australian practice not least through the document that guides Australian thinking and practice, the *Burra Charter* with its reference to social and spiritual values. Integral is the recognition of associations between people and places where '*Associations mean the special connections that exist between people and a place*' (Article 1.15, p.3) and the explanatory note '*Associations may include social or spiritual values and cultural responsibilities for a place*'.[34]

Building on these numerous documents, Australian heritage practice focuses, *inter alia*, on guiding the central platform of values. For example *Steps to sustainable tourism. Planning a sustainable future for Tourism, Heritage and the Environment* of the Commonwealth of Australia, in particular its ten steps diagram.[35] This has been developed in *Stepping Stones for Heritage* (Stepwise Heritage & Tourism[36]) into the concept of a ten steps guide for heritage as a way of guiding people through the heritage management process:

1. Vision for the future?
2. Who is involved?
3. What we know?
4. What is important?
5. What the issues are?
6. Strengths and weaknesses?
7. What the ideas are?
8. What the objectives are?
9. Action plans?
10. Making it happen.

Notably it is a guide that speaks to people, not just experts. It is reflective of Jackson's aphorism that '*We should never tinker with the landscape without thinking of those who live in the midst of it.*'[37]

The shift in thinking about heritage places since 1992 that has accompanied the instigation of the three World Heritage cultural landscape categories is one that links people, places and events through time, and not just the rich and famous people. It has led to the increasing appreciation of intangible values that inhere in places because of the association of ideas between people and place. In this way we can appreciate that all cultural landscapes have associative values. Julian Smith expresses this as follows:

> ... it is useful to think of cultural landscapes as ideas embedded in a place, and to consider the recording of cultural landscapes as an exercise in cognitive mapping rather that physical mapping. The challenge of this approach is that a cultural landscape cannot be observed, it must be experienced. And it must be experienced within the cultural framework of those who have created and sustained it ... some would argue that this find of cultural landscape is an associative cultural landscape.[38]

The discussion on associative value suggests that it is time to review and alter the definition of Associative Cultural Landscape with its emphasis, mistakenly, on '*powerful religious, artistic or cultural associations of the natural element*'. It is intended to relate to indigenous aboriginal cultures, but fails to appreciate that for aboriginal cultures, natural elements of their world are part of the made-world of people and ancestors as in the Australian Aboriginal concept of the Dreaming.[39]

---

[34] Australia ICOMOS (1999) *The Burra Charter: the Australia ICOMOS Charter for Places of Cultural Significance*. Australia ICOMOS: 3.

[35] Commonwealth of Australia, Department of the Environment and Heritage (2004) *Steps to sustainable tourism. Planning a sustainable future for Tourism, Heritage and the Environment*: 7.

[36] www.stepwise.net.au

[37] Jackson, J. B. (1963) Goodbye to Evolution. *Landscape* 13/1: 2.

[38] Smith, J. (2010) Marrying the Old with the New in Historic Urban Landscapes. In Van Oers, R. and Haraguchi, S. (eds) *Managing Historic Cities*. World Heritage Papers 27. Paris: UNESCO World Heritage Centre: 46.

[39] I argued this point in 1993 when Australia was given the task of commenting on the Associative category. In my view the reference to natural element is a Eurocentric Romantic notion of the deep relationship between Aboriginal Australians and their country (landscape).

Urban Identity, Plurality, Sustainable Development Tools for Urban Landscape Planning and Conservation Practice

In dealing with urban heritage conservation it is vital that those involved – whether they be government urban planning/urban design agencies, politicians, NGOs, or inhabitants of cities – understand that historic cities consist of a plurality of communities. These shift and change through time imposing different values, thereby contributing to the layering of the city. With this imagery in mind, the idea of a circumscribed inner area of an historic city, immutable in time, where rigid conservation of the architectural fabric may be enforced as a way of attempting to stamp a sense of local identity irrespective of how social values and ways of living change is not necessarily the best model to follow. Such an approach is in danger of leading to a product that is nothing more (or less) than the driver of heritage tourism or enveloping gentrification. In connection with the latter phenomenon Justin Davidson, *New York Magazine* architectural critic, speculated on WYNC Radio (New York), 15 October 2012,[40] that preserving urban districts architecturally does not prevent change, it accelerates it in effect because it leads to gentrification. It preserves architectural integrity but not communities. This is not a blanket dismissal of gentrification, but rather an emphasis that there will be examples where local community voices need to be heard and allowed to participate in change that does not disenfranchise them: the example of Zhuijiajiao (near Shanghai) given below is a case in point.

Additionally for historic cities there is the curse or benefit – depending on your point of view – of tourism. We may also question whether the circumscribed historic city as driver of tourism is truly commensurate with the cultural landscape idea, notwithstanding the latter's incorporation of change into its intellectual menu. It is how change is understood and handled that is critical. Even so, it has to be acknowledged that the tourism cityscape is a cultural landscape form: but so is Disneyland. In this vein Ashworth and Graham argue in relation to the post-war European city that if it '*exists as an idea, then it is composed of conserved urban forms and the idealised urban form that these contain*'. They place this within the context of vernacularism '*viewed as a self-conscious and deliberate expression of localism [where] the conserved historic city has adopted many vernacular elements drawn from the folk museum*', suggesting then that '*it is a short step from the deliberately assembled museum town to the vernacular "museumification" of existing towns and districts*'.[41] This for me is a chilling thought.

That identity is grounded in heritage is well established. It is part of an inclusive sense of belonging that is communal and embracing, but it might also be exclusive. For tourism purposes for example, inclusivity is central to interpreting and presenting places for outsiders where, from this knowledge, they could imagine being involved in creating what it is that constitutes the identity of the place. Alternatively they might evoke an interest in a place and want to know more about it and to explore it. The hustle and bustle of everyday street scenes with shop houses and markets in Asian cities is a cogent example. The streets are often vibrant, living entities where everyday life – real vernacular as opposed to ersatz vernacular – and sense of living history are palpable. For urban communities identity grounded in heritage is central to sense of place, a sense of being at home, '*something that we ourselves create…It is the result of habit or custom*'.[42]

---

[40] See website: http://www.wnyc.org/story/243306-urban-development-bloomberg-years/

[41] Ashworth, G. J. and Graham, B. (2012) Heritage and the Reconceptualization of the Postwar European City. In Stone, D. (ed) *The Oxford Handbook of Postwar European History*. Oxford: Oxford University Press: 594 and 591.

[42] Jackson, J. B. (1963) Goodbye to Evolution. *Landscape* 13/1: 151.

One of the dangers inherent in urban areas of what is termed 'heritagisation' (ghastly word) is an historic city brand image with replicable heritage items, bric-a-brac, and standard 'off the peg' heritage. It is, according to Ashworth and Graham, reflective of the hallmark of some European cities typified as 'catalogue heritage.' Within this overall classification are separate categories or possibilities to market heritage distinctiveness. Ashworth and Graham refer to these as 'popular optional 'add-ons' ... 'tourist-historic waterfront', medieval old town, 'ethnic' district, festival calendar, sanitised 'red light', and gentrified 'urban village'; 'all devised to be different but ultimately becoming the same.'[43] The cultural landscape model would suggest that

wherever possible this catalogue list approach to urban heritage conservation is best avoided, or at least restricted in extent and reproduction. Instead should be a systematic approach where the socio-cultural and political context of the cultural landscape as process by which identities are formed is applied.

In contrast to these examples are some from Shanghai. The first concerns the changes taking place along the Suzhou River in Shanghai where older brick warehouse buildings are being adaptively reused as professional studios, workshops making such things as furniture, and shops selling the furniture or designed goods. Figure 7.4 shows a group of old warehouse buildings, adaptively reused as studios and workshops and

Figure 7.4 The Suzhou River in Shanghai.

[43] Ashworth, G. J. and Graham, B. (2012) Heritage and the Reconceptualization of the Postwar European City. In Stone, D. (ed) *The Oxford Handbook of Postwar European History*. Oxford: Oxford University Press: 595 and 598.

seen against a background of modern multi-storey apartment developments. This example represents a model of how changing values and economic change have led to changing use of the buildings whilst retaining their structures, sense of place and functional link with the past: another layer in the urban cultural landscape. The second is the famous example of the major waterfront of The Bund, Shanghai, China, where, although a huge tourist attraction and changes have occurred, there is a sense of the stream of time engendered in the conserved *art deco* waterfront buildings and the sheer exhilarating, kaleidoscopic view of the river with its constant movement of boats of all kinds – working boats, tourist boats, ferries – plying the water. It is a layered Historic Urban Landscape where the ultra-modern city development of Pudong heightens the sense of exhilaration.

A third example is the canal towns of Shanghai such as Zhuijiajiao. Taking Zhuijiajiao we see changes have taken place, but they are changes that can be seen not to be simply tourist fashionable vernacularism. Such towns have rich histories, traditional architecture, and daily life that make them distinctly and unmistakably Chinese (Figures 7.5 a,b).

Notably the local community consists of people who have traditionally lived here for generations; people who want to continue to live here because it is a community, not merely a population. It is a cogent example of changing social values where tourism now substantially helps the local economy, but where changes have not destroyed the place from the point of view of tangible values (traditional buildings and canal setting), and from the point of view of intangible values (people's lives, community feeling and sense of place). Significantly, the place still belongs to them and they belong to it. In one building you may catch a glimpse of a local aged persons' group playing *mahjong*. Heritage conservation planning addressed the views and feelings of local people who wanted to stay in their community: here is the essence of the city as cultural landscape. The sense of authenticity and that of integrity are palpable.

Tools

The synergy between a cultural landscape approach and the Historic Urban Landscape is inherent in an overview of the changing socio-economic and political conditions affecting the holy Indian city of Varanasi. Singh establishes the need for urban heritage to be seen as a sustainable resource for future development of the city where heritage is protected and monitored continuously; impact of heritage protection constantly evaluated; development follows specific heritage guidelines; where there is active participation of residents and stakeholders and where heritage brings sustainable economic benefits to the local population.[44]

The cultural landscape is an approach convincingly expressed by Punekar in the following comments:

A cultural landscape approach enables diverse communities to be seen as part of that landscape. That is, cultural, historical, and political conditions affecting contemporary communities are part of the process of human engagement with the place. The cultural landscape approach can be a means of reuniting fragmented approaches to valuing and constructing the environments we inhabit, a means of overcoming distinctions between historic environment and new development, nature and culture, built heritage and context.[45]

---

[44] Singh, Rana P.B. (2011) Varanasi, India's Cultural Heritage City: Contestation, Conservation and Planning. Singh, Rana P.B. (ed) *Heritagescapes and Cultural Landscapes*. Planet Earth & Cultural Understanding, Series Pub 6. Newcastle–upon-Tyne UK: Cambridge Scholars Publishing: 209–250.

[45] Punekar, A. (2006) Value-led Heritage and Sustainable Development: The Case of Bijapur, India. In Zetter, R. and Watson G.B. *Designing Sustainable Cities in the Developing World*. Aldershot, UK and Burlington, VT: Ashgate: 111.

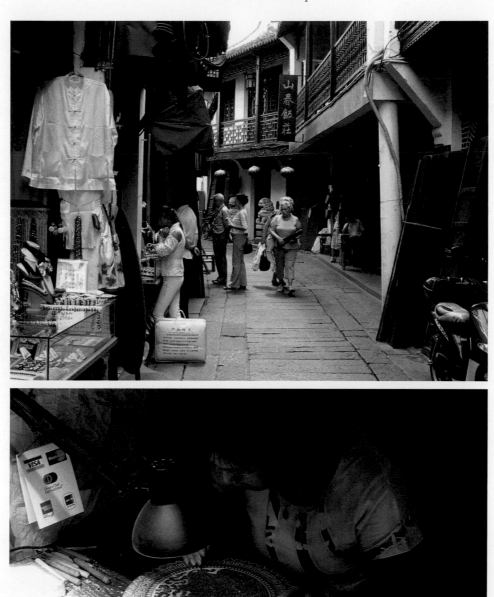

Figure 7.5 (a) The old town of Zhuijiajiao near Shanghai. (b) Local craftsperson displaying traditional skills to the public.

There is a fundamental need to initiate a dialogue with city planners and urban designers on the Historic Urban Landscape paradigm. It is important in this dialogue that it is understood that the concept of urban conservation and the reality of city development and expansion are not mutually exclusive, acceding that change to city form will be inevitable. Inherent in this process are some or all of the following considerations:

1. Understanding the city as an evolving process – living entity – not merely a series of objects (buildings). The idea of process embraces intangible cultural heritage values, *genius loci*, and interaction between culture and nature, as well as tangible heritage aspects.

2. Addressing the overall urban morphology of the city in its landscape setting so that future development does not overwhelm (a) the landscape physically, or (b) its intangible meanings and values, thereby maintaining the link between culture and nature and intangible connotations. The case of the World Heritage Dresden Elbe Valley cultural landscape delisted in 2009 is a case in point where the construction of a four lane bridge was deemed to destroy the values of the landscape setting of the city.

3. Urban landscape under the banner of visual and physical integrity is not just a matter of quantitative visual attributes where management is nothing more than dealing with views and skylines as seen objects. Rather it is to do with the cultural context of the way of seeing dynamic urban form and its relationship with the wider setting: who has been involved and why are some views

important. Of note in this connection is the following reference by Moggridge:

> *During the Ming and Qing Dynasty (XIIIth and XIVth century) it was the custom of every city or town in China to select generally eight [sometimes a few more] best landscape scenes in the vicinity that best represented the local character. The selected scenes were normally those preferred by the local people and were the most popular destinations.*[46]

From this we may see as Moggridge posits *'special views are a major cultural resource of cities.'*

In the category of special resources forming the setting for cities are, *inter alia*, views of hills and mountains visible above or between buildings. Some cities have taken the line of constructing cones of view circumscribing views which are identified as needing protection, for example Vancouver. The urban morphology management regime of Paris achieves a similar outcome by protecting identified significant inner urban zones from overshadowing by high-rise development. By this mechanism the significant morphology rooted in history of the planning of Paris is recognised. Others such as George Town, Penang, look at how to control heights for new buildings to safeguard sense of place and landscape settings.[47] There are lessons and conclusions for the city of Hangzhou and World Heritage listed West Lake, China (Box 7.2), which are critical to a Historic Urban Landscape approach to interaction between the city and the lake setting.

---

[46] See: Moggridge, H. (2010) Visual Analysis: Tools for Conservation of Urban Views During Development. In Van Oers, R. and Haraguchi, S. (eds). *Managing Historic Cities*. World Heritage Papers 27. Paris: UNESCO World Heritage Centre: 66 (Moggridge is quoting I Kairan, University of Sheffield PhD dissertation).

[47] UNESCO (2009) *State of Conservation (SOC). Melaka and George Town Cities of the Straits of Malacca*. WHC 33COM7B.28. Paris: UNESCO World Heritage Centre. See also: http://whc.unesco.org/en/soc/757.

Box 7.1 City as evolving process: Need for a Historic Urban Landscape approach for Canberra, Australia's national capital

Figure 7.6 Nature and culture in the city setting.

Canberra was conceived and planned as a city not like any other for the first 75 years of its conception from the Griffin plan of 1912 with planning guidelines specifically modelled to maintain it as the city in the landscape. It was a cumulative approach over the years that may be seen to be a forecast of the Historic Urban Landscape paradigm (Figure 7.6, 7.7, 7.8 and 7.9). Since self -government in 1988 planning has been governed increasingly by the mantra of increasing urban densification, urban consolidation and high-rise buildings without regard for the fundamental significance of its historic landscape ethos leading to loss of landscape space and trees and blocking views of the surrounding hills that are quintessential to the setting and ambience of the city. Action needs to be focused on the following to reflect the Historic Urban Landscape:

- Special nature of the city as with its vision of a planned ideal city – a city not like any other;
- *Genius loci* of the city inherently centred on the culture-nature interaction;
- Preparation of metropolitan plan for the whole city rather than separate piecemeal plans for separate suburbs or groups of suburbs;[48]
- Within metropolitan plan need for precinct plans for suburbs that relate to specific character of the suburb(s) and where local residents are consulted; currently the model is the development of precinct codes that are then incorporated into the Territory Plan as technical amendments without local community input; planning for cones of view and protecting significant vistas;
- Need to establish appropriate partnership between planning authority and residents for local area (suburb) planning.

(Continued)

---

[48] Canberra's underlying plan is a grouping of six towns linked by open space corridors. Each town is subdivided into suburbs with groups of suburbs forming districts.

Figure 7.7 Nature absent in Canberra new medium density development.

Figure 7.8 Academy of Science National Heritage listed building, 2011.

Figure 7.9 New high-rise building impact on the setting and context of the Academy of Science, 2013.

Box 7.2 Hangzhou and West Lake, China: maintaining visual and physical integrity

Hangzhou and World Heritage listed West Lake (Figure 7.10), China: critical aspects of Historic Urban Landscape approach to interaction between the city and the lake setting:

- Maintain special qualities of the meaning of West Lake as the city grows in the future by not allowing city form to overwhelm the scale and ephemeral poetry of landscape qualities of the Lake. Expansion zones for the city and high rise development are kept away from the eastern side of the lake (Figures 7.11 and 7.12).
- Protect hill and mountain view sheds as backdrop to vistas from the lake using cones of views.
- Engage with local people and tourists through such tools as cultural mapping to research and understand how people see and appreciate the special character, *genius loci*, of place. In cultural mapping terms this is known as local distinctiveness.[49] Engage with Chinese domestic tourists and their perceptions from an Asian perspective. Discussions in tourism often give attention to marketing, facility management or growth statistics.
- Think of the cultural landscape of West Lake and Hangzhou in terms of the idea of an *associative cultural landscape*. Here the critical question on determining future urban form should focus on whether new development 'disrupts existing aesthetics and valued rituals or whether it respects them'[50] and where it may add to the dynamic quality of the urban form and its relationship with the landscape of West Lake with all its accumulated meanings (Figure 7.13).

(Continued)

[49] Taylor, K. (2013) Cultural Mapping: Intangible Values and Engaging with Communities with Some Reference to Asia. *The Historic Environment*, 4/1: 50–61.
[50] Smith, J. (2010) Marrying the Old with the New in Historic Urban Landscapes. In Van Oers, R. and Haraguchi, S. (eds) *Managing Historic Cities*. World Heritage Papers 27. Paris: UNESCO World Heritage Centre: 51.

Figure 7.10 West Lake looking east towards protected hills. Hangzhou city is on the right.

Figure 7.11 Model looking south showing relationship of city with the lake and protection of the eastern, western and southern hills along the lake.

Figure 7.12 Protected low density urban area on east side of lake.

Figure 7.13 Enjoying a lake park near the city.

4. Organising regional workshops, through NGOs and agencies such as UNESCO and ICOMOS, to propel the idea of thematic studies that critically examine urban development and conservation practice. Such studies open the opportunity to focus not only on major cities, but secondary towns/cities and their settings, and vernacular settlements as suggested in *World Heritage: Challenges for the Millennium*.[51] The role of NGOs should not be underestimated. The Penang Heritage Trust (PHT), for example, has been active since 1986 in promoting conservation needs for the historic urban area of George Town, Penang, and is a leading example. Its conservation efforts have been recognised by seats on various state heritage committees, local government committees and the UNESCO World Heritage Committee. PHT publishes a regular newsletter and heritage trail brochures, and offers school education programmes.

5. Use of Urban Heritage Zoning as described by Yuen in Singapore where a shift in the mid-1980s from a demolish and rebuild approach to city planning has seen a greater effort to reinforce and integrate past heritage with present developments, with a major turning point being a 1989 planning act amendment.[52] This saw the appointment of a conservation authority, designation of conservation areas with associated conservation requirements and guidelines. The number of identified conservation areas has increased to more than twenty (total area 751 ha). Many of these are interpreted and presented for tourism purposes through attractive, informative trail brochures such as for Jalan Besar, Balestier and Bukit Timah. Involving historic shop house areas being saved from demolition and specific restoration guidelines with information for owners to help protect authenticity, these Singapore exemplars demonstrate how change and adaptation towards improved environmental character shows how the past should serve the future as Yuen neatly expresses it. Architecturally, old and new combine to present a vigorous lively sense of socially vibrant urban life, rather than simple preservation of old areas. The variety of old and new buildings, including high-rise framing skyline views, adds diversity and interest. A prominent example of treating a city through the Historic Urban Landscape approach and using complementary cultural mapping exercises is Vigan, Philippines (see Box 7.3). As Eric Zarrudo observes '*Vigan was declared a World Heritage city in 1999. It is the most intact nineteenth century district with the fusion of Asian and European architectural and artistic expressions in the Philippines. It has a rich fabric of heritage resources, natural, built, intangible and movable. The local government has developed a Vigan Master Plan and the Vigan Municipal Ordinance N.4 and has embarked on various community programs that protect and enliven the heritage precinct.*'[53]

---

[51] UNESCO (2007) *Challenges for the Millennium*. Paris: UNESCO World Heritage Centre.

[52] Yuen, B. (2005) Searching for place identity in Singapore. *Habitat International*, 29: 197–214.

[53] Zarrudo, E. B. (2008) *The Cultural Mapping Project of the Heritage City of Vigan: Towards Building a Framework for Heritage Conservation and Sustainable Development*. Paper presented to *Third International Conference of the UNESCO Memory of the World Programme*, 19/22 Feb 2008, Canberra, Australia. Quoted in Cook, I. and Taylor, K. 2013. *A Contemporary Guide to Cultural Mapping An ASEAN-Australia Perspective*. Jakarta: ASEAN Jakarta: 184.

Box 7.3 Vigan

Heritage mapping project of the heritage city of Vigan, Ilocos Province, the Philippines. Transforming heritage resources for economic and societal development.

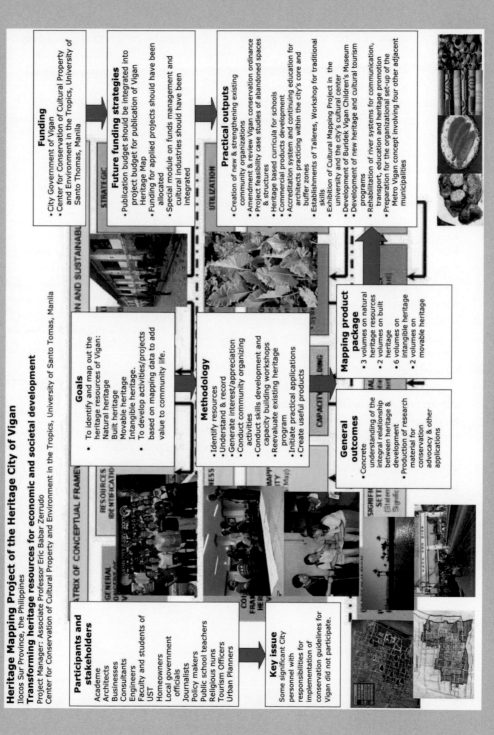

Heritage Mapping Project of the Heritage City of Vigan
Ilocos Sur Province, the Philippines
Transforming heritage resources for economic and societal development
Project Manager: Associate Professor Eric Babar Zerrudo
Center for Conservation of Cultural Property and Environment in the Tropics, University of Santo Tomas, Manila

Participants and stakeholders
Academe
Architects
Businesses
Consultants
Engineers
Faculty and students of UST
Homeowners
Local government officials
Journalists
Policy makers
Public school teachers
Religious nuns
Tourism Officers
Urban Planners

Key issue
Some significant City personnel with responsibilities for implementation of conservation guidelines for Vigan did not participate.

Goals
- To identify and map out the heritage resources of Vigan:
  Natural heritage
  Built heritage
  Movable heritage
  Intangible heritage.
- To develop activities/projects based on mapping data to add value to community life.

Methodology
- Identify resources
- Understand & record
- Generate interest/appreciation
- Conduct community organizing activities
- Conduct skills development and capacity building workshops
- Reevaluate existing heritage program
- Initiate practical applications
- Create useful products

General outcomes
- Concrete understanding of the integral relationship between heritage & development
- Production of research material for conservation advocacy & other applications

Mapping product package
- 3 volumes on natural heritage resources
- 2 volumes on built heritage
- 6 volumes on intangible heritage
- 2 volumes on movable heritage

Funding
- City Government of Vigan
- Center for Conservation of Cultural Property and Environment in the Tropics, University of Santo Thomas, Manila

Future funding strategies
- Publication budget should be integrated into project budget for publication of Vigan Heritage Map
- Funding for applied projects should have been allocated
- Special module on funds management and cultural industries should have been integrated

Practical outputs
- Creation of new & strengthening existing community organizations
- Amendment & review Vigan conservation ordinance
- Project feasibility case studies of abandoned spaces & structures
- Heritage based curricula for schools
- Commercial products development
- Accreditation system and continuing education for architects practicing within the city's core and buffer zones
- Establishments of Talleres, Workshop for traditional skills
- Exhibition of Cultural Mapping Project in the university and the city's cultural center
- Development of Buridek Vigan Children's Museum
- Development of new heritage and cultural tourism programs
- Rehabilitation of river systems for communication, transport, education and heritage promotion
- Preparation for the organizational set-up of the Metro Vigan concept involving four other adjacent municipalities

Source: University of Santo Tomas. In Cook, I & Taylor, K. (2013).[54]

6. Educating planners to understand the importance of social significance of urban heritage places and the need for conservation management plans that address identified zones.

7. Understanding the role of communities of people and stakeholders and the need to organise for their participation. In this part of the process it is essential that differing communities are identified from a range of socio-economic and cultural backgrounds. Central to this aspect is that urban communities are not cohesive single units, but pluralistic and changing. A technique available to deciding on community groups and keeping dialogue flowing is found in REAP methods (Rapid Ethnographic Assessment Procedure). REAP can assist in revealing conflicts and differences as well as similarities in community opinions and values.

8. Understanding by planners of vernacularism, as an expression of localism alongside the existence of regional differences in urban areas, as for example in the Zhuijiajiao case.

9. Understanding the role and importance of cultural mapping and cognitive mapping in urban conservation work. Cultural mapping is the term used to describe the set of activities and processes for exploring, discovering, documenting, examining, analysing, interpreting, presenting and sharing information related to people, communities, societies, places and the material products and practices associated with those people and places. A cultural map may be created as an end in itself or provide an input into another endeavour. The cultural mapping process may focus on the past, the present and also the future. In this respect cultural mapping can be used to monitor change in material culture as well as intangible cultural practices.[54] Cultural mapping is to do with teasing out local distinctiveness and identity.

Conclusion

In 1933 Fernand Léger at the Fourth International Congress of Modern Architecture opined that:

> There are some essential qualities to which the average person is attached and which he insists on having. If you destroy those qualities, then you have to replace them. The problem is an essentially human one. Put your plans back in your pocket, go out to the street and listen to the people breathe; you have to be in touch with them, steep yourself in the raw material, and walk in the same mud and the same dust.[55]

It is therefore fundamentally important to listen to communities and learn how to communicate findings to planners, politicians and developers who will be influential in making land-use policy decisions. A model for listening to community voices of ordinary people in an historic urban setting is found in the work of Chin and Jorge in Malacca:

> We began by listening: Listening to the young; listening to the old. Listening to shopkeepers and craftsmen, traders and fishermen (...) We began listening to those who are often not heard. And to countless more who often dare not speak.[56]

---

54 Cook, I. and Taylor, K. (2013) *A Contemporary Guide to Cultural Mapping An ASEAN-Australia Perspective*. Jakarta: ASEAN Jakarta.

55 UNESCO (2008) *Historic Districts for All: A Social and Human Approach for Sustainable Revitalization*. Paris: UNESCO Division of Social Sciences: 2.

56 Chin, L. H. and Jorge, F. (2006) *Malacca, Voices from the Street*. Melaka: Lim Huk Chin: 10.

SECTION 2
Building the Toolkit

8 Evolution of the Normative Framework

Jukka Jokilehto
Special Advisor to Director-General of ICCROM,
Professor at University of Nova Gorica, Slovenia,
and Honorary Visiting Professor at University of York, UK

What strange phenomena we find in a great city, all we need do is stroll about with our eyes open. Life swarms with innocent monsters.

Charles Baudelaire

Introduction

The adoption of the normative framework for the urban environment is associated with the general socio-economic development trends. From the heritage point of view, it is also associated with the recognition of the heritage significance of specified territories and associated attributes. While the 2011 UNESCO *Recommendation on the Historic Urban Landscape* is not proposing another heritage category, the identification and recognition of relevant qualities and resources in historic areas must be based on a critical process. The recognition of what is important or not important in the historicised territory has had its ups and downs. Indeed, an early appreciation of historic areas can already be detected in the nineteenth century, which however also saw drastic changes due to industrialisation. This resulted in the development of modern planning guidelines in the early twenti-

eth century, which normally required Urban Master Plans and centralised planning control. This process tended to be undermined by the emerging private sector towards the end of the century, giving place to more ad-hoc developments and strategic planning. By the early twenty first century, there is a new phase aiming at raising awareness and stressing the informed involvement of communities in the management not only of specifically protected zones, but also in the integrated conservation and sustainable development of the historic urban landscape at large.

Early Appreciation of Historic Townscape

The roots of the modern appreciation of historic urban areas and their recognition as cultural heritage go back to the nineteenth century and

Reconnecting the City: The Historic Urban Landscape Approach and the Future of Urban Heritage, First Edition. Edited by Francesco Bandarin and Ron van Oers.
© 2015 John Wiley & Sons, Ltd. Published 2015 by John Wiley & Sons, Ltd.

even beyond. An early example of the appreciation of the traditional urban environment was seen in Nuremberg, in Germany, which was perceived as the product of the creative spirit of the Fatherland at the end of the eighteenth century. It was also part of the beginnings of German Romanticism.[1] In the second half of the nineteenth century, John Ruskin, the renowned English art critic, was keen to make people learn to see the quality of good architecture. He did not only focus on individual buildings, but also saw them in their context. Advising on a visit to a historic Cathedral city such as Amiens in France, Ruskin recommended that:

> ... the really right thing to do is to walk down the main street of the old town, and across the river, and quite out to the chalk hill out of which the citadel is half quarried – half walled; – and walk to the top of that, and look down into the citadel's dry 'ditch,' – ... and thence across to the cathedral and ascending slopes of the city; so, you will understand the real height and relation of tower and town.[2]

In the same spirit, in his *Notre-Dame de Paris*, Victor Hugo included a chapter on the history of the town: 'A Bird's-Eye View of Paris'. Here he imagined the historic urban landscape as it would have been when viewed from the top of the towers of Notre-Dame in 1482:

> The spectator, on arriving, out of breath, upon this summit, was first of all struck by a dazzling confusion of roofs, chimneys, streets, bridges, squares, spires, steeples. All burst upon the eye at once – the formally-cut gable – the acute-angled roofing – the hanging turret at the angles of the walls – the stone pyramids of the eleventh century – the slate obelisk of the fifteenth – the donjon tower, round and bare – the church tower, square and decorated – the large and the small, the massive, and the airy. The gaze

> was for some time utterly bewildered by this labyrinth, in which there was nothing but proceeded from art – from the most inconsiderable carved and painted house front, with external timbers, low doorway, and stories projecting each above each, up to the royal Louvre itself, which, at that time, had a colonnade of towers.[3]

These two views on historic urban areas had a nostalgic undertone already anticipating the forthcoming developments that were to drastically change the historic urban landscape. From the second half of the eighteenth century to the mid nineteenth century, there was a period of the early phase of the Industrial Revolution. This involved a transition from hand production to new mechanised production methods, involving waterpower, steam, and coal as fuel. Industry developed into systems of mass production. The introduction of steam ships and railways completely transformed the communication systems. These developments 'liberated' people from the land and farming activities causing a major shift towards industrial production. Due to developments in science and medicine, the population increased; in Britain it passed from about eight million in 1801 to over 30 million by 1901 and continued to expand in the twentieth century. Consequently, urban areas became subject to unprecedented growth and expansion.

The Development and Impact of Modern City Planning

The nineteenth century urban development caused drastic changes to traditional urban environments and the related historic urban landscapes. This was also met by increasing criticism

---

[1] In 1796, an anonymously published small book of essays, *Herzensergiessungen eines kunstliebenden Klosterbruders* (*Confessions from the Heart of an Art Loving Friar*, 1797) by Wilhelm Heinrich Wackenroder (1773–98), aroused the enthusiasm of a wider public with its description of the old curved streets and historic buildings of Nuremberg associated with German cultural history. It was an early example of the romantic admiration of the Middle Ages, which then became a real basis for the development of the modern world from the early nineteenth century.

[2] Ruskin, J. (1897) *The Bible of Amiens, Our Fathers have Told Us, Sketches of the History of Christendom for Boys and Girls who have been held at its Fonts*, London (quote: IV: 6); first published 1880–85.

[3] Hugo, V. (1953) *Notre-Dame de Paris*. London: J.M. Dent & Sons Ltd: 114. (first published in 1831).

by those involved in the emerging conservation movement. In 1841, A.W.N. Pugin published his *Contrasts: or, a Parallel between the Noble Edifices of the Middle Ages, and Corresponding Buildings of the Present Day showing the Present Decay of Taste* (London 1841). The purpose of this polemical book was to alarm about the conflicting changes that were taking place in traditional historic buildings and areas. In the book was included a pair of pictures, showing 'A Catholic Town in 1440', and 'The Same Town in 1840'. The pictures showed an imaginary medieval town with its glorious church spires in 1440, and the same town transformed due to an increasingly industrialised urban landscape where the ruined medieval churches could be distinguished between the new multi storey buildings. The growing criticism was not only addressing larger buildings; it was also about the poor conditions that were faced by people living in evermore crowded urban areas.

Modern movement in architecture and town planning became international particularly after the First World War. The origins are, as usual, much earlier. The 'Urban Historic Centre of Cienfuegos' (Cuba), founded in 1819, is considered the first example of the new ideas of modernity, hygiene and order in Latin American urban planning. One of the counter reactions to the nineteenth century slums was the development of international socialism. This also resulted in the attempt to create healthier environments for factory workers and for urban areas. The ideal industrial village of Saltaire, in West Yorkshire, with textile mills, public buildings and workers' housing is an outstanding model for a harmonious style of high architectural standards and an urban plan reflecting the Victorian philanthropy of the second half of the nineteenth century. The Garden City Movement that developed in England from the end of the nineteenth century, was inspired, for example, by the utopian models of green cities in contrast to the evermore polluted and crowded industrial towns.[4] Other early modernist trends came from various European countries, such as the Netherlands, for example H.P. Berlage (1856–1934).[5]

The growing interest in historic areas is illustrated for example by the writings of Camillo Sitte at the end of the nineteenth century,[6] and of Gustavo Giovannoni later in the twentieth century. After the First World War, in the 1920s and 1930s, the modernist concepts became the real slogan of the time, and had consequences for decades after in the impact that these had on urban design and on historic urban landscapes. In 1923, the Finnish architect Gustaf Strengell (1878–1937) published a book called: 'City as Artistic Creation',[7] where he analysed the morphology of a city, referring to it as a 'closed plastic ensemble' and an independent, architectural form of art. He took note of urban elements, such as the street and the public square, and referred to the city as an artistic organism. It is this sort of holistic approach that also became a guiding principle in early modernism, and was reflected in the Bauhaus School (Weimar, Dessau and Berlin, 1919–1933), directed by Walter Gropius, Hannes Meyer and Ludwig Mies van der Rohe.[8] The new trends aiming at coherence and integrity are found in the new architecture of the 1920s in Berlin, in the so-called Siedlungen ('housing estates'), designed by Bruno Taut, Martin Wagner and Walter Gropius.[9]

---

[4] See for example, Morris, W. (1977) *Three Works: News from Nowhere, The Pilgrims of Hope, A Dream of John Ball*, London: Lawrence & Wishart, (first published in 1888).

[5] Several important properties representing modern architecture have been inscribed on the World Heritage List, see: Jokilehto, J. (2002) 'Great Sites of Modern Architecture'. UNESCO, 2002. *World Heritage*, N.25: 4–21.

[6] Sitte, C. (1909) (4th edition). *Der Städtebau nach seinen künstlerischen Grundsätzen*, Vienna (first printing in 1889).

[7] Strengell, G. (1923) *Kaupunki taideluomana* ('City as Artistic Creation'). Helsinki: Otava.

[8] Gropius, W. (1956) *The New Architecture and the Bauhaus*, London: Faber and Faber Ltd (first published in 1935).

[9] Hajos, E.M. and Zahn, L. (1928) *Berliner Architektur der Nachkriegszeit*, Berlin: Albertus Verlag. Six Berlin Modernism Housing Estates were inscribed on the World Heritage List in 2008.

An interesting early example of the impact of modernist ideas in architecture and urban planning is the case of Asmara.[10] It developed from early settlements of the Tigre people. It was occupied by Italy in 1889, and made the capital city of Eritrea. In the 1930s, it was largely developed by the Italian modernist architects and represents one of the most concentrated and intact results of modern town planning of that period especially in Africa. The development of the town followed the initial ideas with new parts added to form a real historic urban landscape with its embedded stratification. The town also includes small industrial areas and vernacular housing, all part of the integrity of the urban landscape. Notwithstanding the claim that modernist architecture represents an 'international style',[11] as also referred to by DoCoMoMo,[12] it is interesting to note that in Asmara the modernist buildings of the 1930s represented different stylistic currents, not only reflecting the ideas of Bauhaus but also recalling various stylistic currents from the past – even though these were later often condemned by more purist modernists. Indeed, it has also been noted that 'modernism is not a style', but rather a 'loose collection of ideas'.

The *Congrès internationaux d'architecture moderne* – CIAM (International Congresses of Modern Architecture), founded in 1928, organised a series of congresses on specific themes regarding contemporary concerns. Its members included some of the leading architects of the time, such as Le Corbusier and Sigfried Giedion. The fourth CIAM conference took place onboard a ship, SS Patris II, sailing from Marseille to Athens. The declaration of this conference, the Athens Charter of 1933, discussed the functionality of the city, and became a frequently quoted

document, also commented by Le Corbusier in a later publication.[13] The Charter was critical about the social and physical situation in many cities, and recommended more air and space, more hygienic conditions and green areas. It also noted that historic 'architectural assets' should be protected, but that the earlier styles should not be tolerated in new architecture. Regarding the historic areas, it was stated:

> They will be protected if they are the expression of a former culture and if they respond to universal interest and if their preservation does not entail the sacrifice of keeping people in unhealthy conditions and if it is possible to remedy their detrimental presence by means of radical measures, such as detouring vital elements of the traffic system or even displacing centres hitherto regarded as immutable.[14]

The Second World War with its devastating destruction gave a new impulse to the implementation of the modern town planning ideas and objectives. Notwithstanding the above statement of the Athens Charter, modern planning tended to aim at unification and standardisation, and existing earlier building stock could become an obstacle to this aim if not harmonious with the new planning ideas and regulations. For example in Finland, like elsewhere in Europe, the traditional urban areas (here mainly built in timber) had suffered much destruction during the war. In the 1950s and 1960s there was a major reconstruction period. In this process, the focus was on creating a homogenous townscape, and older buildings had to be sacrificed. Indeed, the legislation even sustained this trend and a municipality had the possibility to oblige landowners to upgrade land use to meet the standards of the new town plan. It was considered that, apart from a few legally protected build-

[10] Denison, E. et al. (2003) *Asmara, Africa's Secret Modernist City.* London-New York: Merrell.

[11] In 1932, Henry-Russell Hitchcock and Philip Johnson published a book, called *The International Style: Architecture Since 1922*, reprinted with new introductions: Hitchcock, H. R. and Johnson, P. (1995) *The International Style.* New York: W.W. Norton.

[12] DoCoMoMo Journal n. 28, March 2003. *Modern Heritage in Africa.* 7th DOCOMOMO Council Meeting.

[13] Le Corbusier (1957) *La Charte d'Athènes*, Paris: Éditions de Minuit (first printing in 1943).

[14] La Charte d'Athènes. (1933) pp. 66–70.

ings, any structure older than 50 years would have come to the end of its 'useful life' and would be mature for 'Sanierung', that is redevelopment, literally: 'sanitation' or improving the sanitary conditions. The results were seen in the rapid modernisation of existing urban areas, and the loss of the historic fabric. Even in the best cases, the 'historic centre' remained an island within the contemporary, often anonymous, building stock. In the 1980s, with the increasing role of the private sector and the gradual loss of the centralised control, the more systematic urban master plans regulating land use tended to be replaced by 'strategic plans' that focused on development.

Development of Instruments for Urban Conservation

One of the positive impacts of the Second World War, nevertheless, was to encourage local communities to recognise the interest and the memories of their destroyed neighbourhoods. The 1950s and 1960s became crucial in the development of elements that led to a more systematic methodology in the 1970s.[15] One of such elements related to the visual analysis of the urban landscape or townscape. Thomas Gordon Cullen (1914–1994), a British planner who worked for the journal *Architectural Review* in the 1940s, produced a large number of influential editorials and case studies on the theory of urban planning. In 1961, these ideas resulted in the publication of *Townscape*. Here he proposed a visual analysis of spatial relationships in the urban landscape. Kevin Lynch (1918–1984), an American urban planner, published his analyses on the urban fabric in: *The Image of the City* (1960) and *What Time is This Place?* (1972). Another related question concerned the social issues. Henri Lefebvre (1901–1991), an influential French Marxist philosopher and sociologist, was a critique of everyday life. He introduced the concepts of the right to the city and the production of social space. His principal work in this regard was: *La production de l'espace* (1974) (The Production of Space), which further contributed to the sensitisation of professionals and the international community towards the importance of social and cultural dimensions in the management of historic urban areas.

From the point of view of the Historic Urban Landscape concept, important were some research developments in the analysis of the urbanised territory, starting from the late nineteenth and early twentieth centuries. The studies were conducted especially by geographers in various countries. Particularly interesting were the studies in Germany on 'cultural landscape' (*Kulturlandschaft*), object of research in 'cultural geography' (*Kulturgeographie*). While the geographers focused on landscape in general; the human geographers introduced research in historic urban areas. One of the earliest was Otto Schlüter, who wrote about settlement geography (*Siedlungsgeographie*) already in the early years of the twentieth century.[16] This experience was brought to Britain by M.R.G. Conzen (1907–2000), whose early writings date from the 1930s. In the post-war period, he studied particularly the small market town of Alnwick, but also referred to many other urban areas in Britain. He considered that historical townscapes presented three systematic aspects to morphological analysis expressed by the distinct form categories of town plan, building fabric, and land utilisation. Beyond these static links, he identified the

---

[15] There was an increasing literature regarding town planning, which took note of the social-economic as well as psychological conditions in urban areas, such as Simmie, J.M. (1974) *Citizens in Conflict, The Sociology of Town Planning*, London: Hutchinson Educational. In Blowers, A. *et al.* (eds) (1974) *The Future of Cities*. London: Hutchinson Educational & Open University Press.

[16] See: Whitehand, J.W.R. (1981) *Background to the Urban Morphogenetic Tradition*. Whitehand, J.W.R. (ed) (1981) *The Urban Landscape: Historical Development and Management; Papers by M.R.G. Conzen*. London, New York: Institute of British Geographers, Special Publication, N. 13, Academic Press.

dynamic morphology that depended on the differential time response of the three form categories to changing functional needs. He recalled that the town plan is not merely a 'street plan' but consisted of four complexes: site, street system, plot pattern, and building arrangement, all physically interrelated. The principles of planning could vary from one period to another, but he noted that the British market towns generally rested on the survival of medieval plans and a stock of traditional buildings. Conversely, the Victorian industrial society had failed to develop its own architectural form language. Therefore, their designs had resulted in confusion and devastation of historic urban areas.[17]

In Italy, the study of urbanised territories has its origins in the research and teaching of Saverio Muratori (1910–1973), who taught particularly at the school of architecture in Rome.[18] His early writings date from the period following the Second World War, when he discussed the life and history of towns,[19] and especially studied the case of Venice.[20] The ideas were inherited and further developed by his students, particularly by his assistant Gianfranco Caniggia (1933–1987), who taught at the University of Florence.[21] Muratori had his architectural and philosophical background in the early twentieth century, including Benedetto Croce's philosophy, but he developed his own thinking starting from the technique and language of architecture. He defined architectural organism as: *'formal unity of cooperating, cohesive and conspiring structures, subject to transformation in space-time'.*[22] Mura-

tori also played a key role in the development of the concept of 'type' in relation to town as well as individual buildings. To him, the concept of 'type' was associated with the integration of matter and idea, the tangible and intangible dimensions in architecture.

While Caniggia had his own personality, which differed from Muratori's, he continued to work on the same issues, giving major attention to the concept of 'typology'. To him, this notion referred to *'the relationships spontaneously codified between the environment and the work of an individual, by the community'.* And, the community was *'part of humanity in a particular location, where over time it has conditioned related structures with specific, individual and codified characteristics'.* He saw typology as a system of concepts on the basis of which the existing objects had been matched (conformed) in historical, temporal-spatial succession.[23] To Caniggia, the concept of 'typology' was strictly associated with project development in the planning and design of historic urban areas. History was not static but consisted of a dynamic process. It implied a constant redesign in reference to associated models and concepts. This was necessary in order to allow for a confrontation with existing structures in view of future development. There was need for a clear methodology for the analysis and interpretation of the 'typology', not as a fixed model but as a cultural approach. Indeed, it was not possible to identify a type if not understood in the context of a 'typological process'. It was necessary to understand

[17] Conzen, M.R.G. (1981) Historical Townscapes in Britain: a Problem in Applied Geography (first published in 1966). Whitehand, J.W.R. (ed) (1981) *The Urban Landscapes: Historical Development and Management; Papers by M.R.G. Conzen,* London, New York: Institute of British Geographers, Special Publication, no. 13, Academic Press.

[18] Cataldi, G., Maffei, G. and Vaccaro, P. (2002) Saverio Muratori and the Italian School of Planning Typology. *Urban Morphology,* 2002, 6 (1): 3–14.

[19] Muratori, S. (1950) Vita e storia delle città. *Rassegna critica d'architettura* 11–12: 3–52.

[20] Muratori, S. (1959) Studi per una operante storia urbana di Venezia I–II, *Palladio* 3–4, Rome: Istituto Poligrafico dello Stato.

[21] Cataldi, G. (2003) From Muratori to Caniggia: the Origins and Development of the Italian School of Design Typology. *Urban Morphology,* 2003, 7 (1): 19–34.

[22] Muratori, S. (1963) *Architettura e civiltà in crisi,* Rome: Centro Studi di Storia Urbanistica; quoted from Cataldi, G, (2003) From Muratori to Caniggia: the Origins and Development of the Italian School of Design Typology. *Urban Morphology,* 2003, 7 (1): 23.

[23] Caniggia, G. (1976) *Strutture dello spazio antropico: studi e note,* Florence: UNIEDIT, Biblioteca di Architettura, Saggi e Documenti 3; Caniggia, G. and Maffei, G. L. 2001. *Architectural Composition and Building Typology.* Florence: Alinea Editrice.

the origins from where successive types had derived in order to capture their meaning and the significance of the different phases also in view of eventual restoration. A 'typological process' was guided by its intrinsic laws developed through historical transformations. A re-design in a particular moment in time should be autochthon in its character, it should be associated with its native origins. On the other hand, while re-design was based on continuous verification it also depended on the evolving requirements in community. Consequently, it also involved a creative process.

It is interesting to compare the approaches of Conzen and Caniggia. Indeed, they lived in different contexts and never had any direct contact. While Conzen developed his practice on mediaeval sites, Caniggia's reference was in the ancient Roman and then mediaeval land-use, which continue to be present not only in the individual settlements, but also in the entire Italian territory. Apparently, Caniggia was completely unaware of Conzen's work, who also only learnt about the other later in his life. Whitehand has indicated that the single most important feature in common with the two was the stress they placed on historical continuity.[24] Conzen worked as a human geographer, while Caniggia was an architect. However, Caniggia starting from the buildings worked also towards geographical scale. In some way, Conzen and Caniggia could be seen as complementary in a multidisciplinary approach to the planning and management of the historic urban landscape.

International Recognition of Historic Urban Areas

When the Second International Congress of Architects and Technicians of Historic Monuments met in Venice, in May 1964, under the auspices of UNESCO and the participation of ICCROM, the question of the protection and rehabilitation of historic centres was adopted in the form of a motion (document 4). This motion called for appropriate legislation, and that the rehabilitation of historic centres should be taken into the programmes of national and international bodies. It is however only in the 1970s that there is a more consolidated approach to the recognition of historic urban areas. The year 1975 was declared by the Council of Europe 'The European Architectural Heritage Year', and it resulted in a series of case studies in several European countries, including UK, France, the Netherlands, Germany, Denmark, and Italy. This became a real turning point regarding urban conservation. In October of the same year, the Council of Europe adopted the European Charter of the Architectural Heritage. Here it was declared that: 'The European architectural heritage consists not only of our most important monuments: it also includes the groups of lesser buildings in our old towns and characteristic villages in their natural or manmade settings'.[25]

It was warned that this heritage was in danger, but that 'integrated conservation averts these dangers'. The question of 'integrated conservation' was again a theme introduced in the closing conference of the Heritage Year, 'The Declaration of Amsterdam'. The question here was still more about traditional architecture than about historic towns, even though the aim was to integrate conservation into planning norms.

In the following year, November 1976, UNESCO adopted the *Recommendation concerning the Safeguarding and Contemporary Role of Historic Areas*. The 1976 Recommendation was really the first international recommendation that clearly framed the issues related to the urban heritage seen in the broader context. *'Historic areas and their surroundings should be regarded as forming an irreplaceable universal heritage. The governments and the citizens of the States in whose territory they are situated should*

---

[24] Whitehand, J.W.R. (2003) Editorial Comment. *Urban Morphology, Journal Online*, 2003, vol. 7.2; http://www.urbanform.org/online_public/2003_2_editorial.shtml.

[25] Council of Europe (1975) *The European Charter of the Architectural Heritage*. Art. 1.

deem it their duty to safeguard this heritage and integrate it into the social life of our times' (art. 2). The question here is about areas that could now be interpreted as Historic Urban Landscapes. As a matter of fact, the 1976 Recommendation was taken as the principal reference for the Vienna Memorandum in 2005, where the notion of Historic Urban Landscape was first defined. This Recommendation lists a series of legal and administrative measures to be taken into account for safeguarding historic areas. These include the specific conditions and restrictions for areas identified for protection, and conditions for new constructions, technical, economic and social measures, relevant surveys and specifications of actions. The Recommendation stresses research, education and information, as well as international cooperation.

The International Council on Monuments and Sites, ICOMOS, created in 1965 in Krakow with Piero Gazzola as its first President, started soon a series of national and international conferences, which often touched the question of urban conservation. It is however only several years later that an ICOMOS *Charter for the Conservation of Historic Towns and Urban Areas* appeared. This came as a result of some twelve years of study by specialists, following the 1975 Heritage Year. The study was coordinated by the Hungarian ICOMOS Committee, and text was revised and finally adopted by the ICOMOS General Assembly meeting in Washington in 1987. The same year, the Brazilian ICOMOS Committee came forward with the 'Carta de Petropolis'.[26] Here it was stated: *'Urban historical sites are part of a wider totality, comprising the natural and the built environment and the every-day living experience of their dwellers as well. Within this wider space, enriched with values of remote or recent origin and permanently undergoing a dynamic process of successive transformations, new urban spaces may be considered as*

environmental evidences in their formative stages.' The Brazilian Charter again emphasises that the conservation of historic urban areas is only feasible if understood in the broader context of forces of development that are having an impact on a particular city. Indeed, this means not only in a city in the strict sense but also its relevant socio-economic context.

In the early 1990s, there were two parallel processes, one with the Council of Europe and the other with the World Heritage Committee, both aiming at the definition of the concept of 'cultural landscape'. The World Heritage Committee adopted the notion in 1992, and the Council of Europe in 1995. This can be considered a crucial period, when the broader territorial context is at last entered into the realm of heritage. Until that time, historic towns nominated to the World Heritage List were normally understood as 'groups of buildings'. It is only after the adoption of the 2011 *Recommendation on the Historic Urban Landscape* that this definition may be broadened and associated with the category of 'site'. This is also brings into discussion the need for a broader cultural approach to identify the significance of the contributions of the different generations over time within the various social-cultural contexts in the world. The 2005 *Convention on the Protection and Promotion of the Diversity of Cultural Expressions* indeed goes in this direction. It affirms that *'cultural diversity is a defining characteristic of humanity'*, and that it *'forms a common heritage of humanity and should be cherished and preserved for the benefit of all'*.[27] While the creative capacity of humanity has thus produced an important heritage, it is also noted that the processes of globalisation, characteristic of the twentieth and twenty first centuries, tend to standardise the built environment. This is obvious when one compares any historic urban centre dating from pre-industrial age to what is being built at the present. It is also

---

[26] Adopted by the first Brazilian Seminar concerning the Preservation and Revitalisation of Historic Centres, in Itaipava, July 1987.
[27] UNESCO (2005) *Convention on the Protection and Promotion of the Diversity of Cultural Expressions*. Preface.

obvious that the question is not only about diversity in its own right, but rather of the quality that accompanies the creative product of a particular community.

The 2011 *Recommendation on the Historic Urban Landscape*, adopted by UNESCO, recognises that *'historic urban areas are among the most abundant and diverse manifestations of our common cultural heritage, shaped by generations and constituting a key testimony to humankind's endeavours and aspirations through space and time'.*[28]

Indeed, the reason for this Recommendation has been motivated by increasing problems that threaten the integrity and the traditional continuity of such historic urban areas from the outside. However, the Historic Urban Landscape approach is not proposed to stop in the immediate surroundings; it looks beyond the notion of 'historic centre' or 'ensemble' to include the broader urban context and its geographical setting. The problem of controlling development in historic cities, in fact, is not so much related to the historic centre, where the protection tools are usually sufficiently appropriate, but in particular to the areas outside it. Town planning tools often insufficiently control these areas, and they can be beyond UNESCO protection (in the case of World Heritage property). It is here that we often see an uncontrolled development in construction and infrastructural works giving negative impacts on the protected areas of the territory.

This way, the fundamental goal of the Historic Urban Landscape approach is to guarantee the proper conservation and development of historic urban areas that are an integral part of the overall urban landscape as it has evolved over time (the city as a 'whole'). The notion of Historic Urban Landscape is aimed at the recognition of the qualities and character of such wider urban landscapes, identifying their vulnerability, and proposing suitable remedies and/or control

mechanisms for appropriate management and planning regimes. However, it is still necessary to recognise the qualities of such historic urban landscapes, because it is only on the basis of the recognition that one can decide about appropriate policies and strategies.

The resources and qualities are referred to the tangible and intangible aspects of heritage. All these qualities, together, constitute the common resources that define the character of the historic city, whose significance is the outcome of a series of natural and anthropic events linked to the conformation of the place and the continuous interaction of man and nature with the place itself. Indeed, it should also be repeated that Historic Urban Landscape is not intended as another heritage category. The concept of Historic Urban Landscape is not new, because the concepts characterising it can be found in old European theories, international documents and old plans. What is new is the changed perception and the huge potential of this concept, the possibility of treating these urban areas not as static objects of admiration but as living spaces for *sustainable communities.*

> *The perception of its potential pivotal relevance today, however, is. It is a concept that draws from experience in urban conservation and cultural landscapes and seeks to encompass values relating to natural elements, intangible heritage, authenticity and integrity, and genius loci. Genius loci embraces key components of the sustainability agenda such as sense of place and community belonging, cultural identity and cultural diversity, and – alongside intangible cultural heritage – it subsumes associative values.*[29]

How Normative Frameworks Respond to the Challenges of Change Caused by Urban Development

In the mid twentieth century, Giovanni Astengo (1915–1990), Italian urban planner and director

---

[28] UNESCO (2011) *Recommendation on the Historic Urban Landscape.* Preamble.
[29] UNESCO (2007) *Summary Report of St. Petersburg Conference.* Paris: UNESCO World Heritage Centre: 2.

of the periodical *Urbanistica*, described the urban landscape of Assisi as seen from various points of view. Seen from the plain, the 'visage of the city' appeared as a whole, as a great constructed scenery, articulated by long walls that formed steps converging in the massive fortress on the summit, all embedded in the greenery of the mountain slopes. This view was complemented by various others from the outside and from within the walled city itself. He summarised this in 'The Visage of Assisi' (*'Il Volto di Assisi'*), concluding:

> *Landscape, light, colour, houses, mediaeval towers, squares and renowned monuments; countless mutual views from the plain and from the Hill and, within the city, in the large basins of this built space; a sense of widespread tranquillity and sweetness. These are the elements that combine to form the visage of this exceptional city.*[30]

The ancient city of Assisi, home of San Francesco in the Middle Ages, was one of a hundred Italian municipalities to be invited in 1954 by a Ministerial Circular to prepare a General Master Plan in accordance with the Urban Planning Law of 1942 (n. 1150). In 1955 the municipal administration appointed Astengo for this project. Within two years of intense work, he prepared a complex and innovative instrument both in terms of its method and the proposed norms. Astengo appreciated the historically consolidated image of the city within its rural setting. Indeed, his plan was:

> *complete in all its parts: from the general town plan of its overall structure, also including, for the vast agricultural territory, plains, hills and mountains, an economic development plan based on reforestation and irrigation, with a detailed plan for the historic centre and the expansion to the east of the walls.*[31]

The general master plan was integrated with a detailed historical and morphological analysis of the entire historic urban landscape. This included the entire municipal territory as a basis for precise coding of the regulations controlling the conservation and the limits of transformation of individual elements. The plan included protection of the natural and rural environment with a possibility of limited growth outside the city walls.

The work of Astengo in Assisi was an incentive for the creation of ANCSA, *Associazione Nazionale Centri Storici e Artistici* (National Association Historic and Artistic Urban Centres). It was established in 1960 in Gubbio on the occasion of a National Congress for the Recovery and Safeguarding of Historic Urban Centres (*Risanamento e la Salvaguardia dei Centri Storici*). The conference was organised by Astengo for the presentation of his master plan, and it grouped a number of municipalities, politicians and academic people. It adopted a declaration, called *La Carta di Gubbio* (1960). The charter called for urgent recognition of urban conservation on the national level, and the identification of areas to be safeguarded and recovered, as an essential prerequisite for the development of the modern city. This required appropriate urban planning and relevant norms.

The normative framework, subject of this paper, tends to remain a set of principles at the international level. Taking into account cultural and heritage diversities, which are the real objective of conservation, such principles must necessarily remain more like incentive to be understood and applied in each specific case. The most practical advice concerning historic areas has been offered in the 1976 UNESCO Recommendation. This set of principles certainly continues to be valid even though the governance strategies have since continued to evolve. The relatively centralised control mechanisms that were in place in the 1970s as part of the modern movement in town planning have been increasingly replaced by a shift to decentralised management and more emphasis to the private sector.

---

[30] Astengo, G. (1958) *Il Volto di Assisi*. Urbanistica XXVII, N.24–25: 9–132.
[31] Description of the Master Plan, quoted from the PhD thesis of: Martini, Viviana. 2013. *The Conservation of Historic Urban Landscapes: an Approach*. Slovenia: University of Nova Gorica.

The ideals of harmonious integrity in urban areas, characteristic of the early modern movement, expressed for example in the design of Le Havre,[32] were sometimes considered 'boring'. By the 1990s, they tend to have lost their earlier appeal in favour of more adventurous and personalised architectural expressions, such as those in Dubai or other fast developing economic centres on the globe. These trends have resulted in threats to many World Heritage cities, just to mention Vienna, Graz, Liverpool, or Baku. It is such tendencies that have also called for the adoption of the 2011 *Recommendation on the Historic Urban Landscape*. The Historic Urban Landscape approach is based on a holistic and qualitative analysis of the meanings in the functional and visual integrity of the dynamism of the urbanising territory in relation to its environmental setting. This is particularly relevant now when the population increase in the world is focusing especially on urban areas. The urban expansion can take two forms, or a combination of these. One concerns the occupation and transformation of the rural and natural surroundings of existing urban areas; the other is the increasing density of existing built areas, resulting in the escalation of high-rise buildings. In both cases, the historic urban landscape is at risk, either internally or through outward expansion. At the same time, decision-making in large urban areas is gradually shifting from the public towards the private sector. Consequently, the aim of the management of recognised historic urban territories is to maintain continuity while keeping and integrating recognised meanings and qualities. This is a dynamic process, which implies knowledge-based involvement and empowerment of the local community, guided in concert with competent professionals and administrators. The problems that result can easily be seen in historic towns.

The historic city of Rome is an excellent illustration of an historic urban landscape that preserves its historical stratigraphy and overall visual integrity better than most. With nearly three million inhabitants in 2013, it covers ca. 1300 sq km, out of which the ancient walled city is one percent. Modern town planning started from 1873. In 1887, the principal archaeological area was already protected. The Master Plan of 1931, at the time of the Fascist Regime, marked some drastic changes. However, the entire historic centre was considered a conservation area. After the 1962 Master Plan protection was extended outside the walled city, with precise norms for rehabilitation and conservation of the urban fabric. In 2008, the Municipal Council adopted a new Master Plan, Roma 2000, where the entire municipal area, including built and natural areas, was subject to the same control mechanisms. The city was open to the territory, and it was made more accessible internally and externally, a polycentric system with new central nuclei based on existing social identities linked with a system of mobility and the hierarchy of exchange points. The plan was based on systems of land use, environment, and mobility. Furthermore, a series of strategic orientations focuses on specific problem areas, such as the archaeological park and the relationship with the river. Another excellent example in present-day Italy is Assisi, where the 2011 Structural Plan and Landscape Guidelines follow the planning tradition started by Giovanni Astengo 60 years earlier.

There are other World Heritage cities that have well retained their historic urban landscape. These include the city of Bath, UK, where the urban conservation area covers not only the historic town itself, with the famous circuses, but also the surrounding landscape.[33] It is indeed this entire area that was inscribed on the World

---

[32] An example of the systematic implementation of such planning criteria in modern town planning is seen in the reconstruction of the city of Le Havre after the destruction during the 1940s war. The rebuilt town, designed by a team led by Auguste Perret, was inscribed in the World Heritage List in 2005.

[33] See: Worskett, R. and Redston, R.V. (1978) *Saving Bath, A Programme for Conservation*. Conservation Study, Stage 2, Final Report. Bath: Bath City Council Bath, Department of Architecture and Planning.

Heritage List. The instruments to guarantee the continuity of the qualities and the integrity of the historic urban landscape are embedded in the planning norms of its Urban Master Plan. In 1997, the Municipality adopted the 'Bath Manifesto', which stressed the role of the City Council to direct pressures so as to preserve and enhance the fabric and landscape of Bath within a balanced social, cultural and economic structure.

Many large metropolises, such as Rio de Janeiro in Brazil, clearly illustrate the phenomenon of twentieth century urbanism. The question is whether or not such urban growth should be taken as part of the Historic Urban Landscape. In the case of Rio, the recent World Heritage nomination (2012) was limited to the 'Carioca Landscapes between the Mountain and the Sea', including some of the famous natural features, and the creative parks and landscaping of the shorelines to the design by Roberto Burle Marx. At the same time, the typical Latin American high-rise city was set aside. The phenomenon of high-rise buildings tends to spread even to rural areas, where it is not justified. The small Municipality of Antônio Prado, in southern Brazil, founded by Italian immigrants, has retained its original rural character with wooden buildings within a natural setting. However, only these wooden buildings were protected as 'national monuments', in the 1970s, and there was no concern for the context. By 2012 it had become clear that there was a risk of transformation in the form of unsuitable additions and even multi-storey buildings resulting in the destruction of the visual integrity of the place. A Management Plan guided and implemented by the local authority in concert with the local community had become urgent.[34]

Kyoto is perhaps the most important historic town in Japan, and various parts have been legally protected as conservation areas, and also inscribed on the World Heritage List. Like many towns in Japan, Kyoto has been subject to excessive high-rise development after the Second World War, which tends to hide the still existing traditional urban fabric. Nevertheless, since the 1990s, there have been various initiatives at the neighbourhood level to sensitise the local population about traditional housing, the 'Machinami'.[35] The results are already seen in the growing interest to live in a traditional house, and to improve the townscape in those areas. In particular, the Municipality has worked systematically to lower the height of multi-storey buildings, clear electric wirings and redesign commercial labels in a more harmonious way. All this is gradually improving and partly 'reconstructing' the historic urban landscape.

New Tools for the Management of the Historic Urban Landscape

The former Imperial Palace and some other properties of Beijing are already inscribed on the World Heritage List. Nevertheless, in 2012, there was a new proposal taking into account the overall planning scheme, and proposing to nominate 'The Central Axis of Beijing (including Behai)' to the World Heritage List. The Central Axis was taken as a symbolic planning reference already from the thirteenth century, when the old city of Beijing was first built. It has remained a reference up to the twentieth century, and it also refers to the best-preserved parts of Beijing. It includes not only some of the major public monuments but also traditional *hutong* areas, presently threatened by demolition in many parts of the city. The recognition of the Historic Urban Landscape would offer a new framework and a new motivation for the conservation and

---

[34] Prado has been discussed by Piccoli, C. (2011) *Historic Urban Landscape of Antônio Prado, Brazil: Formation, Recognition and Preservation*. Slovenia: University of Nova Gorica.

[35] The Japanese *Machinami* Association adopted the *Machinami* Charter in 2000, calling for respect of traditional housing and its cultural and physical context.

eventual rehabilitation of associated historic areas, which so far have remained outside official protection policies. In this case, the buffer zone is proposed to be divided into five different levels: from no construction zone to areas with one to three stories and to higher buildings; in all cases the form, scale and colour should be compatible with the character of the historic area. In China too, like in many parts of the world, the real question is awareness raising and consciousness about the improvement of living and working standards without destroying the traditional urban fabric. In the case of Beijing, the use of virtual models and GIS techniques has been found useful tools to help in the sensitisation process. Recognising the significance and value of heritage remains one of the biggest challenges in contemporary China.

A particular problem regarding traditional historic urban landscape is the relationship of traditional continuity and impact of modernism. In the 1970s, UNESCO promoted the development of pilot planning projects, for example, in Kathmandu Valley (Nepal) and Ahmedabad (India). In either case, these initiatives were not properly implemented. In Kathmandu, the boundaries of the World Heritage property had to be revised some 20 years later. In Ahmedabad, the urban fabric is subject to continuous transformation, losing gradually the integrity of its traditional fabric, even though maintaining its traditional continuity, which has always been based on change.[36] Indeed, in both cases, a holistic look, a Historic Urban Landscape approach, should be based on a system of communication and informed involvement of the local community in the management of the historic town and its broader setting.

This is the case in Zabid (Yemen), which has been placed on the *List of World Heritage in Danger*[37] in 2000 due to serious deterioration of the built heritage, removal of market activities outside the old town, and loss of the traditional economic role of the city. In 2011, the city has developed an Urban Conservation and Development Plan, prepared by an international team.[38] This Master Plan has been discussed in detail with the population and the authorities of the city so as to inform them about the policies and strategies to conserve the World Heritage area of the city, and the historic urban landscape that it is part of. Indeed, it is essential that the mistakes made in the past be corrected, but it is equally important to make sure that the historically established relationship of the urban area with its rural surroundings be maintained. Much has been achieved in the management of the old town, instructing the building owners in maintenance and conservation. After the tumultuous period of 2011–2012 in Yemen, it is necessary to again properly involve and convince the local decision makers to continue with the policies and strategies already initiated.

Generally speaking, the principal management tools already exist, and they are based on modern structural and spatial planning. These include the methodologies of study aiming at understanding the urban morphology and typology, as indicated by Muratori and Caniggia in Italy, and Conzen in UK. Such analyses are fundamental for understanding the meaning and signification of the territories relevant to the definition of the Historic Urban Landscape in each case. At the same time, for many people, such analyses may be too 'scientific' to be easily digested. A more 'accessible' approach concerns the visual integrity of a protected area. Indeed, this is one of the key issues often discussed by the World Heritage Committee. In March 2013, an expert meeting took place to discuss the question of visual integrity in relation to World Heritage properties. The meeting recommended

---

[36] Desai, J. (2012) Conservation – in Context of Sustainability. *Journal of Architecture*, India.

[37] See website: http://whc.unesco.org/en/list/.

[38] MEDINA Project (2011) *Zabid 2020: Urban Conservation and Development Plan*. MEDINA Project.

visual impact studies and assessments to be developed and integrated into spatial planning processes as part of the control instruments to retain the overall integrity of the place. It also called on university networks to engage in relevant assessments and identification of 'best practices' taking into account the specificity of each place.[39]

Almost as an anticipated response to the invitation by the Agra meeting, Viviana Martini has concluded in April 2013 her doctoral thesis concerning precisely the analysis of the different aspects of the Historic Urban Landscape and the development and implementation of pertinent strategies for planning control and management.[40] Martini started her research from the Action Plan proposed in the 2010 'Draft Recommendation on the Historic Urban Landscape' of UNESCO.[41] More specifically, this Action Plan proposes the identification of resources, their significance and vulnerability in view of possible change within the overall context. There are different issues that need to be examined in the process, including the characterisation of qualities that range from static to dynamic, as well as the tangible and intangible dimensions of heritage in relation to cultural diversity and identity represented by the specificity of each place.

Martini's analysis has included the verification of the structural and visual integrity of the urbanised territory in reference to all the important elements seen within their context. Next, she has analysed the intensity of resources having relevance for heritage, seen from different angles, including protected areas, historical and cultural resources, landscape resources, physical-natural resources, socio-economic resources, pilgrimage and tourism resources, etc. Such resources are identified area by area, and the intensity is given a numeric value, on the basis of which it has been possible to prepare an Intensity Map. The same elements are then verified in terms of their vulnerability for change, depending on the existence or lack of planning norms and other legal measures, resulting in a Vulnerability Model. It is on this basis that it is then possible to update the policies and strategies and assess the relevance and efficacy of required control instruments in order to maintain the overall balance of Historic Urban Landscape, monitor developments and prioritise actions in view of potential changes and developments in the future. As part of her thesis Martini has made full use of digital techniques illustrating the results of the analyses in 3D models.

The Historic Urban Landscape approach should consider aspects connected to the managerial, economic and sociological scope of the subject. Historic Urban Landscape is not seen as a new town plan, but rather as a general management tool, which should be integrated in tools already existing. This requires the involvement of the entities in charge of managing the territory, of the public and private stakeholders, of the public administrations and, above all, of the population, which must be informed, educated and involved. The key question seems to be in the recognition of the Historic Urban Landscape by the community and relevant decision-making bodies, and in the need to develop and implement required planning and management instruments at the micro and macro scales. The key point is in article 5 of the 2011 Recommendation, which addresses: 'the need to better integrate and frame urban heritage conservation strategies within the larger goals of overall sustainable development, in order to support public and private actions aimed at preserving and enhancing the quality of the human

---

[39] UNESCO Report of the International Expert Meeting on Visual Integrity, Agra, India, 6–9 March 2013.

[40] Martini, V. (2013) *The Conservation of Historic Urban Landscapes: an Approach*. Slovenia: University of Nova Gorica; the thesis includes three case studies: Assisi, Urbino and Ferrara, each exemplary in its way of Italian experience in urban conservation, ranging from the 1950s to 2011.

[41] UNESCO Executive Board: 'Invitations to the intergovernmental meeting of experts (category ii) related to a draft recommendation on the conservation of the historic urban landscape', PARIS, 7 October 2010, 185 EX/46.

environment' (Art. 5). The World Heritage Convention itself stresses that one of its aims is *'to give heritage a role in the life of the community'.* This demand was taken as one of the principal points to be highlighted in the Kyoto Vision, adopted by the conference in Kyoto celebrating the 40th anniversary of the World Heritage Convention in November 2012.

The Council of Europe Framework Convention on the Value of Cultural Heritage for Society (Faro 2005) stresses the same issue. It is noted that a *'heritage community consists of people who value specific aspects of cultural heritage which they wish, within the framework of public action, to sustain and transmit to future generations'* (Art. 2). It is the right and responsibility of everyone alone and collectively as part of this community to identify and recognise heritage, and in our case, the Historic Urban Landscape approach. The Faro Convention also stresses the importance of relationship of culture with economics.

In order to make full use of the potential of the cultural heritage as a factor in sustainable economic development, the Parties undertake to: raise awareness and utilise the economic potential of the cultural heritage; take into account the specific character and *interests of the cultural heritage when devising economic policies; and ensure that these policies respect the integrity of the cultural heritage without compromising its inherent values (Art. 10).*

The Historic Urban Landscape approach depends on the involvement and empowerment of the community's capacity to recognise and understand the qualities of the historic urban landscape. Consequently, as has been stressed by various international organisations, capacity building at all levels is one of the fundamental instruments for guaranteeing the required continuity in the planning and management of the historic urban landscape. This should involve property owners, the administration and all stakeholders in decision-making. At the individual level, capacity building should aim at creating the conditions for the individual members of the community to engage in learning processes. The existing institutions should be involved in a process of verification of their roles and competencies, and their mutual communication systems, for the benefit of guiding and supporting the integrated conservation and sustainable development of the historic urban landscape.

9 Civic Engagement Tools for Urban Conservation

Julian Smith

Architect, Planner and Educator and currently President of ICOMOS Canada

> *Cities have the capability of providing something for everybody, only because, and only when, they are created by everybody.*
>
> Jane Jacobs

Introduction

The 2011 UNESCO *Recommendation on the Historic Urban Landscape* represents a significant shift in how we frame our approach to urban conservation issues and their relationship to urban development more generally. It moves us from a primarily commemorative and aesthetic interest in the historic city to a broader ecological perspective. In doing so, it creates the basis for a synergy between three major perspectives that to date have operated somewhat independently – the contemporary design and development perspective, the environmental perspective, and the heritage conservation perspective.

Where the heritage conservation community can make its strongest contribution is in the area of reading and interpreting the existing urban landscape. This is fundamental to developing interventions that strengthen existing values and re-calibrate traditional and current patterns. The legibility must be related not only to the visual morphology of the city, but to the cultural practices that have created that morphology and that continue to animate it. And, in a world of increasing immigration and movement, it is important that the heritage conservation community understand the city as a layered reality. It is a layering that is often both temporal and cultural – the overlay of different periods in the evolution of a place, but equally the different realities that co-exist in the same place from a variety of cultural perspectives.

This reading of the historic urban landscape was largely lost in the planning and design activities of the modernist period. There was a deliberate decision, as part of the modernist enterprise, to reinvent the world as part of rescuing us from the horrors of the great wars of the first half of the twentieth century. But that utopian vision of refashioning the world for 'the New Man' and the new society has run its course. There is now a realisation, far deeper and more widespread

Reconnecting the City: The Historic Urban Landscape Approach and the Future of Urban Heritage, First Edition. Edited by Francesco Bandarin and Ron van Oers.

than at any time in the last fifty years, that true sustainability involves continuity between past, present and future.

There is a young and global generation that has grown up with the concept of ecology, the first generation to do so in a self-conscious way. There is also a growing realisation that ecology in its broadest sense – the creation of self-sustaining communities that live within a balance of natural and cultural resources – is closely tied to traditional knowledge and the long experience in many communities of living with limited resources. Ecological initiatives gain increasing priority as we realise that unsustainable patterns endanger our very survival.

The Historic Urban Landscape approach is dynamic rather than static. That is an inevitable consequence of shifting one's attention from the value of objects to the value of relationships. Relationships among objects, and between tangible objects and intangible cultural practices, are fundamental to any ecological understanding.

To read a dynamic environment involves understanding the forces at work that create conditions of equilibrium or disequilibrium. It is for this reason that the Recommendation speaks of social, cultural and economic processes as well as architectural evidence and distinctiveness.

The following are some of the tools available to researchers in what is essentially a cultural landscape approach to the urban situation.

Ways of Seeing

The emphasis in understanding historic urban districts has tended to be on the documentation of physical form. Photographs, drawings and cartography have been central to this exercise, which emerges out of the aesthetic bias in conservation. Photographs have been particularly important, continuing an emphasis from within the European art historical tradition of understanding place by translating it from three dimensions to two dimensions through the use of one-point, two-point and three-point perspective. To these physical records are added the historical biographies of individual buildings, streetscapes and open spaces to provide a commemorative narrative to deepen or shape the artistic appreciation.

At the scale of the city, the urban landscape is often understood in a similar way. There has been considerable study of view planes from key vantage points in and around the city. To these are added the study of monuments as organising principles within the visual landscape of specific areas.

To address the 2011 Recommendation's broader definitions of the historic urban landscape, as *a historic layering of cultural and natural values and attributes*, one must move beyond the outward morphology of the city to understand how it is experienced from within. A physical form is experienced through the rituals of inhabitation, and it is these rituals that create a cumulative experience of the urban landscape that is essential to both its understanding and its sustainability. These rituals include not only the daily and seasonal lives of its inhabitants and visitors, but the rituals of shaping and reshaping the physical form itself, over time (Figure 9.1). In a number of cities there have been proposals to remove residents from a portion of the historic urban landscape to create hotel complexes for tourists. The visual form may remain intact, but the experience of the city would change radically.

Monuments restored as visual museums, in cities around the world, continue to attract wide interest and visitation, but that interest is declining. The visual imperative of the conservation field was part of a larger twentieth century approach to design and development. Architecture in the modernist sense was understood as an exercise in visual sculpture, and the photograph became the dominant mode of transmitting its intent and evaluating its quality. The conservation field adopted some of the same assumptions, happy to put the sculptural and aesthetic quality of historic monuments and sites against anything the modern movement had to offer.

Figure 9.1 Skyline, Cairo.

But the thrust of cultural tourism has already shifted from observation to experience, with increasing emphasis on understanding place through participation. This can sometimes be a problem for host cultures. In New York City, for example, churches in Harlem have seen a large increase in tourist attendance at the gospel services on Sundays, with visitors sometimes outnumbering church members. Nonetheless, this illustrates a broader shift towards appreciating cultural experience as basic to sense of place and sense of identity.

The documentation of experience rather than observation cannot be so easily captured through more and more exact recording and documentation. Although laser technologies and other advances have seemed to open new ground for collecting and sorting massive amounts of data, this data is still about the observable form of the monument or the city. One can exhaust oneself with measurement and still not approach the recording of experience.

The documentation of experience relies on a broader set of media, and a different set of organising frameworks. In terms of media, it benefits from video and film, which add the dimension of time and the experience of movement, and allow narrative and artefact to coalesce. It also benefits from the insights of poets and novelists who translate the urban experience

into frames that capture an emotional as well as intellectual response. It benefits from storytelling and narrative in the oral tradition, and an appreciation for the insights of what is sometimes referred to as 'traditional knowledge' but can be understood more simply as the knowledge born of experience. And it can benefit from recording sacred or secular rituals that interpret the urban landscape through stylised and heightened powers of movement. One of the wonderful discoveries of such documentation is the importance of other senses beyond the visual – sound, smell, touch, taste. The ringing of bells, the penetrating sound of *nagaswarams* in a moving procession, the smell from open markets, the feel of different pavements, the echoes of children's voices in narrow alleys – all of these are part of delineating the urban landscape.

There are many examples of these insights – for Istanbul, the recent writings of Ohran Pamuk; for New York City, the impact of Jules Dassin's 1948 film, *Naked City*; for Paris, Laurence Cossé's recent *Au bon roman*; for Toronto, the poetry of Dionne Brand or Srinivas Krishna's powerful 1991 film, *Masala*; for Detroit, Julien Temple's documentary *Detroit, Urban Requiem?* There are insights in these works that reveal the patterns of inhabitation that bring these places alive and ultimately that sustain them. It is surprising how often these alternative sources of understanding are bypassed, by those documenting the historic urban landscape, in favour of more quantitative measures. As T.S. Eliot once remarked, where is the wisdom we have lost in knowledge, and the knowledge we have lost in information?

In terms of UNESCO's Cultural Landscape definitions – designed, evolved, and associative – the recording of experience is really about applying the category of associative cultural landscapes to the urban setting. When UNESCO first introduced it, the associative cultural landscape category was primarily associated with aboriginal sites. The definition talks about landscapes where there may be very little evidence of human intervention, but where there exists rich natural landscapes that have been invested with cultural meaning.

However, the 2011 Historic Urban Landscape Recommendation is in part applying the concept of associative cultural landscapes to the city. It turns out that the insights of many aboriginal communities around the world, about integrating nature and culture, are fundamental to developing a more sustainable and ecological approach to the historic urban landscape. The 'historic layering of cultural and natural values and attributes' is embedded in the very definition of the historic urban landscape, and explains in part the shift from historic urban 'districts' to historic urban 'landscapes' as a way of suggesting what is essentially a more ecological and dynamic approach to the city as an organic entity. Within this framework, experience is essential to understanding.

Cultural Mapping

If the experiential quality of the historic urban landscape is as important as the visual reading of that landscape, there remains the question of what form the documentation of the city can take, in order to be useful to conservators and interveners.

Once again, modernist assumptions about mapping and documentation can get in the way of what could be considered more accurate or useful recordings of the urban experience. The

reduction of geography to Cartesian cartography accompanied the colonial expansions of both Europe in the west and China in the east. Both civilisations also embraced new styles of perspective-based landscape painting as ways of understanding and controlling territory.

But other realities exist that take different forms. Thongchai Winichakul's 1994 book *Siam Mapped* is an important exploration of the impact of the colonial imperative on the cultural identity of Thailand, including the perceived necessity for maps with boundaries. In North America, W.H. New's 1997 book *Land Sliding: Imagining Space, Presence & Power in Canadian Writing* is a related analysis of the shift in understanding of landscape with the introduction of Cartesian geo-politics. Anita Harper, a prominent Canadian First Nations spokeswoman, has pointed out that a First Nations archaeological site loses some of its cultural identity when the archaeologists set out the grid to begin their excavation and documentation. There are far more assumptions embedded in styles of documentation than we care to admit.

These critiques of mapping have to be remembered as we invest increasingly more resources into the documentation of the historic urban landscape. Geographic information system (GIS)-based mapping is useful for coordinating visual data and capturing the visual morphology of the city, but it has its limitations when recording the urban experience.

The answer is to supplement these maps and photographs and drawings with images that may seem to distort the city but in fact often come closer to its truths. These ideas were explored in Kevin Lynch's classic 1960 book, *The Image of the City*. They are reflected in images such as Saul Steinberg's famous New Yorker cover of 1976, showing a New York City resident's view of the world with familiar cultural landmarks in great detail and less familiar landscapes receding in both scale and articulation. Such maps reveal important cultural insights.

In India, the representation of urban landscape in the *mandala* can be directly correlated to the ritual mapping of the city through a seasonal round of religious festivals, particularly in

Figure 9.2 Kolam exploration by a young girl in south India.

South India. The elaborations of the *mandala*, not only in their formal expressions but in the informal *kolams* of everyday life, can reveal an experience of the urban landscape that is truer to the experience of the city than any Cartesian map (Figure 9.2). The field of cognitive mapping has been developed by environmental psychologists and others to elucidate the urban experience, and it plays a key role in expanding the recording of the city beyond observation.

It is also important, in experiential mapping, to break down assumptions about public and private space, and about exterior and interior space. It is astonishing that Nolli's beautiful eighteenth century map of Rome remains one of the few examples of mapping the cultural landscape of a city beyond its pure physical evidence (Figure 9.3). He indicated private spaces in black and public spaces in white, but went beyond the simple solid/void dichotomy that has been used so many times before and since. Instead, the white represents public space whether in the open void of the street or in the interiors of 'public buildings' – the churches, ancient temples and other monuments that were cultural extensions of the public realm.

Stanford Anderson in his collection of essays and mappings in the 1978 book *On Streets* is one of the few people since Nolli, to attempt a true mapping of public and private space, in both its exterior and interior extensions. His mappings of Paris and Cambridge, Massachusetts, are supplemented by essays from anthropologists and others who understand the concept of experience.

The most significant advance from Kevin Lynch's mapping of the imagined city is the recognition that multiple mappings may occur in the same space. This happens when different communities or cultures share all or part of an urban landscape, and may each have their own experience of the place. It is their different rituals of inhabitation that define their separate realities. The great cities of the world have often been those where this mix of cultures produces the rich layering that the 2011 Historic Urban Landscape Recommendation refers to in its definition.

In this case, each cultural reality needs to be mapped separately, and then the combination mapped again to show the interaction. It is through this layering of experience that one can understand and evaluate the cultural richness of a place.

These mappings of the urban reality can be traced onto standard GIS base maps. In many cases, however, it is important to supplement these with the more free-form maps that reflect both memory and imagination. One of the impressive aspects of cognitive maps is that they exist at the intersection of these two central concepts – memory and imagination. Neither one can be found in the simple physical documentation of the urban landscape. It is not possible, no matter how detailed the observation, to record how people remember the place, nor how they imagine it yesterday, today or tomorrow. In neuroscience terms, memory and imagination are

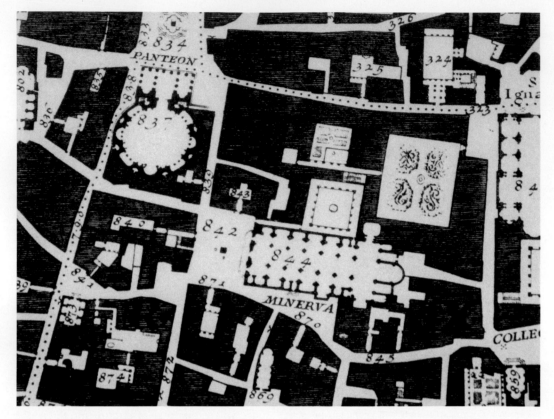

Figure 9.3 Part of Giambattista Nolli's map of Rome, 1748.

strikingly similar brain functions, and these cognitive maps, by calling both into play, create the framework for a dynamic understanding of the city over time. Kevin Lynch's last major work, *What Time is this Place?*, published in 1972, touches on this important intersection of the spatial and temporal dimensions of the historic urban landscape.

One final aspect of cultural mapping is not related to normal forms of documentation at all, but to knowledge that is instinctive rather than self-conscious. This is the understanding of the historic urban landscape by form-givers working within a particular place and a particular tradition. These people – simultaneously designers and builders – are part of the cultural and natural environment they inhabit. They are masons, carpenters, adobe-builders, glassworkers, and metal smiths, who carry design traditions and instincts within their hands. They both repair the old and

build the new, and their particular blend of memory and imagination is fundamental to the ecology of place. The results of their work can be photographed and measured, but the design-build skills themselves take a lifetime to understand. So part of the cultural mapping is simply recognising their presence, recording their names and their skills and their ways of working, and giving them a place of honour within the urban landscape.

The Concepts of Equilibrium and Resilience

Central to the concept of mapping the city of memory and imagination is the political importance of such activity. In many situations, the various cultural layers are the result of political power and control. If multiple realities are not explored, then the dominant view becomes the *de facto* basis for understanding and for intervention. As suggested in the 2011 Historic Urban Landscape Recommendation, the city must be understood as embodying:

> the site's topography, geomorphology, hydrology and natural features; its built environment, both historic and contemporary; its infrastructures above and below ground; its open spaces and gardens, its land use patterns and spatial organisation; perceptions and visual relationships; as well as all other elements of the urban structure. It also includes social and cultural practices and values, economic processes and the intangible dimensions of heritage as related to diversity and identity.[1]

The urban landscape as a reflection of economic, political and cultural practices is central to the investigations of urban sociologists and cultural geographers and others interested in the city as a cultural instrument. Books such as Henri Lefebvre's *The Production of Space* (1974) have helped illuminate the multiple realities that can exist within a single form. It was in part this political reshaping of the urban environment that drove Jane Jacobs to mount her protests against Robert Moses and his modernist visions for New York City.

Dolores Hayden, in her 1995 book *Power of Place: Urban Landscapes as Public History*, reflects this same commitment to understanding and mapping the urban perspective of communities that are often marginalised. A similar strand runs through Lynne DiStefano and Richard Engelhardt's cultural mapping handbook developed for their courses in the Architectural Conservation Programmes at the University of Hong Kong.

The move towards an ecological understanding of the city is central to this shift, a shift from the monoculture of modern utopias to a more organic view of the city as an accumulation of layers. Diversity is healthy, in an ecological perspective. And yet even the heritage conservation movement in the twentieth century often adopted the assumptions of its modernist antagonists and looked for urban areas that had a maximum amount of uniformity. There became a desire, through legislation and regulation, to maintain the static uniformity of these idealised places, often with a strong reference to a particular cultural interpretation. Restoration became a way to remove unsightly changes and revert back to something more pristine and aesthetically pleasing. It was the combination of aesthetic and commemorative interests that created, in many cities around the world, the carefully delineated 'old city centre' as the locus of conservation activity.

The 2011 Historic Urban Landscape Recommendation takes direct aim at this concept, in its very definition of the historic urban landscape. As it says, 'the historic urban landscape is the urban area understood as the result of a historic layering of cultural and natural values and attributes, extending beyond the notion of 'historic centre' or 'ensemble' to include the broader urban context and its geographical setting.'[2] In cities as different as Quebec, Venice, and Baku, the problem of historic centres is increasingly evident as they become monocultures defined exclusively by the cultural landscapes of tourism.

Even more remarkable is article 12, which states in full:

> The Historic Urban Landscape approach considers cultural diversity and creativity as key assets for human, social and economic development, and provides tools to manage physical and social transformations and to ensure that contemporary interventions are harmoniously integrated with heritage in a historic setting and take into account regional contexts.[3]

Both the definition and this reference to the essential role of diversity and creativity shift the

---

[1] UNESCO (2011) *Recommendation on the Historic Urban Landscape*: para. 9.
[2] UNESCO (2011) *Recommendation on the Historic Urban Landscape*: para. 8.
[3] UNESCO (2011) *Recommendation on the Historic Urban Landscape*: para. 12.

paradigm for understanding the city from static to dynamic, from utopian to organic. And this is as much a necessity to rescue heritage conservation from an increasingly irrelevant sidebar position of isolated monuments, as it is a necessity to rescue contemporary city planning from its own static and utopian goals.

The shift in this Historic Urban Landscape document is towards an ecological view of the city, an ecology in which humans are an integral part of the ecological equations and variables. Some understand urban ecology as a study of natural resources within the city – how they are consumed, recycled and managed. But a true ecological framework includes human activity itself as part of the natural ecosystem, and puts cultural behaviour as a central ecological variable.

What is clear from the emerging interest in studying urban ecology more holistically, is that historic urban landscapes are incredibly rich sources of ecological knowledge and expertise. The Historic Urban Landscape Recommendation speaks of cultural diversity and creativity as two key elements of a sustainable urban condition – the coming together of these two qualities is not unrelated to the coming together of memory (related to diversity) and imagination (related to creativity). The Centre for Urban Ecology in Montreal, as an example, has shown the enormous resourcefulness of the traditional Montreal triplex in creating sustainable dense urban centres (Figure 9.4).

Even slums have valuable lessons in the complex uses of limited natural and cultural resources in an urban setting, as pointed out by

Figure 9.4 Montreal triplex: a low-rise high-density approach to ecological urban living.

Arish Dastur and others in their 2008 World Bank publication on *Approaches to Urban Slums*. In many historic urban landscapes, it is often the simple forms and the unassuming craftsmen who created them that best represent an ecological model. The resurgent interest in urban agriculture is another example of learning from traditional locally-based practices. Many urban centres lost their food production capacity due to twentieth century bylaws and regulations, but are now realising they can recover some of those patterns of sustainability.

The key in understanding the city is to map this diversity of practices and skills, and to have the humility to understand that marginalised communities often carry valuable insights not readily visible in the dominant forms and facades.

Sustainable Diversity

It is fine to say that more creative mapping of the city is possible, using a variety of voices, but there are still questions about how such information can be condensed and articulated in ways that become useful for design and development.

One of the answers is to move beyond the twentieth century reliance on legal boundaries and definitions to a more fluid definition of boundary conditions and a shift to mediation from legal confrontation as ways to develop true community-based design and development. First Nations communities in a number of countries have pioneered both this concept of more fluid boundaries, and the related interest in mediation and dispute resolution. It is essential that these new tools become more integrated into the planning framework.

The most important step is to displace the utopian imperatives of twentieth century modernism with the more flexible organic ideas of urban growth and transformation. The utopian city relies on zoning as its primary tool for decision-making, and zoning is a tool that favours monoculture. It can only comprehend one memory and one imagination. It is necessary to summarise the findings of historic urban landscape research in ways that do not rely on this end product.

Cultural mapping, for example, can provide multiple boundaries for overlapping cultural landscapes (Figure 9.5 a, b and c). Various subcultures may inhabit the same physical space in different ways, and their various realities can be mapped on top of each other. This is essential to prevent interventions that benefit one community while destroying key patterns that benefit another. This is not a ghettoising approach, which is the unfortunate result of using a zoning framework for the city; rather, it celebrates the advantages of true diversity that respects difference and allows multiple realities to coexist (Figure 9.6).

If the multiple realities of the city are not recognised and documented, it will always be the dominant reality that shapes the on-going evolution. Tourism, for example, is often a contemporary intrusion into a historic context. If one simply assumes that the host community and the tourist community are engaging with the same reality, then there will be one design solution for both that will inevitably displace historic patterns with new patterns – even though the physical form appears to remain relatively constant. In this situation, the international weight of the tourism industry will often override the local interests, and eventually the host community will feel alienated. If, on the other hand, it is understood that the cultural landscape of the tourist community and the cultural landscape of the host communities are very different, then it becomes possible to modify the physical form to satisfy both sets of rituals. This requires that very careful mapping of experience, a mapping that recognises both cultural landscapes individually, as well as a shared cultural landscape of overlap and intersection. And tourists can be treated as simply one more complicating layer within the political, economic and cultural layers that already share the city's form.

The fact that the multiple boundaries of memory and imagination will spread beyond the

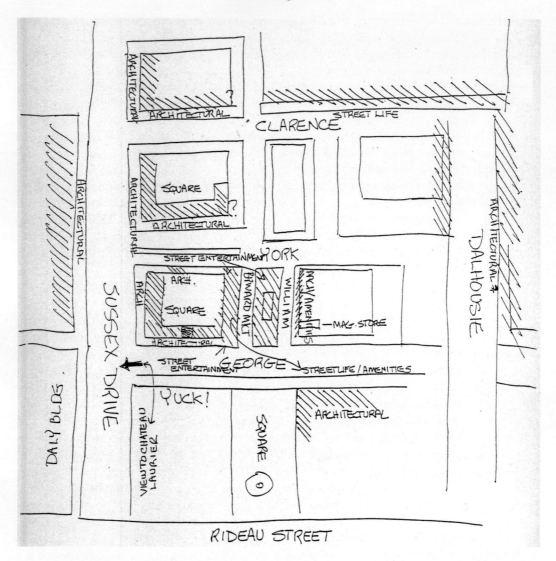

Figure 9.5 (a) Three overlapping cultural landscapes in the Byward Market area of Ottawa, Canada, (b) the aesthetic landscape of tourism, the market landscape of the rural farmer participants, (c) the late-night bar landscape of university students and others.

notion of 'historic centre' to encompass the broader urban landscape is the *quid pro quo* of the Historic Urban Landscape Recommendation. Yes, there will be a more dynamic interpretation of the historic centre, and the possibility of a contemporary layer within the existing layered setting. But at the same time, those layers and those patterns will influence the evolution of the city around them. It is essential that the reading

of the historic urban landscape be a tool not only to understand the past and the present, but as a necessary narrative for framing the future.

There has already been a shift in accepting the environmental imperative as a new and necessary part of contemporary urban planning. The cultural imperative is following – it is showing early signs of success in cities such as Hanoi, which are choosing to use the historic

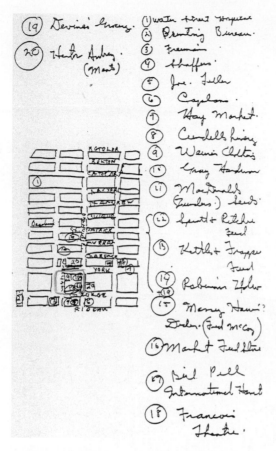

Figure 9.5 (*Continued*)

urban landscape as a framework within which to understand a distinctive and rich future (Figure 9.7).

This is still a kind of planning in its infancy, but there is a young generation that is going to insist on it. Universities around the world, for example, are finding that the best way to attract these young people is to locate in historic urban areas and to design their campuses and their buildings in ways that reflect the cultural patterns of those areas. The more enlightened participants in the tourism industry are similarly looking for designers who understand local culture and ritual and can insert their facilities either in existing buildings or in contemporary surroundings that mirror their environment. And as with the heritage conservation field, these contemporary designers are moving beyond aesthetic and commemorative connections to try to understand how the experiences of the new can be specific to a particular sense of place.

Influences of Civic Engagement: Towards Community-Based Design and Development

The discussion in the previous section has spoken of the importance of documenting the experience of the city, and then mapping these realities in ways that can inform decision-making.

In addition to the shift from a modernist and utopian viewpoint, to a more layered and organic understanding of the city, there is an equally significant transformation at work. That is the re-evaluation of the role of the 'expert'. Just as the shift from a static to a dynamic view of the historic urban landscape is seen by some as a threat, so too is the displacement of the role of the expert to an increased role for civic engagement. Experience is impossible for the expert to document, other than through the knowledge of those who inhabit the city. And in order to understand diversity and layering, it is necessary to experience the city through the knowledge of multiple cultures and subcultures.

This requires significant humility on the part of experts, who are more accustomed to having full control of both the documentation and the analysis. The architect and the architectural historian can do their work by gathering photographs and undertaking recording activities of various kinds. The cultural historian can use historical sources and secondary materials to craft a reasonable narrative. But the ecological complexity of the city is not reachable through these media alone.

This means that the process of producing cognitive maps and recording traditional knowledge and understanding embedded rituals requires a different level of community engagement.

Successful engagement of this kind will almost always lead to expectations by the community

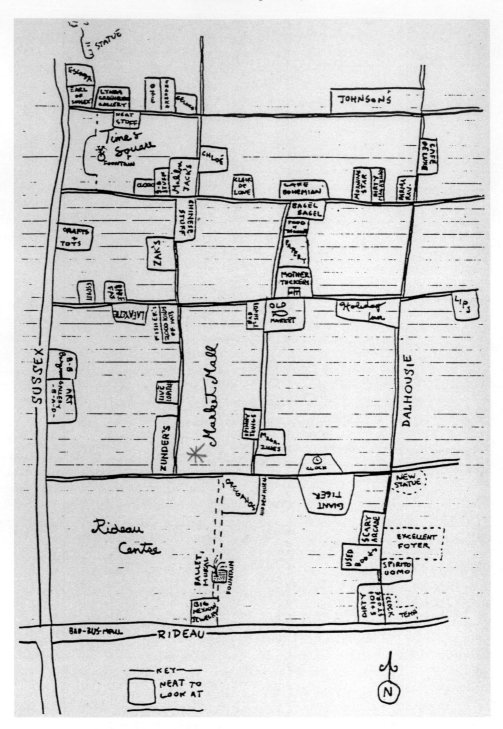

Figure 9.5 (*Continued*)

Figure 9.6 The Byward Market area today, managed as an area of layered experiences.

Figure 9.7 Hanoi: the layered urban landscape.

– expectations that its involvement will extend beyond documenting the place to identifying values to framing future design and development. Cultural mapping can be a very empowering exercise, and the knowledge and even wisdom that results is never fully removed from the voices that make it accessible.

Community-based design and development is the traditional way that communities have created a sustainable environment for themselves. It is a paradoxical but important fact that a community that relies on sustainability 'experts' is not a sustainable community. Sustainability has to be grounded within the community, and this begins by understanding the past interactions between nature, culture, memory and imagination.

The goal of sustainable community-based design and development is not to advance towards a certain vision. This utopian approach requires defining the future, and then moving towards it. The organic approach, or what one might call a true ecological approach, requires defining the present and then choosing interventions that strengthen its positives and weaken its negatives. The future simply emerges as a sum total of these interventions. The point of reference is the present, not the future.

Some people see the dynamic rather than static quality of the Historic Urban Landscape Recommendation and fear that it will lead to nothing more than the 'management of change'. This would indeed be a sorry result. The phrase assumes that change is a given, and that good management skills are the only requisite ability.

Instead, it is better to frame the Historic Urban Landscape Recommendation as seeking ways to establish equilibrium and resilience. In an ecological sense, equilibrium is achieved when all the various parts are healthy in themselves and also in healthy relationships with what surrounds them. That is what urban ecology is about: finding the healthy equilibrium that thrives on diversity but achieves a larger whole that has a certain stability and sustainability. In the right circumstances, change may be relatively minor. If change is abrupt or drastic, it is that much more important that the key variables in the urban landscape are understood and that each one can be adjusted to once again arrive at a healthy equilibrium.

Civic engagement within an ecological framework is a necessary part of sustainability. It is the community that understands its experience, that maps this experience onto the urban environment, and that can identify where the relationships are fragile and need to be strengthened or changed. It is the community that can identify the true landmarks, not always the most highly visible points in the landscape. And it is the community that can respond most creatively to threats.

The role of the expert is to work with the community to develop ways in which the reading of the historic urban landscape can frame the review and approval for future development. Because the boundaries of cultural landscapes can be understood and imagined across large areas, it is not a problem to take traditional knowledge and extend it to new development. Unfortunately, the second half of the twentieth century saw the removal of many of these civic engagement tools, replaced by the advice of an outside expert. Even where the intentions were honourable, the result has tended to be monocultures where diversity and creativity are difficult to sustain.

In terms of sustaining and evolving the physical morphology of the historic urban landscape, there is a required role in community-based design and development for the master-builder tradition, the skilled craftsman-designer who for millennia has been the primary creator of the urban landscape. The rise of professional architects and planners in the modernist framework of the twentieth century has not only led to an unfortunate separation between design and build, but also to a tendency to standardise rather than differentiate. Community-based design-build is an essential component of sustaining and reviving the traditions of vernacular architecture and urbanism. This is part of paying attention to the cultural, social and economic processes that created many of these environments in the first place. Box 9.1 in the following pages illustrates some of these civic engagement tools at work in the context of a historic urban landscape.

Conclusion

The 2011 Historic Urban Landscape Recommendation is a starting point for a new set of principles that applies equally to conserving the past as to planning for the future. The heritage conservation field has a long and proud tradition of working with local communities and understanding the significance of civic engagement. As the field became more professionalised in the late twentieth century, however, it began to adopt some of the patterns of its fellow disciplines in the contemporary design and development communities. It tended to privilege the role of the expert and focus on the object – the building or artefact – as the basis for organising its systems of documentation and regulation. This made it compatible with what architects and planners were doing but did not help develop a more ecological approach.

In the twenty first century, there is an impatience to change these systems and create a more ecological and sustainable perspective. For some, this is a brave new world, but for those in the heritage conservation field it is also a return to some of the fundamental values that have inspired previous generations to value the historic urban landscape. It is not a remnant of a past that is quickly disappearing; it is increasingly a model for a future that is reconnecting with past practices. It is a more dynamic view, but also a more sustainable view.

The greatest strength of the heritage conservation field is that it can still read the past, and because of this it can understand the present. These skills have never been more important. They require the engagement of the communities involved, and from this shared understanding comes the basis for consensus on how to move forward.

Box 9.1 Kensington Market – Civic Engagement Tools at Work

The Kensington Market area is a small part of downtown Toronto, which is Canada's most multicultural and fastest-growing city (Figure 9.8). Almost 50% of Toronto's population is made up of first-generation immigrants, arriving from all over the world.

Kensington Market has for 150 years been a kind of refuge in the city for those who faced discrimination or difficult economic times. In the late nineteenth century it became home to a Jewish community facing discrimination in the city; it later housed a large Portuguese community, and now it is home to significant numbers of Canadians from Italy, Eastern Europe, the Caribbean, China and South Asia. It was originally outside the city limits and saw unregulated lot subdivision and complex residential and mixed use building developments. This somewhat independent attitude has survived to the present – it is generally accepted as the traditional centre of Toronto's punk-rock music scene, and is well known as a creative source of alternative cultural expressions.

The process of trying to understand Kensington Market invariably moves beyond the physical form to the spirit of the place. It is impossible to understand the delicate and shifting balances between public and private space, between residential and commercial occupancies, between past and current realities, without engaging directly with the community that lives and works in the area. And even then, it quickly becomes clear that there are at least three Kensington Markets – the complex retail environment that thrives on owner-operated vintage clothing stores, alternative local food sources, specialised eating establishments, public, semi-public, and private cultural spaces; the larger retail plus residential neighbourhood that embraces the full urban area demarcated by four major city arteries around the perimeter; and the music scene that is centred on Kensington Market but includes venues outside the physical boundaries of the neighbourhood (Figure 9.9 a–e).

(Continued)

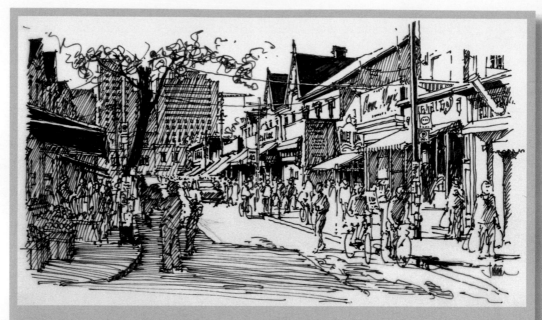

Figure 9.8 Kensington Market, Toronto.

Figure 9.9 (a, b, c, d and e) Five cognitive maps of Kensington Market: a satisfied retail consumer, an uncomfortable visitor, a resident, a merchant, a musician. These maps are part of restructuring the governance model for this neighbourhood to recognise and preserve its diversity.

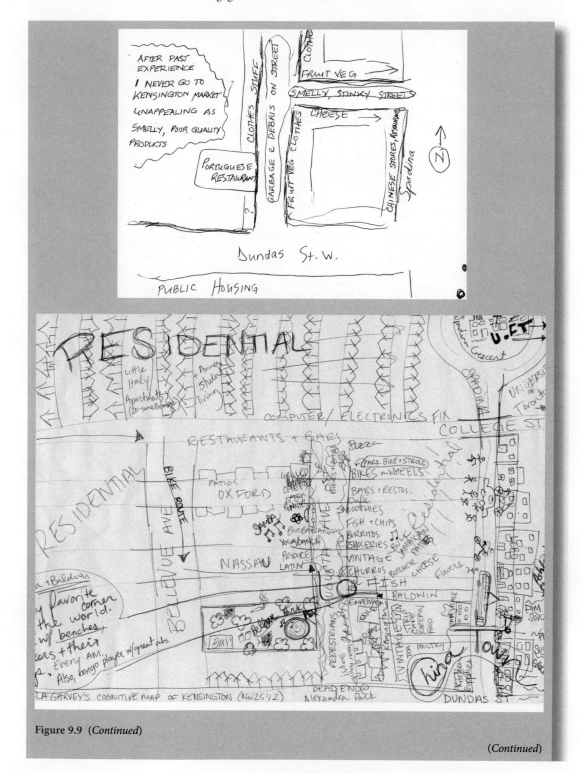

Figure 9.9 (*Continued*)

(*Continued*)

Figure 9.9 (*Continued*)

Today the area faces the pressures of franchise outlets, big-box stores, high-rise condos and all the other characteristics of a quickly developing urban area. The local councillor has recognised the limitations of existing heritage conservation legislation, which can address the tangible form of the area but not its intangible qualities. And yet it is clear that it is the combination of the two that defines the area and gives it a sense of place. The planning structure is not imaginative enough to deal with the fine-grained patterns of occupancy – in fact, their regulations run directly counter to much of what gives the neighbourhood its life. The contemporary design and development community has not been trained to read the existing landscape and design contemporary interventions that speak to its uniqueness.

The response has been to develop a set of narratives and mappings that address the experience of Kensington Market from several perspectives simultaneously. These include the physical mapping of its nineteenth century urban form with complex layers of additive intensification; a large enough collection of cognitive mappings by both insiders and outsiders to define its key pathways, nodes, landmarks and conflicts; a collection of interpretations of the area by authors, musicians, street performers and others to provide some glimpse into its past and present urban reality; a set of discussions with social services and other interveners to gain their sense of the workings of the area; and examination of its relationships to neighbourhoods around it, which further highlight its ability to absorb an astonishing array of people from different cultural groups, economic strata, and political leanings.

There has also been an examination of the economic and social and cultural processes at work in the area historically and at present. From this analysis, it became clear that small lot development, vernacular design traditions, and incremental growth were intrinsic to the physical form. Local control of economic processes was and continues to be critical – as an example, this is one of the few areas in the world where both Starbucks and Nike established a presence, and were later ejected by sustained and imaginative community pressure. The form of the place has been the result of community self-regulation – so much of the land division and building construction was done without any city approvals that a blanket grandfathering of existing conditions was carried out in the 1970s to simply update the city's files. This research activity reflects the Historic Urban Landscape Recommendation's interest not just in the end result, but in the process of getting there.

The next step is recommending ways of moving forward. This is a case where political will outstrips the tools available to planning staff. The 2011 Historic Urban Landscape Recommendation is useful in this context as representing a new paradigm that can allow places like this to thrive. Part of the process is loosening the grip on this area of both the planning regulations and the heritage regulations, and putting in place a cultural landscape listing – one with multiple boundaries to confound any overly simplistic response. The intent of the cultural landscape association is in part to remove the area from oversight by a provincial planning review board that tends to homogenise communities rather than welcome their uniqueness. There is also a necessary process of creating a committee based in the area, working with them to create a consensus set of principles, and giving them the power within the political process to speak directly to the decision-makers. Other recommendations are limiting building permits to developments proposed within historic lot boundaries, to avoid the problem of land assembly and out-of-size developments, and examining regulations to prevent franchises, in favour of locally-owned and operated establishments. There is still exploration of how to encourage locally sourced materials for both building and commerce, and how to re-engage a master builder tradition that has largely been replaced by a separation between design professionals and building contractors. The intention of all of the above is to maintain both the physical intricacy of the place, and its sense of independence and local control. But the assumption is that evolution will continue – changes will continue to occur, creativity will remain an essential part of its character.

As with many cities, Kensington Market is important to more than its own residents and vendors. It is a point of reference for the city as a whole, providing a sense of place and a sense of identity. It is still an open question as to whether it will survive with its spirit intact. But it is worth trying, because what it represents is a healthy ecology and the idea of urban sustainability in its most basic form.

Interview with His Highness the Aga Khan
Listening to the People, Promoting Quality of Life

Founder and Chairman of the Aga Khan Development Network and 49th hereditary Imam of the Shia Imami Ismaili Muslims

October 29, 2013, Aiglemont, France.

Question: *Your Highness, let me ask you in the first place when and why you got involved with urban heritage? As the main private organisation in the world that works on urban heritage, how did you develop your approach to urban heritage preservation?*

Answer: You have to think that in 1957, when I succeeded my Grand-Father, many countries were still colonies or had just obtained independence. Physical planning of cities in the developing world was done in London and Paris, and in any case it was not handled by local technical experts, but by expatriates that consulted their capital cities whenever needed. At that time, planning was done by a small number of specialists, whose responsibility was to design cities that could grow rationally and in an orderly manner. That was true mostly in Africa, as Asia had already achieved a much greater level of independence. A number of cities had suffered from tragedies, such as the division of India and Pakistan, with heavy consequences due to the sudden flow of millions of people (for instance Karachi or other cities in India).

The first lesson I learned, and that I retained until today, is that the process of urban planning cannot be driven only by one force, it has to take into account the multiple forces at play. There are political forces, economic forces, and then technical forces, those managing the planning process. But there is another force that has become more important, one that I consider fundamental for what we are doing. It is connected with a simple question: whom are we trying to help? And once we know whom we are trying to help, we must try to define what is *their* list of priorities, not ours! We have to understand how they define their quality of life. We have spent a lot of time in recent years, in the Aga Khan Development Network, trying to work out how you define quality of life for: an established urban population, a newly migrated population, a rural population, a population in transit, a population in a civil war situation or in a post-conflict situation. I have come to the conclusion that you cannot be really successful in this field unless you are willing, first of all, to listen to the people. Listen and listen and listen! Once you have listened, once you have heard, once you have imbued yourself with their value systems, then you can start thinking with them, on how to achieve the outcomes you want. So, my sense is that planning should be a people-driven exercise, and that the role of the specialists is to try to respond to the people's views and requirements. These are not always rational, feasible, bankable, but you must first get a sense of what people want and of the nature of the environment in which they are living. This has been very important for me and for us. When I was much younger, economics was much more dogmatic, as it had to fit into dogmatic contexts. Over the past 50 years, we have seen that when a dogma is applied to economics, it does not necessarily produce good results. And I ask myself, whether in the field of physical transformation we are facing the same sort of dogmatic approach, one that tries to ring-fence our imagination and desire to think outside of the box!

Question: *How do you define a 'dogmatic' approach in the field of urban planning?*

Answer: A dogmatic approach refers to physical planning based on theoretical standards, on correlations between standards and needs, all within a top-down vision on what you think the population needs, the way they should be organised, the way in which they should relate to their environment. In fact, I do not believe that you can reinvent the social fabric of communities; it has to be understood and respected. This structure will change over time, without any doubt, but a city planning process has to take into account the social dimension, the fact that human beings are social individuals, that they function within a society. Many modern cities can be defined as 'soul-less' places. If you look at these new cities, you realise that it may take decades before they start having civil dynamics, the dynamics of people. This is often a result of abstract planning, like when you plan the size of an infrastructure to serve a certain volume of users, or you create a grid, and the buildings then have to fit into the grid. Very often the grid system over 50 years proves incorrect, because there are new requirements, but the process of change is impeded by a rigid planning process. In my life time, I have seen changes in just about every type of building I have known! Take for instance a hospital, a university, a housing estate, a bank. All these building types are changing in response to the changes of the way people function. I remember looking at a hospital design in 1958: the notion of day surgery did not exist at that time. The average length of stay was 8–10 days; today it is 2–3 days. So it is these sort of situations that I have seen becoming problematic.

Another situation where 'dogmatic' planning does not work has to do with uncontrolled urbanisation, which I have experienced in conflict zones. I saw it for instance in Karachi after the war, but we have seen it in Kabul and in many other places in the world. The 'normal' management process of urbanisation, done with the tools of city planning cannot function in these cases. Look at what is happening in Cairo, for instance, after the Arab Spring, with new constructions springing up everywhere. I have to say, very honestly, that I do not have a clear vision on how cities should or could react to these situations, it is really not easy. Within our Award for Architecture, for example, we did a study on the situation of the city of Dakar in Senegal, where we also studied the squatter settlements. Nobody within the group of researchers which did the investigation seemed to have a clear set of solutions. We also looked at other cases in Turkey, we looked at the legalisation process, as a way to deal with these situations and none of the solutions that were proposed and implemented seemed credible in the long term, they were all short term responses. The conclusion was that out of ten urban situations, you can perhaps generate the solution for one, but all the other cases end up turning into slums. So, you may solve an immediate problem with a short-term solution, but then a medium and long-term problem is created. Quality of life is the victim of this process.

Question: *With respect to these issues, you have seen that the Historic Urban Landscape Recommendation points to the fact that the historic city has a unique quality, and that it should be used as a model to develop an effective urban rehabilitation and expansion process.*

Answer: I agree. One of the most important characteristics of the historic city is its social construct. The city itself was designed around a system of social relationships, and even if those social relationships have changed, and people no longer live in the same area, they keep their own 'soul'. Most historic cities have a soul, even hard-nosed architects know that!

Question: *Many of your projects are developed around important monuments. Why do you choose these points of attack? It is because of their link to people's identity? How can working on a monument bring change in a community?*

Answer: We look at the historic city as we think that they are important drivers for the future of the urban population. They carry cultural continuity, represent symbols, and sustain a mix of relationships between different groups and generations that is very important for social balance. The historic city tends to develop around a focal point, be it a church tower, or a minaret, that represent a 'sign post' of the centre of the city. In our work, we gradually understood that we could not achieve the improvement of life of a community by dealing with only one building. If you deal only with one building, and particularly if it is a building linked to faith, it is likely not to be used by the entire population on a permanent basis. What we want to do is to work on the elements that constitute the social identification of the population. In implementing many projects, we found that the population of the historic cities where we worked, were very poor, recently urbanised, had no property rights. This was essentially a transient population, which did not see itself as permanent users of these places, and so they were in fact accelerating their decay. We concluded that we could not achieve significant results in the improvement of the social situation through an intervention based only on one building, but that we had to deal with a larger area. In fact, we had tried the single building exercise, and it was a failure, it did not work. We moved away from this model to an approach focused on the intervention on an entire community, which could be defined as living within an urban perimeter, as the target of our intervention. In the case of the Al-Azhar Park project in Cairo, we had a community of 200 000 people, based in the area to the west of the Ayyubid Wall, Darb-el-Ahmar. We started by doing a poverty survey, one of the most remarkable surveys I have ever seen. The survey showed us that the poverty level was extreme for the people living near the Wall, while people became less poor as you moved away from the Wall. On the basis of the survey, we oriented our investment, to improve the physical environment. But we also realised that this form of development alone was not sufficient to help people improving their quality of life. That's when the notion of quality of life comes in. We asked ourselves what were the causes of this extreme poverty? We looked at a variety of factors, education, health care, access to credit, malpractice, corruption, and we tried to understand all the causes of social marginalisation. After we did that, we asked ourselves how it was possible to improve the quality of life, and we realised that we could not limit ourselves to clean up public spaces, or to improve historic buildings or to green an area, as much as these things are important. We introduced the notion of 'multi-input', and we defined a process called Multi Input Area Development (MIAD). When we try to improve the quality of life of a particular population, located for instance in a historic city, we have to consider all the different aspects of life. We look at ways to improve education, health care; we look at how to bring in microcredit, commercial activities, in order to expand economic opportunities for the people.

We also look at generational changes, because in many of these environments, we cannot bring change without mobilising the new generation, and I think that this is absolutely central to what we do. And then, obviously, there are unknowns. One of the unknowns is the relationship between the local authorities and us. At the beginning, we cannot know the true willingness of the local authorities to cooperate, the speed with which they take decisions, what they will be seeking from us or from the change process. There are some risks, as we do not know the history of the urban environment, the way in which it is structured, who are the effective leaders. It is important to gradually address these unknowns and to turn the local agents into partners. Once you succeed in doing that, than change becomes possible, although it really becomes implementable only if we can prove that there are low-lying plants that can be planted. You must create the possibility for some immediate results, so that the local population can see the benefits of change. This benefit can be the renovation of a square, of a public space, of a public well; it can be the sewer system or the public toilets. But generally speaking it cannot be limited only to one intervention; it has to be made of a variety of interventions. And these interventions have to be perceived by the community as a benefit for all, not just for a select group.

That's a long explanation of why we do not work only on one building!

Question: *The interventions you do are indeed very different from simple conservation operations. What are your criteria to evaluate success?*

Answer: The first proof of success is when the community says please do not go away, please stay with us. The second one is when the neighboring community says: we have seen what you have done for our neighbors, please come here and do it for us too! And this happens in the rural environment, from one village to the next....

Question: *How do Governments react to the interventions? Here is a case of success, but it does not belong to the Government....*

Answer: In the early steps of the project we try to bring onboard the key governmental decision makers and to explain that we are not trying to take anything away from them, that we are trying to develop a change process that can turn beneficial to them in due course. Originally, we started with what I would call an open-minded relationship, based on good will, but then we realised that there was a need of some form of management of this process, because these projects generally have a life of 10 to 20 years, and they continue through several administrations, several political cycles. So we developed a model of Public Private Partnership agreements, we normally do one with the City and one with the Ministry of Culture. These agreements are immensely important. At the beginning, in the earlier projects, we had made the mistake to transfer the project to the public body in charge of the area, and it all fell apart, things went back to square one in almost no time. So what we now try to do is to establish a partnership relationship with the local community and the decision makers. We establish a sort of joint venture, under which all the agents engaged in the process are responsible to keep the area in good conditions after the end of the intervention.

At the beginning, we had no projects in India. Manmohan Singh was the Prime Minister at that time, and in fact he is still there. At one of the events we had organised in India, he made a very important speech and said that the Government of India needed support in many areas, and within those there was the protection of Indian cultural heritage. As we needed to define priorities, we told him that we could help in the area of urban rehabilitation and restoration, but that we needed to operate within Public Private Partnership schemes. He opened the door to this notion in the country and the PPP process was extensively used in India, as in many other developing countries. In fact, this notion of PPP was new for the cultural domain, while it had been used in other domains such as industry and finance. And you have to consider that this notion had a bad reputation in many contexts, as it was seen as a way for the private sector to exploit government assets, particularly in countries with high levels of corruption. Therefore, it had never been used for culture. The attitude of Prime Minister Singh was fundamental in opening these possibilities in India and in the developing world. He brought to this domain a sense of pragmatic realism that was enormously important.

Question: *Your Highness, let me now ask you a question about future trends. Trends seem today negative for historic areas in many fast developing countries in the world. We need to understand what is happening and what can be done, and what are the acceptable limits of change? What is your assessment of these processes?*

Answer: I think that the question we have always to ask is whether the initiatives we take are necessary or can they become necessary to the people. So the human dimension, what we can do for people, remains very important. I tend to divide what I would call restoration of the historic city from archaeology. I am not against archaeology, but I am saying that if we have to make a choice between two options we will give preference to the one that impacts on the quality of life of the people. The second question I would ask is whether the transformation process is just a flash in the pan or is it a long-lasting one? Are we making a transformation that will remain or something that in a few years will be gone? The longevity of the outcome would be critical for me in assessing a project.

The third important element I would like to mention is the notion that cultural continuity is part of the richness of people around the world. They may not sense it on a daily basis, but when it is taken away from them, they feel it. Cultural continuity is for me very important and I think that it is one of the elements that define the soul of an urban environment. Most new cities are soul-less because they do not have cultural continuity. However, I am not saying that this always applies, we need to look at the specific situations. Some societies do not have a history of a tradition of a high quality built environment, we have to be able to recognise it and act accordingly.

Question: *In the past 50 years, you managed urban projects in different regions of the world. Which is the dimension of the development process that struck you most?*

Answer: Many years ago I worked with the World Bank and I remember discussing with the President of the Bank what the key issues were related to the quality of life of the poor in different environments. We looked at many analyses, and we concluded that the main factor influencing the quality of life of the poorest segment of society, all over the world, is fear. Neither he nor I had this clear perception before. Fear concerns many aspects of life: it's a fear of the loss of a child, fear of a bad season, fear of political power, fear of gangsterism, it's a fear of a family member who has problems and pulls down the whole family, etc … fear in many societies is omnipresent. I think that if you compare this situation with the one of societies in the developed world, if you made a survey in Europe or America among the poor, the notion of fear would not be the dominant one, as there are better systems of social protection, of legality. We then have to ask ourselves: what can we do to protect people from fear? We know that this is a complex issue, but in the end I believe that the answer rests in a multiple input approach, like the one we called MIAD. And I believe that culture is a very important entry point for the development of this approach. To convince the Development Agencies, which very often do not see the importance of culture, we need to develop a new definition of culture and explain why this is important for many dimensions of development, like the perception of personal security, food security, housing security, financial security, heath security, to name but a few. With the World Bank we were able to develop a major study that showed us very clearly that there is a need to innovate in the classic development approaches, to go beyond the 'bankers' vision of development. Culture is the key.

Case Study *Valuing Cultural Diversity*

Richard A. Engelhardt

Consultant in cultural policy and heritage management to governments and former UNESCO
Regional Advisor for Culture in Asia and the Pacific

Introducing the Cultural Diversity Lens

The preamble to the 2011 *UNESCO Recommendation on the Historic Urban Landscape* refers directly
to the issue of engaging the development process in the safeguarding of cultural diversity. Cultural
diversity is a reality. Respecting and promoting cultural diversity is part of an international commitment, as highlighted in the 2001 *Universal Declaration on Cultural Diversity*; in Conventions for the
safeguarding of cultural heritage including the 1972 *Convention concerning the Protection of the
World Cultural and Natural Heritage* and the 2003 *Convention for the Safeguarding of Intangible
Cultural Heritage*; as well as in the many UNESCO Recommendations and International Charters for
the safeguarding of different forms and expressions of culture.

To translate the objectives of these idealist international conventions and recommendations into
reality at all stages of development policy-making and programme implementation, UNESCO has
elaborated an applied methodology: the *cultural diversity lens*. The cultural diversity lens is a planning
tool designed to systematically analyse and evaluate whether programmes, policies and practices
incorporate and promote the principles enshrined in the 2001 Declaration and related guidance
documents. The cultural diversity lens guides decision-makers and programme planners in understanding the cultural context in which they operate, identifying gaps and shortcomings of interventions, and devising alternative programme designs.

Like any lens, a cultural diversity lens is first and foremost a tool. It enables the user to see more
clearly by improving focus on specific details – in this case on the details of culture – as these details
impact policy formulation and programme design, in both the short and long terms. A cultural diversity lens also reveals how the implementation of a programme will impact a community's culture, for
better or for worse. Fundamentally, the purpose of a cultural diversity lens is to **raise awareness** of
the cultural dimension of development and promote knowledge-based decision-making. It opens new
ways of thinking, exposes potential flaws in programmes, and points the way to new solutions.

Fundamentally, a cultural diversity lens is a **tool to evaluate** whether programmes, policies, proposals and practices promote and safeguard cultural diversity and intercultural dialogue, while respecting
human rights, including cultural rights and the right to development. At its most basic, a cultural
diversity lens is a check-list or a list of criteria and questions supplemented by **indicators** and other
means of verification. Most efficient when applied **systematically**, a cultural diversity lens allows planners to see beyond what may be perceived as common sense, to analyse the degree to which cultural
factors have penetrated into programming processes.

A cultural diversity lens is useful at all stages of a programme: planning, implementation, monitoring, and evaluation. It is an interdisciplinary tool relevant for all programmes and activities, not only
those specifically related to culture.

Applied systematically to all programmes, throughout planning, implementation and monitoring
phases, a cultural diversity lens assists policy-makers and programme planners understand the impact
of **cultural factors in determining the success or failure of programmes**, and to make informed
decisions on programme design and (re-) direction. A cultural diversity lens is not simply one possible
alternative to other forms of sectorial analysis (such as gender-sensitive programming) or a substitute
for other programme management tools (such as statistical verification or results-based management
approaches). A cultural diversity lens is a necessary complement to these other development tools,
raising additional questions and utilising different perspectives.

The use of a cultural diversity lens will help to identify new activities or suggest new programme
designs. Policy-makers and managers must then prioritise needs while remain mindful of potential

weaknesses in the project design caused by the failure to take cultural factors sufficiently into account. Rooted in respect for cultural diversity and recognising the overarching imperative of promoting intercultural dialogue, a cultural diversity lens projects a rights-based approach to development, specifically a cultural rights-based approach.

Constructing a Cultural Diversity Lens

There is no single design for a cultural diversity lens. Each programme will need to develop its own lens, tailor-made to suit its purposes. Ideally, a cultural diversity lens is created in a participatory manner by those who use it. Recognising the dynamic nature of culture, a cultural diversity lens should never be so rigid as to constrain effective programme implementation or to stifle response to changing circumstances. A suitably flexible lens will always be in the process of evolving and reconstruction. Continuing intellectual debates and practical experience contribute to a better understanding of the application of a cultural diversity lens in each new circumstance.

However, specifically designed, the application of a cultural diversity lens always comprises two elements in addition to the cultural framework for analysis provided by the lens: a programme to analyse; and options for improving the programme. It is in the dynamic, synergistic interaction of programming realities with the analytical framework of the lens that the value of a cultural diversity lens in improving programme implementation is realised.

At its core, the analytical framework provided by a cultural diversity lens should be designed to probe the extent to which a programme respects cultural rights and safeguards cultural diversity as basic principles of sustainable development. These principles can be further contextualised under three fundamental rubrics:

- Recognising the essential cultural dimensions of development;
- Engaging respectfully with all cultures; and
- Building on culture as source, asset and inspiration for development.

With this framework of reference, a robust cultural diversity lens works by deconstructing the programme under scrutiny according to key themes – themes which have been established as fundamental by the international community through normative instruments. At its most basic, an effective cultural diversity lens will guide inquiry into five major themes of scrutiny:

1. Recognising cultural factors in analysing the larger programme context.
2. Considering diverse cultural perspectives.
3. Ensuring culturally appropriate access and participation.
4. Safeguarding cultural heritage, including intangible expressions of culture.
5. Promoting innovation and creative diversity in the cultural economy.

Each of these thematic areas, in turn, can be further elaborated by a number of questions, guided by key words, designed to probe deeply into the reality of the specific programme, and matched with a set of indicators for the answer to each question.

Through this process, it becomes possible to improve the quality of programming in two important ways: firstly, by increasing the effectiveness of implementation by identifying culturally-appropriate operational mechanisms, as well as any cultural obstacles to success; secondly, by minimising the negative impact of the programme on cultural heritage, while maximising the positive impact in promoting cultural diversity and with it equitable prosperity through rights-based development.

Thematic Structure of a Cultural Diversity Lens

Context of the Programme

The first thematic area of a cultural diversity lens is devoted to culture in the programme's most general context. In this thematic area, the user of the lens grasps the totality of the legal, institutional, customary, and socio-economic dimensions which comprise the frequently complex and cultural diverse background context in which the programme is embedded. Key words guiding programme-specific inquiry include: legislation and regulation; customary law; traditional decision-making and governance systems; enforcement mechanisms; hierarchies in the exercise of authority; distribution of and limitations on power; definitions of which resources constitute a community's capital assets; notions of public vs. private property; markers of wealth.

Diversity of Perspectives

The second thematic area of a cultural diversity lens relates to recognising the diversity of points of view in the conception and implementation of programmes, including the difference in perception of who stands to benefit the most and who stands to lose from the programme. An examination of this theme helps the user of the lens to understand underlying assumptions, perceptions, and ideas from the perspectives of various parties, thereby anticipating potential points of tension. The recognition of diverse perspectives will also contribute to problem-solving dialogue throughout the programme's implementation phase. Key words guiding programme-specific inquiry include: collective memory; knowledge holders and arbiters; hierarchies of leadership and influence; stakeholders and other interested parties; age, gender and other demographic factors; dissention and minority perspectives; local politics.

Access and Participation

The third thematic area of a cultural diversity lens relates to the extent to which all members of the community are able to participate in, and the measures in place to ensure broadest access to benefits provided by, the programme under analysis. Specific consideration of this theme not only promotes cultural diversity by providing equitable access, it also helps to identify cultural factors that might limit access to the programme, as well as culturally appropriate decision-making mechanisms to expand and ensure participation by all. Key words guiding programme-specific inquiry include: language(s); age/ gender /caste /ethnic /religious barriers and other forms of discrimination; challenges for physically and otherwise handicapped; cost constraints; time/distance limitations; other factors affecting access/restriction to resources, goods, and services.

Cultural Heritage

The fourth thematic area of a cultural diversity lens is dedicated to understanding the programme's impact on and links with the cultural heritage – tangible, intangible, documentary, and living – of the population concerned, including the cultural significance of the natural environment and its ecology. Whatever the focus of the programme under analysis, assessing any potential negative impact on the cultural heritage of the people concerned is a necessary component of programme analysis. Potentially negative impacts identified should be avoided, through changes in programme design, or at the very least mitigated in a way acceptable to the local population and in accordance with established norms and safeguards. At the same time, opportunities for building upon existing cultural heritage and strengthening cultural heritage assets should be recognised and harnessed to reinforce programme success. Key words guiding programme-specific inquiry in this theme include: local knowledge; indigenous typologies of nature, space, and society; knowledge of animals and plants, including local pharmacology; hunting and agricultural cycles/practices; food preparation, preservation and eating habits; music and dance; artistic and handicraft traditions; writing systems; notions about the nature of man and the universe; religious and ritual practices; symbolic, social and material values of natural, built, and intangible expressions of heritage including that of associated artefacts, tools, costumes, utensils and documents; traditional safeguarding practices including taboos and other restrictions.

Economics of Culture and Creativity

The last thematic area of a cultural diversity lens is dedicated to creativity and to the economic dimension of cultural activities, goods and services (Figure 9.10). This theme focuses attention on the power of the programme under analysis to promote a diversity of possible economic activities that are linked to it while respecting intellectual property rights of individuals and of the collective group.

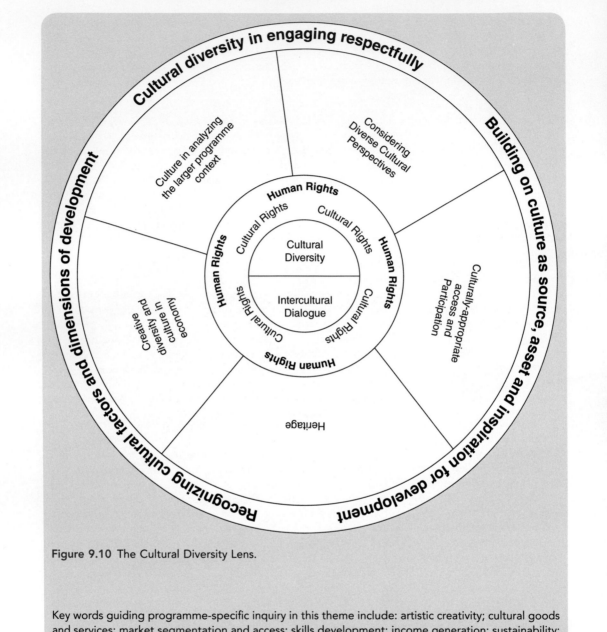

Figure 9.10 The Cultural Diversity Lens.

Key words guiding programme-specific inquiry in this theme include: artistic creativity; cultural goods and services; market segmentation and access; skills development; income generation; sustainability; safeguarding intellectual property.

Structured in this way with five interlocking themes centred on a solid core emphasising the pre-eminence of a cultural rights-based approach to development, a cultural diversity lens guides the user from the most general of themes concerning the context of the programme, to specific themes concerning equity and access, the safeguarding of heritage, the economics of cultural production, and the promotion of innovation and creativity. The way we treat cultural diversity in the development paradigm affects all economic, social, ideological, and political issues. A society that squanders, abuses, or misuses its cultural resources, in all their diversity, cannot prosper. However, if we understand and manage the power of cultural diversity, the well-being of individuals will be enhanced and societies will thrive.

10 Knowledge and Planning Tools

Jyoti Hosagrahar

Director of Sustainable Urbanism International at GSAPP, Columbia University, New York and Bangalore; Chair of the PhD programme at Srishti School of Art, Design, and Technology, Bangalore, India

Maybe we can show government how to operate better as a result of better architecture.

Frank Lloyd Wright

Introduction

Objectives of Developing Tools for Managing Urban Heritage

Paul Ricoeur observed the paradox of becoming modern and returning to sources; reviving an old dormant civilisation and becoming part of a universal civilisation.[1] These are paradoxes that confront historic urban areas in the contemporary context. The tools we seek are those that will help to weave together and complicate the simplistic binaries between modern and historical, new and old. Rather than expecting to see simple temporal sequences of uncontested flows from the old to the new, we need tools that help to identify the essential attributes of historic areas that must be retained to give it identity and those that may be allowed to change to make room for new. What tools would help to conserve the forms, spaces, and meanings that communities choose to remember collectively?[2]

Tools that help to identify the processes that have resulted in the particular historic built environment are also critical. As Walter Benjamin has reminded us, '*The uniqueness of a work of art is inseparable from its being imbedded in the fabric of tradition*'.[3] Although he has said this in the context of the authenticity of reproduction of works of art, this can be seen as valuable also in thinking about reproducing or retaining the relevance and meaning of cultural heritage.

---

[1] Ricoeur, P. (2007) *History and Truth*. Chicago, Illinois: Northwestern University.

[2] Halbwachs, M. (1992) *On Collective Memory*. Chicago: University of Chicago Press.

[3] Benjamin, W. (2011) The Work of Art in the Age of Mechanical Reproduction, *Illuminations*, with an introduction by Hannah Arendt. New York: Random House. First published in 1968.

Reconnecting the City: The Historic Urban Landscape Approach and the Future of Urban Heritage, First Edition. Edited by Francesco Bandarin and Ron van Oers.

The Historic Urban Landscape approach is about managing urban change and integrating contemporary needs and services positively with continuities in cultural heritage. Thus a variety of interdisciplinary and innovative tools, mechanisms, and instruments are necessary to guide and manage change and development to be compatible with heritage identity and values. The tools and instruments have to be adapted to the local contexts in that some places may be able to have access to the latest digital technologies and the availability of precise data while in other places, the decision making is through negotiations between different community groups.

By emphasising locality, context, historical continuities, and identity, a heritage-sensitive approach to development promises greater opportunities for achieving goals of equity and social justice. Thus, the tools are intended to offer opportunities for a plural, gentler, and more contextually responsive mode of development. The tools would also aim to recognise cultural significance and diversity while providing for the monitoring and management of change to improve the quality of life and of urban space.

In writing about the tools for planning in the Historic Urban Landscape approach, the selection and modification of the tools have been adapted to the circumstances of the town and the formal and informal processes of planning operating in the locale. Two of the most significant local influences are first in the emphasis on socio-economic development of the local communities and second in the prevalence of 'living heritage.' Other local conditions include the poor availability of reliable data, the relative cost of tools, instruments, and evaluations, and the paradoxical availability of high quality expertise and inadequate capacities to manage the sites. The absence or minimal availability of written and textual documentation combined with continuing intangible practices including building crafts is also significant in its influence on restoration and conservation.

The tools outlined in this chapter are discussed in three main sections. The first section addresses mapping, measuring, and visualisation tools for knowledge and data gathering. The tools in the next section are focused on reading, interpreting, and analysing the urban landscape. The final section includes tools to help with planning and regulating interventions in historic areas and making decisions and choices regarding the protection, changes and new development. A brief discussion on engaging the stakeholders and the community concludes the chapter. The premise is that the sequence of activities would follow these sections and that some tools are necessary from each of the sections to make up the final tool kit for a particular urban area.

Mapping, Measuring, and Visualising the Urban Landscape

Mapping, inventorying and documentation would need to go beyond privileging the architectural and material attributes of urban heritage, and recognise cultural significance and diversity, as well as help monitor and manage change. A key objective of the data gathering tools, beyond those necessary for routine documentation, is to help identify the attributes that give a place its unique identity or identities.

Surveying and documenting has to be done at several different scales from the city and region to the interior of small buildings and architectural elements. Depending on the context, the survey and documentation work may be carried out with more or less use of electronic devices and systems. Usually it is useful to corroborate data from one kind of survey methodology with that obtained from another. The selection of the tools is influenced by the existing data, funds available, the level of accuracy and precision desired, and the time estimated for the completion of the work.

At the Scale of City and Region

An important first step is to establish a reasonably accurate base map on which to record as much information as possible about the place. Tools for this include total station surveys, Lidar surveys, geospatial mapping of topography, digital elevation models, mobile mapping, and aerial surveys.

Total Station Surveys are perhaps the most common survey technique today. It consists of an electronic-optical instrument that measures angles and distances from the instruments to the points being surveyed. Coordinates to the points, angles, and elevation levels are calculated relative to the instrument's location using trigonometry and triangulation. Total Station requires line-of-sight observations and needs to be referenced with at least two points whose coordinates are known. The data downloaded to a computer and fed into software can generate digital survey maps. Lidar Surveys using laser technology is also useful for mapping a variety of physical features and structures.

Digital elevation models prepared by using remote sensing or direct surveys are especially useful to understand the relationship of the topography of the town in relation natural features in the area such as water bodies or ridges. Geospatial mapping of topography, walking with a hand-held GPS device is useful for surveying the location of scattered heritage structures, fortifications, or natural features including lakes, ponds, or wells.

Aerial maps are valuable for marking the footprints of buildings, water-bodies, and roads when they are reconciled with the reality on the ground.[4] The level of accuracy is co-related with the resolution of the aerial photographs. The higher the resolution the more details that are visible and the more expensive they are. At higher resolutions cartographic distortions due to tilt need to be rectified. One drawback of using aerial surveys is the availability of high-resolution aerial photos and another is the expense of such images.

At the Scale of Streets, Urban Spaces, Monuments, Buildings, Interiors, and Architectural Elements

Measured drawings prepared from on-site measurements of historic structures, streets, squares, and open spaces are invaluable in documentation of a place. The data so recorded may be used to construct hand-drawn or digital drawings. Similar methods can be applied to create plans, sections, and elevations of buildings, streetscapes and street sections, as well as detailed layouts of building interiors. Once plans, sections, and elevations have been digitally constructed, 3D models of the neighbourhood or urban space may be easily created at the level of detailed desired. Remote sensing and Lidar techniques could be used to generate such 3D models (Figure 10.1 a,b).

Panoramic photographs can help to construct street elevations or to document a 360 degrees panoramic sweep from a single point in an urban space. These are especially useful for recording views from or of structures. Photogrammetry helps to determine the shape and geometric properties of objects from photographs obtaining reliable information about physical objects and forms as well as the environment through measuring and interpreting photographic images. Photogrammetry can be useful in producing two-dimensional street and building elevations, as well as three-dimensional views of urban spaces and buildings, interiors, and architectural details.

Reading and Interpreting the Urban Landscape

Inventory and Database

A detailed inventory of a wide variety of cultural resources is crucial. An inventory of existing

---

[4] Dolores Hayden's paper is illuminating regarding aerial surveys. Hayden, D. (2000) Flying Over Guilford. The *Journal of the American Planning Association*. Vol. 66, 9.

Figure 10.1 (a, b) 3D model of a Hindu temple. Drawing of a temple complex around a public space in Bangalore.

cultural resources and mapping of cultural assets sounds like a mechanical compilation of heritage structures and museums. However, a thorough inventory requires a deeper understanding of place, its history, and society. It also requires an expanded definition of cultural heritage as well as an inventory of services and institutions related to cultural activities. Cultural Asset Maps may represent these resources spatially in and around the urban area. A participatory approach involving local residents, school children, and volunteers ensures that the inputs of a community help identify and include those resources that are valuable to them.

Historical Analysis and Mapping

The purpose of the historical analysis is to capture the key developments, processes, and events that have influenced and shaped the city over time.[5] The idea is not to valorise the royal or dominant figures that may have been influential in the city at one particular moment in its history, but rather to understand the conflicting groups, identities, and values that have been central to forming the identity of the place. The historical development of the city may be marked on overlaid maps of different periods so as to enable visualisation of the growth and change at each phase of the city's development. While histories are narratives that offer insightful interpretations of time period or a historical question, place, or event, historical analysis as a tool in the Historic Urban Landscape approach is a temporal analysis and mapping of the spatial and formal structures and relationships. Many cities have been built and rebuilt numerous times; almost annihilated by war or natural disaster only to rise up again. The remains of each era are layers that need to be identified and represented separately as well as to understand the ways that each preceding city influenced those built upon its remains. Oral histories and interviews with people help to clarify the different positions of the residents over time. Oral histories and historical interviews of residents and artisans are especially valuable in recording and interpreting these. In places and regions such as South Asia where there is a paucity or absence of historical texts, records, drawings, and documents, interviews with the local community and crafts people as well as oral histories are especially valuable.

Visual and Formal Analysis

Visual analysis for historic urban landscapes involves analysing all the visual attributes and visible material elements of the urban area. The survey, measurement, and documentation must be completed before undertaking a thorough visual and formal analysis. The analysis is best carried out separately at different scales and for different aspects. A key objective of visual analysis is to arrive at the visual identity of the urban landscape and its key components. One crucial aspect is to recognise that visual identity of a place is not a static artefact frozen at a particular period of time. Rather, the challenge is to consider ways to manage dynamism in the visual identity of a place.

At the Urban Scale

A skyline or silhouette is the profile of the tops of buildings in a city against the sky. The tallest buildings whether they are minarets or commercial buildings are significant landmarks. A distinctive skyline is an important aspect of a city's visual identity and in many instances forms the iconic image of the city (such as New York or London).[6] A historical perspective of the skyline at different periods of time, influences, and impacts may be crucial to understand the layering of the city over time and also to guide how change in the skyline can be managed. A skyline could be interpreted as a dynamic visual artefact, a composition of vertical and horizontal elements. Skylines of cities are not equally distinctive all around, nor are they as clearly visible from everywhere. As important as safeguarding the distinctive visual skyline is protecting the points within and around an urban area from where the skyline is viewed to greatest advantage.

In every city, certain views and sight lines are more image-able, picturesque, and iconic than others. A visual analysis that identifies such

---

[5] See also: Abbott, C. and Sy, A. (1989) Historical Analysis as a Planning Tool. *The Journal of the American Planning Association*, 55, 4.

[6] Hal Moggridge has described skylines as 'images of buildings often national monuments, seen against open sky (occasionally distant mountains). Sometimes a tree canopy or landform is part of the critical skyline.' Moggridge, H. (2010) Visual Analysis: Tools for Conservation of Urban Views During Development. In Van Oers, R. and Haraguchi, S. (eds) *Managing Historic Cities*. World Heritage Papers N. 27. Paris: UNESCO.

North Elevation -1

Figure 10.2 Drawing of street elevation of northside of Devalaya Beedi (main Temple Street) in the historic core of Belur, India.

views and site lines helps to prioritise regulatory measures to conserve and protect them. Such views may be in the distance such as mountains, water bodies, or the ocean; or may be a vantage point that offers the finest visual presentation of a monument or place. Gateways, the space between buildings, or a placement at the head of a central square are all different ways to frame iconic views of urban ensembles to the best advantage. Uncontrolled building heights, and haphazard development interfere with the sight lines of such iconic views. Hence, a visual analysis of the views and site lines would not only identify the significant views but also include the view points, and view cones for each. Views are important to identify at different scales and to consider them 'looking in' towards and 'looking out' from the cityscape, urban space, or ensemble. In Chartres, France, for instance, the view of the city rising up from the agricultural land around has been identified as important and protected as such.

A 3D model of the urban area, especially a digital one, is helpful in analysing the form in many ways. Such a model is a way to track and analyse spatial development of the city over time both horizontally and vertically. It is also useful to analyse formal geometry including axis and zones. Overlaying the 3D digital model of the buildings and urban area on a digital terrain model provides a useful understanding of the relationship of topography to the built form of the city.

Street elevations that use photogrammetry to visually represent entire streets are useful for identifying street elevations that have the greatest visual integrity and authenticity as well as to evaluate the parts of a valuable street-front that is most in need of conservation or other interventions (Figure 10.2).

Cultural Analysis

Cultural analysis and interpretation includes the values, beliefs, practices, and knowledge that constitute a place, give it identity, and make historical structures and spaces meaningful. These may include intangible heritage such as specific forms of music, dance, theatre, literature, or visual arts particular to a place; crafts including traditional technologies of construction and building crafts; knowledge and practices related to managing the local ecosystem and environment. Practices of extracting natural resources for food and fuel to building materials are forms of cultural knowledge that contribute to making the landscape unique. Urban agriculture (such as fruit trees), cuisine, rituals, festivals, and religious beliefs and practices, all form the complex web of culture of a place. However, since these beliefs and values are dynamic and change over time, the cultural analysis needs to reflect the transformations as well rather than freeze the cultural values and beliefs at a particular instant in time. Obtaining information for such cultural analysis is through direct observation of the society, interactions and interviews with a wide cross-section of residents, ethnographies, interactions with community groups and NGOs, and through community mapping exercises (Figure 10.3).

Mapping Cultural Meanings, Practices and Identities

Beyond mapping assets and resources, cultural mapping represents a spatial interpretation of the meanings, activities, values, relationships, and other less visible aspects of places. It relates the material world of buildings, streets, and

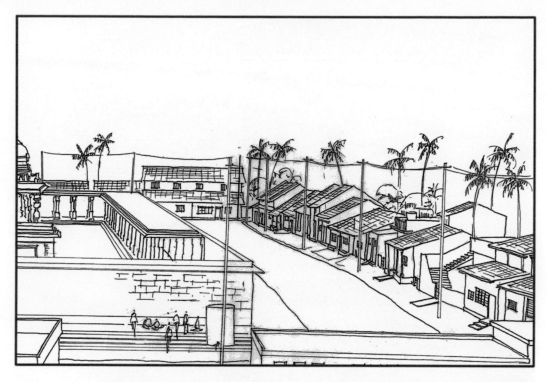

Figure 10.3 Community mapping exercise. Sketch of the Ratha Beedi (Chariot Street) north of the temple in the historic core of Belur, India.

spaces to their meanings and significance. Economic, social, religious, or symbolic meanings and structures can be understood as spatial relationships. It also enables tracking and interpreting cultural changes over time in the built environment. Cultural mapping and analysis of the cultural landscape supplements formal techniques to make the spaces more meaningful both to the present inhabitants and those who created and lived in another era.

Cultural mapping involves:

1. Obtaining cultural information;
2. Representing the information spatially; and
3. Reading the maps to interpret the cultural landscape.

Cultural information may be based on detailed observations both of people and their relationship to the built environment, community mapping, interviews, and other ethnographic techniques. Using archival research, dating buildings and structures, and oral histories can help to create maps that show cultural transformations over time.

Mapping spaces of religious and ritual significance allows the importance of places, as opposed to structures, to become relevant. For instance, in one of the historic towns in southern India the waterfront, its usage and significance, as well as the ritual processional paths were mapped. Processional paths for historically significant rituals, and the use of specific spaces for annual fairs that have been continuing for a few centuries were also part of the cultural mapping exercise.

The mapping of the age of buildings within the Srirangapatna Fort (Figure 10.4 a) captures changes over time and layering of a city where the monuments are of one era and the ordinary fabric of the city reflects the renewal of the vernacular buildings during different time periods

RELIGIOUS STRUCTURES
AND GHATS
FESTIVAL AREA

a

0 - 50 Yrs.
50 - 100 Yrs.
100 - 200 Yrs.
200 - 300 Yrs.
300 << Yrs.

b

Figure 10.4 (a) Map of Srirangapatna Fort. A 3D model of the fort area of the island town of Srirangapatna, India, showing spaces of religious and ritual significance including steps for bathing in the river, and processional routes. (b) Shows the remnants of different layers of fortification, and the spatial distribution of the age of structures.

within a centuries old planning framework (Figure 10.4 b).

Different social groups that live in the different neighbourhoods are also represented in this reading. Similarly, a mapping of water channels helps to reveal not only the flow of water through the town but also enables a spatial interpretation of the significance of water for the island town of Srirangapatna.

Socio-economic Analysis

Socio-economic analysis of the historic areas is essential to understand the inhabitants and users, as well as the development pressures the historic area is under. Densities of population within certain areas would be helpful to understand situations of crowding and congestion if they exist. Such analysis is especially relevant as in many parts of the world, the historic core of cities have become the centres of poverty and deprivation, overcrowded, under-serviced, and with poor living conditions; while in other parts of the world, historic cities are declining both in population and economic activity becoming 'shells' as it were. Such data gathering would also help to reveal the role of cultural activities including music, dance, theatre, literature, arts, festivals, and crafts in providing employment to residents. In Delhi, this was also a way to identify areas dominated by migrant workers from those that had stable and long standing communities.

Morphological Analysis

Figure-ground studies or solid-void studies help to identify the structure of the urban area and to highlight the relationship between solid mass and open space. The hierarchy of the different urban elements and the links connecting them are made apparent by such studies. Categorisation of buildings and spaces based on their formal and visual attributes or typologies is also a valuable interpretation of the landscape. The classification may be based on height, volume, geometry, materials, shape, location, elevation, architectural features, function, roof form, fenestrations, and other features. Numerous scholars such as Gianfranco Caniggia, Peter Hall, Jeremy Whitehand, and Michael Conzen have been influential in this regard.[7]

Infrastructure Analysis

The prevailing systems of infrastructure services available and their adequacy and available technologies need to be carefully assessed. Traditional systems of water provision, and sewage disposal may be still prevalent but inadequate in some places. Circulation patterns and movement are critical with recent emphasis on motor vehicles in a huge shift from earlier designs for pedestrians and horse-drawn carriages.

Geospatial Referencing

Spatialisation of a variety of quantitative and qualitative information on city maps and plans is a valuable way of integrating and correlating a range of data with particular points in the city. All of the information relevant to a particular point in space can be accessed as a database of that location. While such geospatial referencing is most effectively and commonly done with computer technology using Geographical Information System software (GIS), non-computerised system of creating vectorised maps, layers of drawings, and datasets for each coordinates of each point on the map, could be equally effective.

Protecting, Enhancing, and Improving the Urban Landscape

Evaluating the potentials and development pressures on a site and weighing a range of possible actions and interventions, prioritising them, and anticipating and evaluating their impacts are central to the Historic Urban Landscape approach to managing urban areas. While protecting and safeguarding the continuity of the most significant heritage attributes is critical, finding ways to address development needs and potentials in a way that is compatible with the heritage character is equally important in this approach. Tools that help with some aspects of the decision-making include:

---

[7] See for instance: Caniggia, G. and Maffei, G. (2001) *Architectural Composition and Building Typology: Interpreting Basic Building.* Firenze: A. and Conzen, M. R. (2004) *Thinking about Urban Form: Papers on Urban Morphology 1932–1998.* Bern: Peter Lang.

SWOT Analysis

A SWOT analysis is a step towards developing a comprehensive management plan. Such an analysis is usually a matrix and is a structured planning method where strengths and weaknesses inherent in a place are assessed in comparison to other similar places, as well as the threats and opportunities present. The matrix lists factors influencing the place such as cultural resources, access to and quality of infrastructure services, tourism facilities, natural environment, and creative industries. SWOT matrices help analyse the attributes of the place that would contribute positively to the goals and those that are impediments to achieving them. However, the matrix format does not allow us to qualitatively examine all the internal and external factors or to prioritise them.

Visioning and Strategic Planning

Arriving at an overall vision for the historic urban landscape is a crucial component around which a management plan can be built. Ideally, again, such a vision would be arrived at by consensus through a participatory process although the input of experts is necessary to elaborate on the potentials, benefits, and disadvantages of different options. The vision would help to set the direction of actions and priorities. Strategic planning towards such an agreed vision helps to establish strategies, directions, and priorities for action as well as balance conflicting needs, and uses. A strategic plan can also be viewed as a path or a road map to achieving defined goals. It also helps to identify management structures necessary.[8]

Cultural Heritage Impact Assessment

When any development project is proposed, the potential impacts, both positive and negative, of the projects need to be carefully evaluated. Such assessments are also useful to understand the projected impacts of existing development trends. Design of interventions to enhance and support the benefits, and others to mitigate the negative outfalls are necessary to counter balance the proposed projects.

Typically, a cultural heritage impact assessment includes a base-line study of all the existing cultural resources prior to the commencement of a proposed project or at the beginning of the study period.[9] The impacts may be assessed as projected trends or based on previous experience. Direct, indirect, and cumulative impacts need to be assessed on cultural resources and cultural heritage that are tangible and built, intangible, and on cultural knowledge and practices. Proposed interventions may seek to reinforce positive impacts and mitigate negative ones through measures that may be reactive to current or expected trends, or be proactive to safeguard and conserve and give more opportunities for the utilisation of cultural resources.

A Visual Impact Assessment is often an integral component of a cultural heritage impact assessment but may also be carried out as a separate exercise. This is an evaluation of the aesthetic and visual repercussions of proposed development or trend on the surrounding areas especially cultural heritage properties and significant views. For landscapes and areas whose visual attributes are of high value or are otherwise visually sensitive, such assessments help to evaluate the location, siting, nature, scale, and distance from visually significant elements in the

---

[8] In this regard a useful review is: Myers, D. (2000) Constructing the Future in Planning: A Survey of Theories and Tools. *The Journal of Planning Education and Research*, 19, 3.

[9] See for instance, ICOMOS, *Guidance on Heritage Impact Assessments for Cultural World Heritage Properties*, International Council on Monuments and Sites (ICOMOS), 2011. See also: Tarrafa Silva, A. and Pereira Roders, A. (2012) 'Cultural Heritage Management and Heritage (Impact) Assessments.' *Proceedings of the Joint CIB W070, W092 & TG72 International Conference on Facilities Management, Procurement Systems and Public Private Partnership*. Cape Town, South Africa.

urban landscape. Any visual impairment that the proposed project or trend may cause to the views, sightlines, skylines, and visual integrity of the urban area may be mitigated by alterations to the designs. Equally valuable are visualisations of development control guidelines and building regulations that would enhance the visual integrity of historic areas.

Environmental Impact Assessment

An environmental impact assessment (EIA) is an evaluation of the predicted impacts, both positive and negative, that a proposed project, development, or trend might have on the environment of a place. In recent years, the environment is more broadly defined to include the natural environment as well as the social and economic impacts. EIAs are a valuable planning tool as they enable decision makers to, plan a course of action, accept or reject development proposals before they are implemented, and influence changes in the design of interventions to mitigate negative impacts and enhance positive ones. While EIAs are extremely useful in providing a rational basis for decision-making, they can only inform as the decisions themselves are based on the values of the decision makers. Furthermore, the direct impacts on the local environment are more easily measured and predicted than global or regional ones that may be outside the scope of the project to fully evaluate or control.

Zoning

Protective zoning is another important management tool that has been effectively used for managing historic urban landscapes. Zoning is a way of identifying a ringed area around heritage properties that have more stringent development controls than those at some distance. But zoning is also a way of identifying sub-districts within a historic urban landscape that may have different regulations.

World Heritage properties are typically required to identify core and buffer zones. The core zone is the boundary of the inscribed heritage property. Any change or development within the core zone would directly impact the values, authenticity, integrity, and identity of the heritage property. Around each heritage property is a larger buffer zone that is intended to provide 'an additional layer of protection to a World Heritage property'.[10] Any development or change in the buffer zone could significantly impact the values, integrity of the heritage property and the identity of the urban area. In addition to a buffer zone, the World Heritage Committee recommends defining a larger 'area of influence' where development activities may impact the heritage properties.[11]

Zoning is also a way of identifying sub-districts in an urban area where differing levels of change are acceptable. As cities are inherently dynamic, unlike for monuments, the designation of core and buffer zones needs careful consideration for urban landscapes. In our work, we have argued for the use of internal management zones to manage development and change. Several zones may be established within the city with different building by-laws and varying degrees of control. For instance the walled historic part of a city with the oldest streets and neighbourhood we designate as a historic precinct with the most stringent development controls. While the height regulations in the next ring around it may be more lenient. However, the activities and uses may remain more highly regulated in this second ring or zone than the third. Hotels and tourism facilities were only permissible in the third ring and industrial uses and

---

[10] The concept of a buffer zone was first introduced in the *Operational Guidelines for the implementation of the World Heritage Convention* in 1977. In the 2005 version of the Operational Guidelines, buffer zones have been recommended but not made mandatory.

[11] UNESCO (2009) *World Heritage and Buffer Zones*. World Heritage Papers N. 25. Paris: UNESCO: 188–189.

apartment blocks beyond this area.[12] Such an internal zoning plays a more positive role in guiding urban development that is compatible with the values and integrity of the heritage area than merely protective regulations.

Zoning is also useful for regulating activities that in turn influence the urban landscape. While modern city planning principles have generally advocated single use areas in the city such as residential, retail, commercial, and industrial, mixed use areas with residential, commercial, and small workshops were commonplace historically in most urban areas. Thus, carefully designed zoning can enhance and support activities that contribute to the values and identity of the place.

Traditional and Customary Systems of Management

In the Historic Urban Landscape approach, traditional and customary systems of management would be important to recognise and reinforce where possible. While this aspect has had some discussion in the management of cultural landscapes and natural landscapes, much needs to be done in the management of urban areas in this regard. This concept is significant in numerous ways in the South Asian context. In many towns with substantial populations of indigenous and ethnic groups, community or tribal groups are responsible for decision-making about their urban landscape. In other instances, ownership and management of some properties have remained in the hands of old elite families such as erstwhile princes and maharajas. Their authority and benevolence has continued to care for the upkeep of places and kept them meaningful. Numerous palaces, forts, and gardens as well as urban areas within and around them are in this category.

Alternatively, formally appointed non-governmental and non-profit organisations are in charge of some sites. For instance, traditional institutions like Temple Trusts continue to regulate activities within and around the temple area including festivals, processions, and ceremonies adjacent to the temple. Many of the shops, houses, and nearby agricultural fields also often belong to the Temple Trusts. In India, Islamic philanthropists have historically dedicated properties for religious, pious, or charitable purposes that are overseen by formally constituted *Waqf* boards. These include sites of Islamic significance such as mosques and *dargah* and in many instances also the areas immediately around them including shops, open spaces, and residential structures.

The most challenging to recognise and reinforce are the informal community spaces that give urban areas their identity. These may range from occasional transformations such as weekly markets or rituals held by various communities, to everyday community spaces such as courtyards, platforms, wells, and water bodies.

Contextualising the Historic Urban Landscape Approach

As the Historic Urban Landscape approach integrates the goals of urban heritage conservation with social and economic development, adapting the Historic Urban Landscape planning tools to local context begins with engaging local communities and stakeholders as an integral aspect of the entire process. The aim would be to empower a diverse cross-section of stakeholders to identify key values in their urban area and negotiate trade-offs to safeguard heritage and promote sustainable development. Bruno Latour has argued for recognising the connections between nature and culture and also between cultures past and present.[13] As the towns and cities urbanise and globalise, their conditions and contexts transform, so too must the tools and instruments for planning and managing their heritage values and development needs and aspirations.

---

[12] Sustainable Urbanism International. *Site Management Plan for Conservation and Sustainable Development of the Hoysala Heritage Region, Karnataka*, Vols. 1–6. Monograph prepared for 15 towns for the Government of Karnataka, India, 2011.
[13] Latour, B. (2012) *We Have Never Been Modern*. Cambridge: Harvard University. First published in 1991.

Case Study *Reading the City of Tokyo*

Hidenobu Jinnai

Professor of Urban Design at Hosei University in Tokyo, Japan, and since 2004 Director of the Laboratory of Regional Design with Ecology at the Graduate School of Hosei University, Japan

Introduction: Differences in the Landscape between Japan and Italy

In the early 1970s in Venice, I learned the method of 'reading a city'. Back then, design survey of a well-kept, traditional urban landscape was the common method used in architectural research and education in Japan. But the idea of selecting a vibrant real-life city as a subject for historical or design research was extremely remote. It was also very rare to refer to old maps in order to understand a city.

Italy, where I studied, is a country that takes pride in its urban history. Starting in about 1960, a re-examination of its urban history rekindled an interest in historical cities and the interwoven urban fabric interpretation method came to be established as a result. It was indeed refreshing for me to hear the students at the University Institute Architecture of Venice use the words 'reading the city' casually during classes and seminars when they engaged in spatial survey or design work.

It was Saverio Muratori who first proposed a method to decipher the origin of a complex city that consists of layers of historical periods. He had passed away just before I arrived in Venice to study, but I had the opportunity to be acquainted with Paolo Maretto and Gianfranco Caniggia who were both central figures of the Muratori School. I was very fortunate to have been able to acquire the teachings directly from them.

Unlike the urban landscape visual approach method that had developed widely in Europe, including England, I later learned that Italy uses a very distinctive method to read the spatial structure from its layers of history by going all the way back to the ancient period. Architecture and history were carved into the topography, and it was this Italian structural concept that proved to be very helpful to read the Japanese cities that had difficulties inheriting the architecture and townscape due to recurring earthquakes, conflagrations and war disasters.

In Japan, it was the literary scholars who first showed an interest in reading cities, and the works of Toru Shimizu, the French literature scholar, Kouichi Isoda, Ai Maeda and others, were particularly prominent. It was in fact Takashi Hasegawa who discussed the definition of urban space from a literary angle, one that is not so different from an architectural. His leading work was *Urban Corridor* (1975).

Even before these scholars, Takeo Okuno had already used the sensational term, 'archetype landscape' in his *Archetype Landscape in Literature* as early as 1972. Back then, this term had broken away from the modern school of thought and had inspired those in the architectural field who sought to delve into the depths of architecture, urban space and such to discover the cultural identity. From here emerged a series of leading works such as *Archetype Landscape of Tokyo* (1979) by Noboru Kawazoe, *Aesthetics of Townscape* (1979) by Yoshinobu Ashihara and *Appearing and Disappearing City* (1980) by Fumihiko Maki. Each presented his topologic of Tokyo as an attractive place. Ashihara did so based on his urban experience during his Italian architectural pilgrimage. Kawazoe, on the other hand, valued the surroundings of Sugamo known for their numerous scenic flower-viewing spots as his archetype landscape and Maki, the labyrinthine space in the Yamanote area of Minato-ku.

Reading the City of Tokyo by Utilising the Italian Experience

In the fall of 1976 several years before the emergence of this series of leading works, I took the podium at Hosei University as an adjunct professor upon my return from Italy. In January of the following year, I immediately formed the 'research group' at Hosei University with some enthusiastic students to engage in the survey and research of Tokyo. The first thing that I wanted to do in Japan was to directly apply this 'reading the city' method that I had learned in Italy. Unlike cities such as Kyoto that are veiled in history just like Italian cities are, everyone looked at Tokyo as being a city that had lost its history. So, it was both paradoxical and thrilling to choose Tokyo as a subject and to uncover its history.

There are several wards in Tokyo that survived the Kanto Great Earthquake and Fire (1923) and where the old townscape remains. One of them was Shimoya/Negishi district in Taito-ku that I chose as my first research subject. I felt that it was the perfect subject for applying the Italian-style typological method. In the area for the townspeople along the main roads, there were tradesmen's houses. In the alleys were row houses and further back were shrines and temples. A continuous flow of larger samurai residences made up the residential area that extended beyond the shrines and temples. In the surrounding areas sprawled farmers' houses. In this survey, I was able to analyse and observe the historical formation and development of the architecture and cities by weaving the landscape through the urban context. This experience proved that even here in Tokyo, it is possible to carry out an interesting survey from an historical angle. At the same time, I think that I was able to portray a Japanese style area that was somewhat different in nature from an Italian urban space by characterising the relationship between houses and professions of residents, connection between the community and setting and spatial partition of the sacred and worldly in the town.

To the Concept of Spatial Anthropology

Yet, it was not possible for us to delve into dynamic Tokyo simply by passively observing the traditional downtown district that was rich in nostalgia. For this reason, we changed our concept completely for our next research subject and chose the whole community of what used to be the Yamanote area in former Edo (inner ring of Yamanote Line), whose urban structure seemed complex and hard to understand at first. We then challenged ourselves to read the area by graphing the characteristics chronologically. The fact that the area played a central role in the post-Meiji Restoration modern period made it even more attractive. Yet, unlike a European city, it was not as though there were many old buildings left in the city centre. Nevertheless, we assumed that there were likely to be elements that could be inherited.

In order to understand this face of Tokyo, we needed to interpret the Italian method even more loosely, transcend the architecture-centred concept and focus on how the varying topography, complex road network, luscious vegetation, architectural genealogy that has its roots in the modern era (the area for townspeople, residences of *daimyos*, houses of middle-class samurais, houses of lower-class samurais, shrines and temples, etc.) came into formation through interrelationship and how each transformed to shape the identity of the area. Once we began, the work proved to be very interesting. The Edo model clearly uncovered the formation of the modern and contemporary Tokyo (Figure 10.5).

I was encouraged by Muratori's words that once an urban structure is formed, it is not easily lost. Yet, it still came as a surprise to me when the old Edo map matched perfectly against the current map. Compared to Italy, the physical logic that is connected per se to architecture, however, is weak in Tokyo. In the end, this led to the formulation of a Japanese-style applied characteristics analysis method that focused on land and location, including topography, vegetation, sacred places, road network, zoning, etc. The in-depth study proved to be an interesting and intellectual challenge. Feeling that an anthropological and folklore approach would be necessary to decipher the features of Edo Tokyo, I also considered a concept of spatial anthropology (*Tokyo, A Spatial Anthropology*, 1985).

I applied my experiences in Venice for my research of downtown Tokyo after I realised that it was a water city. Furthermore, the perspective of spatial anthropology actually proved to be very effective for this study. Indeed, water played multiple roles in this city, including drinking, farming, boat transport/distribution, production, fishery, religion, religious ceremonies, drama, amusement, entertainment and such. A principle of formation also emerged, showing that the waterside was not merely a

Figure 10.5 Relationship of the mountain ridge road in the Yamanote area and *daimyo* residences.

marketplace but a distinctive place where shrines and temples, amusement districts, famous sightseeing spots, red light districts, etc. had developed.

Suburban Community – Focusing on the Layer of the Ancient Period

Within the inner ring of the Yamanote Line, remains of the city of Edo are scattered throughout and they can be sensed as you walk down the street. However, it was generally thought that the context of a historical community did not play a key role in the outer ring that extended all the way to the old farming villages in the suburbs. Surprisingly, it was my experience in Sardinia that gave me the methodological hint to tackle this new difficult subject. During the sophisticated stone civilisation of the Nuraghe Period (1500 BC–300 BC), people built sacred places and villages in areas where spring water supply was abundant. Occasionally, these locations were inherited by later historical periods as sacred places for the community, and even today, they have retained their importance. We then learned that these historical layers could be found even in the surroundings of Tokyo. Spring water, archeological sites, old roads, sacred places became our keywords. Together with architect Yuji Yanase, who had studied Sardinia for his master's research, we published *Sacred Island in the Mediterranean – Sardinia* (2004). This experience proved very helpful in the study of the Tokyo suburbs. Utilising our Sardinian experience, Yanase and I formulated a hypothesis based on my archetype landscape. Referring to this hypothesis, we carried out repeated in-depth fieldwork on bicycle with my study group students and we succeeded in reading the rich historical framework that had

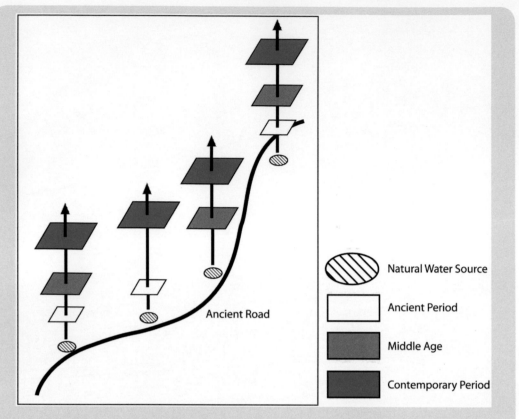

Figure 10.6 Configuration of the historical layers of a suburban community.

been inherited by the deep historical layers of the suburban community (*Tokyo Suburban Topography – Reading History from the Everyday Scenery*, report, 1999).

I myself was raised in Narimune in Suginami ward to where I moved at about the age of two. During the pre-industrial growth period, there were still remains of Musashino in this community; the topography can be recognised clearly, rivers and hills have a major presence while the old villages and locations of shrines and temples narrated such characteristics. Dispersed were fields, thickets, ponds and attractive looking playgrounds for children that had integrated with the topography and vegetation. Needless to say, all of these elements of my archetype landscape gave me important hints for portraying the identity of a suburban area (Figure 10.6).

When we direct our eyes to the suburbs of Tokyo, we are instantly drawn to the old structures from the medieval and ancient periods. We are able to identify with them because *Earth Diver* (2005) by Shinichi Nakazawa that was published at a later date highlights the historical layers of Tokyo from indeed a similar perspective. On the other hand, the area in the inner ring of Yamanote Line that was already a city in the Edo era was veiled in modernism. So the primitive historical layers preceding this period are hidden and emerge only in half-concealed state.

When suburbs are the research subject, we have the advantage of being able to compare the accurately formulated maps to concisely and realistically understand the entire process as the area transforms from a pre-modern day peaceful rural state to modern day residential and urban district. We can easily read the habits, trends, principles and mechanism in this vernacular transformation of a Japanese countryside as it develops into a city. It's interesting to compare this transformation with that of a European city.

Various kinds of fresh discoveries can be made even within the deep historical layers of a particular community. First of all, many shrines and temples are built in locations where there are springs. Omiya Hachiman-Gu Shrine near my house is a typical example. Since the ancient period, people have been living around the shrine. Archaeological sites from Jomon, Yayoi and Kofun eras have revealed that from the olden days, residential districts have been built in slightly elevated areas along the Zenpukuji River. Extending northward from Omiya Shrine is an old road from the medieval period that is connected to Suginami's other religious centre, Asagaya Shinmei Shrine. Midway, we find the important archaeological site of Matsunoki. Saginomiya Hachiman Shrine is located further north on this road.

In fact, Asagaya Station-front shopping centre (current Pearl Center) was built after the construction of Asagaya station. The loose linear arc that extends vertically is a typical example of the route of an old road. Its landmark found at the intersection of another old road is a sacred place, housing the Koushinzuka (roofed rest area for worshippers) and Jizo (Buddhist guardian deity of children).

If Chuo Line Did Not Exist

The majority of people who live in the surroundings of Tokyo will undoubtedly first consider the railway as the spatial axis in relation to their community. The fact is that Seibu Shinjuku Line, JR Chuo Line and Inogashira Line pass radially through Suginami-ku and cultural and commercial zones have been built around them. Yet, we note that these zones are nothing more than those that were built from the end of the Meiji era to early Showa era. They have nothing to do with the community's deep layers of history.

Archaeological findings from the slightly sloped region along the river show that people lived there since ancient times. Occasionally, roads in the ancient and medieval periods extended vertically, connecting key locations such as shrines and temples with springs. Later, during the modern period, the roads spreading radially westward from the Edo centre formed an axis for a wider area. Only today does the railway seem to have a major role but back then, it was a total newcomer.

Under the stimulating hypothesis, 'If Chuo Line did not exist,' I am enjoying my field survey to portray the deep historical layers of the wide area along the Chuo Line, beginning from Nakano-ku and Suginami-ku and going all the way out to Kokubunji. What I do is to observe the entire area by reading the topography as I go along the old road, scrutinising all the elements that constitute the environment and landscape. In addition to shrines and temples, large residences, small shrines, road signs, trees, stone walls, hills, alleys, bridges and underground drains, old-timer establishments such as liquor shops, soba restaurants, eel restaurants and rice shops also serve as key indexes for reading the region.

On a recent visit to Nakano, I was finally able to solve my long-time enigma of the gay geisha quarters that had been established in Nakano Shinbashi (Figure 10.7). There, the conditions are picture-perfect. Kanda River flows horizontally north of the Nakano Shinbashi subway station. Further north of the river extends a gentle downhill slope of favourable conditions. In the area north of this slope are situated Fukujuin Temple and Hikawa Shrine that were most likely built in the medieval period. They are situated side by side, facing the southward downstream of the river. Between the sacred precincts and the Kanda River developed the restaurant–tea house–geisha house quarters (geisha dwellings, Japanese tea house with geisha, classical Japanese style restaurant). In particular, it has been confirmed on the residential map from 1958 that as many as eight classical Japanese-style restaurants dotted this idyllic riverscape. This was also proven through interviews with people.

In such a way, I visited and studied the territory that used to be a suburban farming village during the Edo era. Although I did not discover an Edo urban structure, I was able to experience first-hand the formation principle of a typical Japanese topography that goes back to the ancient and medieval periods and also, the logic behind the formation of a modern city there.

Westerners tend to critically view general suburban districts apart from planned countryside. Undoubtedly, suburbs are inferior both culturally and scenically when compared to a historical charm-filled city. Yet, if this kind of method is used to penetrate into the deep historical layers of the Tokyo suburbs, one begins to see that the suburbs are indeed a charming place. I am confident that Japan

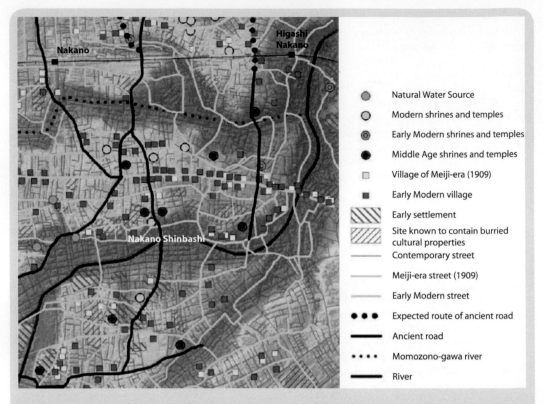

Figure 10.7 Layers of history in the communities surrounding Nakano.

will be able to present its own version of a distinctive evaluation method for a suburb. For that, we cannot merely adhere to methods that look only at the architecture; we need to gather all our knowledge, including that of ecology, geography, folklore and everyday life studies.

Fusion of History and Ecology – Survey of the Rural Scenery of Hino

From my research series that began with my studies on Yamanote, an area of hills, that was followed by my studies on watersides of downtown Tokyo and most recently, on the suburbs, I can see that the city of Tokyo is inherently an eco-city. Fortunately, the Laboratory of Regional Design with Ecology was founded in our Hosei University ten years ago and the centre has been conducting research using the history and ecology-fused community revival method.

The research of Hino-City, a 'water city' in the Tokyo suburbs is an example. Hino has a rich topography. Tama River and Asa River form alluvial plains and a sprawling plateau lies to the west. Archaeological findings have shown that along the foot of the plateau, there are signs of human settlements during Jomon, Yayoi and Kofun eras. For a long time, people have settled in the hills south of Asa River, taking advantage of the springs there. From the ancient to the medieval period, shrines and temples came to be built near the springs at the foot of the plateau that saw an expansion of the residential area further down. Small-scale rice farming developed, utilising the water drawn from water channels. In the Edo era, farmland and residential area expanded, covering a wide region as many water channels were built to draw water from rivers. The water was then carried to the alluvial plains. As a result, Hino became a region known for its rich and varying rural landscape.

However, from about the 1980s, there was a sharp decline in rice fields and the landscape began to change as it became more residential, more urban. In collaboration with the administration and

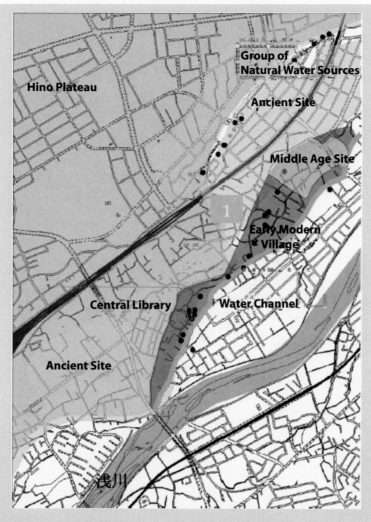

Figure 10.8 Spatial structure of the region in Toyota, Hino-City, Japan.

residents of Hino City, we are studying the formation of a community that enhances the rural landscape of a 'water city' (Figure 10.8). The rural framework has been considerably lost up to about Suginami. Yet, as one moves out from the centre, by the time one reaches Hino, the rural landscape has been inherited considerably, igniting our enthusiasm to develop a new kind of research method.

From the early days, much attention has been drawn to the water channels of Hino, the primary rural characteristic; furthermore, various studies have been conducted and they have led to the enactment of safety regulations. Yet, there is still an infinite abundance of environmental assets that lie dormant in this territory. They are major elements with hidden potential such as plateaus, sloped green tracts and other topographical features, springs, rivers, archaeological remains, shrines and temples, medieval mountain castles, old roads, villages, large residences, stone walls along roads, residential forests, farmland (rice paddies and crop fields), trees, small shrines, big roads and relay stations. This is why we need to understand the comprehensive landscape from the rural landscape concept (Figure 10.9).

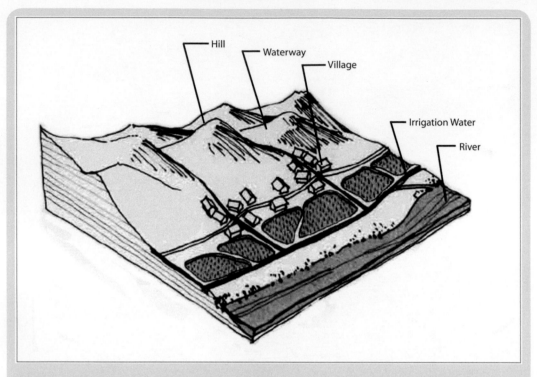

Figure 10.9 Model of a plateau – configuration of the topography and village structure (Hirayama region).

In fact, hints for this kind of concept were obtained from the research method used by Paola Falini, an urban planner at Rome University, for her studies on Val d'Orcia in Tuscany and Assisi in Umbria (both are World Heritage sites). This method analyses the man-embellished natural landscape as layers of distinctive development traces from the prehistoric, ancient, medieval, and modern to contemporary periods. This kind of method is certain to be much more effective in Japan where the topography is much more closely interlinked to nature than in Italy. We should be able to give a greater ecological twist to this method that originates in 'reading the city' to analyse the rural surroundings of Japanese cities, including Hino. By doing so, we should be able to clarify the comprehensive value of the historic urban landscape.

Agricultural landscape inheritance in the surroundings of large cities such as Tokyo and the continuation of agriculture are both major contemporary issues. This is the reason why it seems that the 'reading the territory' method that we tested in Hino will gain importance. Not limited to cities now, it has spread to the rural area.

11 The Role of Regulatory Systems

Patricia O'Donnell

Landscape Architect and Planner, and Founder of Heritage Landscapes LLC,
Preservation Landscape Architects and Planners, based in Charlotte, Vermont
and Norwalk, Connecticut, USA

> *A city is not an accident but the result of coherent visions and aims.*
>
> Leon Krier

Defining Regulatory Systems

Laws and cultural traditions are regulatory systems that address the form, content, uses of and changes to the historic urban landscape of communities worldwide. These controlling systems influence both tangible and intangible outcomes in living human settlements of all types and sizes. Regulatory tools interact with civic and community engagement, knowledge and planning and financial tools, rather than functioning independently.

Urban landscape integrity, the degree to which historic character, features, resources, and traditions remain legible today, is dependent upon the historical, contemporary, and future interactions of humanity and nature that form settlement landscapes. An array of regulatory systems influence the urban landscape, its character, and features. The regulatory tools, customs, practices, and laws that affect the urban landscape are not necessarily urban landscape specific or central to these resources and features. Nonetheless, the guidance or mandates they address influence the continuity, change and management of tangible urban resources and the intangible heritage practices that use and shape them.

Regulatory systems encompass a group of tools that directly or indirectly shape human settlements. Existing laws at the local, regional, and national level and international treaties address the larger scale of the city or settlement and can range down to the small-scale of an individual property. There are also traditional systems and practices to be considered, in terms of public and private land uses, building techniques and materials, and places where customs and practices are

Reconnecting the City: The Historic Urban Landscape Approach and the Future of Urban Heritage, First Edition. Edited by Francesco Bandarin and Ron van Oers.

expressed. Culture is dynamic and over time traditions and legal frameworks remain intact or transmute. Therefore, it is useful to note that not only traditions and legal tools are at hand, new regulatory systems can be created based on prior ones or framed to address the urban imperatives and innovations of the twenty first century. This process is underway as professionals, officials, and citizens work together to shape new regulations and legal tools through collective action. Current regulatory initiatives address topics like green construction, climate change response, dark sky lighting practices, and character of townscape.

The rate and type of urban changes in the twenty first century was the impetus for the UNESCO *Recommendation on the Historic Urban Landscape*, adopted on 10 November 2011. Historic Urban Landscape extends thinking from specific heritage resources to address the broader urban context and its geographical setting. This chapter asks two questions:

1. What are regulatory systems – to preserve, shape, enhance, and innovate within the historic urban landscape or degrade it?
2. Do these traditional practices, guidance, legal mandates, and regulations, that exercise control in the historic urban landscape, enable and support preservation, vitality, and authenticity of urban places or do they fail to do so?

These regulatory systems are enumerated in three lists, and are grouped as traditional systems, and laws and regulations that directly or indirectly impact urban heritage and character.

Legal Regulations Directly Addressing Public and Private Lands

Legal mandates addressing public and private lands that are directly applied to heritage that expresses the character of villages, towns, cities, and city settings include both tried and true tools and emerging ones, such as:

- Applying municipal plans and zoning laws that set forth required land use conditions;
- Following historic preservation laws strictures;
- Directing historic property and district commission review processes;
- Detailing guidance for purchase and transfer of development rights;
- Legislating the protection and stewardship of public parks and open spaces;
- Preparing and legislating urban viewscape controls;
- Applying the new urbanism charter and carrying out new urbanism projects;
- Developing and legislating form-based zoning codes that focuses on urban character; and
- Developing and legislating sustainable zoning codes and green construction laws and mandates to integrate energy efficient, carbon footprint reduction driven development.

First and foremost specific historic preservation legislation directing attention to cultural heritage is globally widespread, although cultural heritage is not afforded specific legal protection in some cultures. Earliest efforts to recognise and protect valued heritage date back centuries in some parts of the world. In the USA the Antiquities Act of 1906 was a national milestone, followed by other preservation focused regulatory tools.[1] The Historic Preservation movement came of age with the passage and application of the National Historic Preservation Act (NHPA) of 1966, which as amended to the present provides clear guidance for the designation and protection of local, regional and national heritage. This Act declares that *'the historical and cultural foundations of the Nation should be preserved as a living part of our community life and development in order to give a sense of orientation to the American people'*, and that irreplaceable

---

[1] American Antiquities Act of 1906, available at www.nps.gov/history/local-law/anti1906.htm.

heritage is a vital legacy to society, among other objectives.[2] The NHPA law provides standards for researching, documenting, and listing historically significant properties, sites, and structures as a baseline. Documentation of heritage resources prior to degradation is required under Section 110 of NHPA and mitigation is directed when negative impacts to designated heritage resources are proposed in federally funded project under Section 106 of NHPA. Related codes provide income tax incentives for the preservation of historic buildings. This national law established a framework for state and local historic preservation legislation that enables the setting up of historic commissions to oversee local and regional cultural resources and directs the work of designation and oversight of changes through a certificate of appropriateness process. These statutes often include delay of demolition clauses that can be employed to rescue or at a minimum record threatened historic property. These state and local regulatory tools bring historic preservation actions to the historic urban landscape level. Preservation of historic resources has overly focused on historic architecture, to the detriment of identification and inscription of urban landscape resources. In the past 30 years a broadening of efforts has been witnessed, but urban landscape heritage is often viewed as secondary, and able to withstand radical change without diminution of heritage. When in modern cities 50% (or more) of the space is landscape, not buildings, the full breadth of urban heritage resources needs to be stewarded effectively to retain village, town, and city identity and uniqueness into the future.

Conservation easements are another type of legal tool that separates certain rights from a land parcel to preserve the overall property character or to focus on the exterior and sometimes interior appearance of historic structures. One source notes that: *'Conservation easements provide landowners with a legal means of protecting their properties' conservation values while maintaining ownership, retaining certain uses of their land and earning significant tax benefits. A conservation easement is a voluntary legal agreement between a landowner and an easement holder by which the landowner imposes permanent restrictions on the way the property is used'.*[3] Such easements address continuity of heritage, and can be applied to entire parcels, requiring that their enumerated and documented character-defining features be stewarded to retain: spatial and visual landscape patterns; building fenestration and massing; access ways; topography; fencing; and small scale features, which contribute to the heritage character of the property. Another aspect of managing components of property rights is the purchase (PDR) and transfer (TDR) of development rights from and between properties. These PDR and TDR laws create frameworks that allow for the level of development on target properties to increase or decrease, often in support of community character preservation and conservation goals.[4] TDR can be used to increase the scale and density within a non-historic quarter while retaining the lower scale of an historic one. PDR is applied to support the purchase and removal of development rights from a property, reimbursing the owner for this loss of rights, so that heritage can remain intact while property value is not diminished. Both TDR and PDR are legal frameworks that employ financial incentives to control level of development.

From the early twentieth century, municipal plans and resulting zoning regulations have shaped urban form in many countries. At that time zoning of land for specific uses was an

---

[2] National Historic Preservation Act of 1966 as amended through 1992, Public Law 102-575, http://www.nps.gov/history/local-law/nhpa1966.htm.

[3] An example of conservation easements is found at the Georgia Land Conservation Program website, http://glcp.georgia.gov/conservation-easements.

[4] The American Farmland Trust offers a summary of Transfer of Development Rights at http://www.farmlandinfo.org/documents/27746/FS_TDR_1-01.pdf.

innovative response to urban overcrowding and degradation of the living environment. Through zoning private land uses are controlled by regulations that address the location of residential, commercial, and industrial uses, and individual lot public access, coverage, building bulk, overall density and other factors. Control over land uses predicts urban form that is shaped by the aggregate of private lot uses, as much as it is by the controls on public lands and rights-of-way. In the USA planning and zoning laws are often enabled at the state level while being shaped, detailed, and enforced at the local level. Planning is an advisory, non-regulatory process, until the recommended plan is approved when it becomes the guiding document for community intent and is codified in zoning mandates. In the USA many municipalities require a ten-year review and revision of their community plans, which is an iterative process. Effectiveness of prior plans, and recent growth or decline of an urban area are taken into account as the plan review cycle iterates.

The outcomes of planning and zoning can encourage or damage traditional and historic settlements. In some cities twentieth century zoning for discrete land uses has directed the segmentation of historic city mixed uses, resulting in sterile empty centre cities after the work day. While in others, mixed uses that blend commercial, residential, and light industrial activities, have been reinforced. City and town zoning has often failed to protect agriculture, scenic beauty and natural resources, encouraging fringe growth and sprawl over the landscape adjacent to historic cities, which compromises their setting. London is a good case study as the legislated green belt surrounding it remains intact and the city within the belt is increasing in density, although the character of the impact to heritage as densification increases is not clear. Zoning is rightfully criticised because if applied

to new settlements these regulations would fail to shape urban areas resembling the valued cities, towns and villages. Another issue with zoning codes is that they are socially and economically stratified and restrictive, and are unsuccessful in integrating diverse people into a community, forming like-type enclaves and generally ignoring need of lower income and poor citizenry. Some applications of planning and zoning regulations demonstrate positive outcomes, as these examples attest. There are also emerging concepts in form-based codes and green sustainable codes that offer promise.

One interesting example of an uncommon state land use law that prescribes land use concepts and details, not simply enables local law, is the 1970 Vermont Act 250.[5] This pioneering development control legislation affects the quality of land use development directing that it will not have adverse impact on scenic and aesthetics quality, natural and historic resources, and wildlife habitat, create an undue impact on water resources or public educational facilities, congest traffic or cause pollution.[6] Act 250 regulates direct development into existing settlements as a measure toward controlling sprawl that degrades agricultural lands and aquatic and forest landscapes. Further, the process requires the developer to meet this array of criteria with proof in order to obtain a permit. Vermont is a north-eastern state well known and highly regarded for its picturesque qualities and recreational tourism is an important income source. There are strong quality-of-life, jobs, and visitor experience incentives to retain the scenic and environmental qualities of Vermont's villages, towns, cities and countryside. The shortfall from Act 250 is that the laws only apply to major development, addressing about 40% of construction, not small-scale lot-by-lot projects that have a considerable cumulative effect. Sprawling development that compromises resources has

---

[5] Vermont Act 250, at http://www.anr.state.vt.us/dec/permit_hb/sheet47.pdf.
[6] Albers, J. (2000) *Hands on the Land: A History of the Vermont Landscape*. Cambridge, Massachusetts: MIT Press for the Orton Family Foundation: 316–317.

occurred quite broadly over the past four decades. Regardless, the 40 year record of managed change that Act 250 has been credited with, to a notable degree, exercised controls that have secured the settlement character. Vermont remains a state characterised by picturesque towns set in lush green valleys along streams, with surrounding hills and forested mountains. The result in Vermont of a state-wide level of development control is unique in much of the USA as development controlled by each municipality fails to consider the larger scale of the character of the land around and between each settlement.

An interesting example of the result of land use controls at the community scale is found in the historic village of Riverside, Illinois, begun in 1869 and shaped by the General Plan of Riverside designed by Frederick Law Olmsted Sr. and Calvert Vaux. In contrast to the usual grid rectangles of many settlements, the pattern of this historic town, as a result of the innovative plan, follows landforms in a curvilinear, naturalistic organisation around the Des Plaines River, and reserves central open spaces as public lands for community benefit. Village residents are passionate about their community and engage in a constant dialogue about the value and retention of its character. The opportunity to amend the National Historic Landmark documentation in 1999 provided a deeper analysis of community patterns.[7] In that project the streetscape was studied and the relationships of alignment of streets, walks, location of open spaces and configuration of lots were all distinctive characteristics of this settlement. Individual building lots vary in size and organisation, however an important characteristic of many properties is a variable frontage setback that reinforces the overall community curving pattern. The retention of these historically variable setbacks withstood a legal challenge, by an owner who wanted to add a front porch, as the clearly delineated setbacks

on a 1930s map have been used as a controlling tool to retain village spatial organisation into the twenty-first century. The combination of land use zoning, and local historic district legislation has to a substantial degree, retained Riverside character while managing change.

The protection of community parks, open spaces, natural resources, woodlands, and clean water supply, is governed by community traditions in some cultures and by legal frameworks in others. The public lands of cities are shaped to meet the needs of the populace applying both traditions and regulations. Legislating the protection and stewardship of public parks, parkways, boulevards, riverways, and other open spaces is a further regulatory tool that shapes the urban landscape. Parks are generally set aside in perpetuity by law, so that generation after generation can benefit from these public open spaces, often valued as designed urban landscapes that are a legacy of common heritage. Recent decades have seen a rise in support through non-governmental organisations in public-private partnerships. In addition to parks, streets, alleys, canals, streams, open spaces, etc., are public assets, most often stewarded by the local government. For example, the unique blue coloured stone paving on the streets of Viejo San Juan, Puerto Rico, is a character-defining feature of this 500 year old urban landscape. Stones were brought to the Americas as ship ballast and employed in street paving. The paving is preserved and repaired as a component of Old San Juan historic district regulations today.

As stated in the Historic Urban Landscape Recommendation,

Urban heritage, including its tangible and intangible components, constitutes a key resource in enhancing the liveability of urban areas and fosters economic development and social cohesion in a changing global environment. As the future of humanity hinges on effective planning and management of resources, conservation has become a strategy to

---

[7] O'Donnell, P. et al. (1999) *Riverside National Historic Landmark Amendment*, for the Village of Riverside.

achieve a balance between urban growth and quality of life on a sustainable basis.[8]

An example of planning and zoning regulations that have effectively preserved a historic urban ensemble is Annapolis, Maryland, where a colonial harbour area is the core of this Maryland capitol city. The Annapolis urban heritage core is a cohesive composition of narrow radial streets, three-storey buildings, limited street space with brick sidewalks, and narrow roads that define the unique character of this urban landscape. Private spaces within building lots are enclosed gardens with large trees, plantings, paths, ponds, sculptural objects, etc. The urban pattern of radial hubs with iconic structures, the Governor's Mansion and churches, shapes view lines and multi-sensory perceptions of this older town within a modern city. At the city scale, Annapolis, Maryland protects its historic core through zoning regulations that control building heights, facade forms and proportions, to retain unique streetscape scale and details.

Capturing iconic city identity is a current challenge. To plan for the shared visual urban character, the notion of 'defining the sky' was coined by landscape architect Hal Moggridge.[9] In his recent study of Edinburgh, Scotland, a series of viewsheds have been mapped to provide public visual access to iconic urban elements into the future. Urban viewscape controls define the sky as a framework for urban growth control. Applying this guidance to development in Edinburgh proposals are required to demonstrate compliance with the protection of identified iconic city view corridors and vistas, a visual commonwealth of Edinburgh dwellers, visitors, and future generations.

There are a number of proven and emerging planning and zoning directions being tested that demonstrate the potential to influence urban settlement character while retaining heritage. As one trend, harking back to traditional townscapes, New Urbanism emerged in the 1980s in the USA advocating settlement form that applies traditional land use patterns with walkable streets with community shops, services and housing, in a socio-economically mixed pattern.

Since the late 1990s there has been a trend toward community vision planning and scenario planning that becomes development and conservation regulation. The publication of a Connecticut River Valley regional land use scenario book in 1988[10] and the emergence of computer models to display images of development build out and city fly-through models, led to a widespread process of community and regional visioning plans that use development build out models for transportation, open space, residential/commercial/industrial land uses to evaluate, dialogue about, approve and implement urban and regional plans. These visioning processes have often raised the issue of harmonious new development, rather than radically different projects. Emerging regulations may be outcomes of these community dialogue processes, reflecting grassroots desire for continuity as growth proceeds. These emerging regulations trends are form based codes and sustainable development codes.

Form-based codes are structured to emphasise the physical form of new development that respond to the streetscape and adjacencies, rather than to the land use. Well-informed citizens are calling for development controls that contribute to retention of community character, as opposed to having contemporary development, insensitive to its context, alter and erode desired community heritage over time. The definition is that:

Form-based codes foster predictable built results and a high-quality public realm by using physical form

[8] UNESCO (2011) *Recommendation on the Historic Urban Landscape*: para. 3.

[9] Moggridge, Hal. (2009) *Edinburgh study*. London: Colvin & Moggridge.

[10] Yaro, R., Arendt, R., Dodson, H. and Brabec, E. (1988) *Dealing with Change in the Connecticut River Valley: A Design Manual for Conservation and Development*. Cambridge, Massachusetts: Lincoln Institute of Land Policy.

(rather than separation of uses) as the organising principle for the code. They are regulations, not mere guidelines, adopted into city or county law. Form-based codes offer a powerful alternative to conventional zoning.[11]

These codes provide a regulating plan and public space and building form standards that are highly prescriptive, and can be seen as overly limiting. The purpose of form-based codes is to reflect the traditional urban fabric and experience of place. This approach to integration of city heritage and new development is based on developing in harmony with the existing, evolved streetscape and building character, rather than land use. In fact this approach fosters mixed uses, and can support socioeconomic integration rather that isolation. Form-based codes are rooted in individual communities, place based and place derived, to retain character and therefore support the evolved heritage of communities.

Developing and legislating sustainable zoning codes and/or green construction laws seek to integrate energy efficient and carbon footprint reduction through development process guidance that reflects twenty first century resource limitations and climate change issues. Sustainable zoning and development codes begin with sustainable land-use plans and evolve into zoning codes, subdivision regulations, and construction codes that target green best practices. Global climate change response and a desire for greater community resilience is one driver of this trend. This approach to municipal regulations, formalising owner driven green development, to foster better use of resources in new development applying green programmes, like the United States Green Building Council (USGBC) programme, Leadership in Energy Efficient Design (LEED), and the Sustainable Sites Initiative (SITES) to address renewable energy sources, lower carbon footprint construction practices,

water management with greywater systems and rainwater harvesting and use, waste management with cleansing wetlands, and stormwater management with retention and infiltration on site.

How do green regulatory systems influence urban form? In addition to the temporary beneficial effects of construction disturbance limitation and improved waste management, there are innovations in building materials, and property adaptations for rainwater harvesting, stormwater infiltration, solar panel installation, and other building performance targets, which are changing built form and land uses in ways that are not traditional. Meeting sustainable targets will generally alter character in heritage areas. Model codes are being tested. To date these codes fail to meaningfully integrate heritage and city character. However, there is a hybrid 'sustainable form-based code' arising, that may be an effective tool to achieve carbon footprint objectives while retaining community character through focusing on current qualities of urban areas while incorporating green updates.[12]

Legal Regulations with Indirect Influence on Urban Heritage

Regulations that address varied topics, often without consideration of urban heritage resources, have an indirect impact on historic villages, towns, and cities. Indirect regulations that address related topics and influence urban character include:

- Requiring clean water and water use controls and regulating stormwater absorption;
- Enforcing public health, safety, access and sanitation laws and guidelines;
- Requiring lighting controls and 'dark sky' compliance for urban properties;
- Directing transportation mandates that influence urban form and character; and

---

[11] Form-Based Codes Institute, http://www.formbasedcodes.org/what-are-form-based-codes.
[12] American Planning Association, seminar on sustainable form-based codes notice, March 2013.

- Disaster preparedness, response and recovery to severe weather events and human conflicts.

Water is a shared resource foundational to all life on Earth. In the USA, the Clean Water Act requires that urban waste water that is discharged into adjacent water bodies be cleansed of city and settlement pollutants. Over recent decades this foundational law toward public health has directed the separation of intricate networks of combined sanitary sewer and rain water piping within cities. When high water flows run through combined pipes unhealthy discharges are deposited in urban water bodies. Large subsurface infrastructure upgrades are a project type that can adversely impact historic cities with significant disturbance. In dense urban areas, high stormwater flows into antiquated combined storm and sanitary sewer systems causing unacceptable contamination. Predominantly impervious surfaces of urban areas gather and flow rain runoff quickly into catch basins and drain pipes, with little pervious ground or roof to absorb water. These overflows are a huge problem causing failure of waterways to meet standards for healthy fishing, swimming, and wildlife. Legal mandates to clean water and improve overall water quality force actions.

The traditional approach to water quality problems has been to build new infrastructure, at high cost, but alternatives and innovations are on the rise. While many cities historically used cisterns to capture rainwater for reuse, few rainwater harvesting features are in use today. That traditional technology is being applied to the renewal of the USA National Mall, where harvested rainwater will be gathered, cleansed, and used to irrigate the turf panels while the historic appearance and soil and turf health is renewed.

There are some new techniques being applied to the public and private landscapes of cities that aid in resolving these high stormwater flows and resulting water pollution. In Chicago, Ann Arbor, Washington DC, Portland and many others, on site stormwater capture, reuse and/or infiltration are on the rise. Home owners are building modest rain gardens and bioswales on their lots and along neighbourhood streets to hold and infiltrate stormwater. Public parks are being used to gather stormwater and use it for public water features. A current example of public-private infrastructure project, mandated by clean water legislation, the stormwater initiative cited above was pursued because of legislated water quality mandates, but the financial aspect is another important facet. New York State supports the New York City green infrastructure initiative that *will eliminate 1.5 billion gallons of sewer overflow annually by 2030, while 12 billion gallons will be kept out of New York's waterways through combined green and grey infrastructure systems, saving the state and city loads of money in the process*.[13] The projected cost of each new bioswales is $13,000 with more than 100 constructed in 2012 applying a standard design. Green infrastructure funding for New York City is $2.4 billion in public and private investment towards implementation over the next 18 years. This project changes urban form, while meeting a clean water law mandate, and has an indirect influence on urban heritage.

A green infrastructure of tree lawn bioswales along streets, green roofs, and other stormwater management techniques is becoming widespread. These methods apply contemporary best practices to take pressure off the antiquated undersized, subsurface stormwater system. Does this intervention change the historic urban landscape? It does. However, researching historic city streets indicates that in many urban neighbourhoods continuous tree lawns or verges, between sidewalk and street, were the historic norm. This application establishes more continuous green

---

[13] The New York Times Green Blog, 4 April 2012.

edges, not only improving water quality, but also enhances the urban environment aligned to the historic patterns.

Enforcing public safety, and access laws and guidelines can require changes to features of the urban landscape. Safety codes require sidewalks with clear passage widths that control signage and lighting location and height, which can be contrary to local traditions. Grand historic buildings with ceremonial steps are generally unused as they block the way for people with disabilities to access. This change in building use often means that visitors are routed along a secondary course through the landscape and into the public building, missing the historic grand entry experience. Federal legislation access codes also control the level of pitch, specifying gradients of not less than 2% cross slope and 5% continuous grade. While applying these gradients to public sidewalks, park and open space paths improves accessibility, it can also alter historic grades and materials. Further, disabled access codes require very slight variations (less than 1/4 inch) in walking surface level. Historic stone and brick pavements with greater surface variation can be replaced with more level but not original materials.

A related aspect is urban growth that increases light pollution as it gains intensity night lighting, which also addresses public safety for night-time movements. Night lighting can be traditional and character defining. Recent interest in energy efficient lighting and in lighting that does not illuminate the night sky is changing the directional qualities of lighting to support urban stargazing. Dark Sky compliance is the movement guiding a drop in urban light levels and some municipalities have added these targets to lighting standards and zoning compliance.

Transportation mandates and practices that influence urban form and character are most often directed to vehicular traffic and seek improved traffic flow and air quality. Projects target performance improvements, not retention of the adjacent urban heritage directing toward street widening and traffic control coordination. These initiatives can be damaging unless the values of adjacent heritage are part of the planning and implementation process. A burgeoning global interest in multi-modal transportation, and approaches like designated bicycle space on road, or trails off-road, shared streets, road diet, and complete streets are all on the rise. Projects addressing transportation equity, such as the Transmilenio and innovative bus rapid transit system and for those who cannot afford the bus a wide bicycle commuting highways in Bogota, Colombia, are also surfacing.[14]

In a recent project in Fort Wayne, Indiana, for Ardmore Avenue a designated commercial trucking corridor, the passage of construction raw materials through the city from source points to development areas where materials were in use traversed a historic neighbourhood.[15] The project effort looked at the history and significance of the Wildwood enclave to the east and a historic Frank Lloyd Wright residence to the west and proceeded to develop plans to mitigate construction traffic noise and vibration. Design and construction integrated planted earth berms, cast stone walls and vertical sound barriers along both sides of the commercial truck corridor to protect the adjacent residential community to both sides. The character of the corridor was wholly altered, while the historic neighbourhoods remained intact with one changed boundary.

Another indirect area influencing cities, towns and villages is disaster preparedness, response and recovery to severe weather events and human conflicts. The legal mandates that govern human conflicts, directing that in war cultural

---

[14] ASLA The Dirt, The Rise of the Endless City, 2/22/2013, by J. Green, http://dirt.asla.org/2013/02/22/the-endless-city-doesnt-have-to-be-a-bad-thing/

[15] City of Fort Wayne, Department of Public Works, Ardmore Avenue, construction and neighbourhood sound barriers, completed July 2010, Heritage Landscapes, Projects Landscape Architect, Crossroads Engineering, Project Engineer.

heritage is protected, occur at the international level with the Hague Convention on armed conflict and the protocols of the Blue Shield.[16] We know that at the local and national level, war often destroys or degrades cultural heritage, as we are currently seeing in Aleppo, a World Heritage city in Syria. However, as severe weather events are increasing in frequency around the world, heritage protection, stabilisation, and restoration to recover from damage are not included in disaster planning. In fact, the heritage of public open spaces is particularly vulnerable as parks, riverfronts and open lands are often pressed into service for dumping rubble, providing temporary housing or other needed but damaging uses. Local disaster response plans regulate how to save human lives, and protect private property with little, if any, attention to places of heritage value. This is another area of regulatory systems that requires consideration of heritage resources to be more effective.

This group of indirect regulations effects changes in urban form and content. Individually these mandates may not appear to be significant, however the cumulative effects of their application, and that of other indirect mandates, can have a notable influence on urban heritage. Particularly damaging results can occur when heritage character and features of the city are overlooked in the application of these important laws, protocols, regulations and traditions.

Conclusion

The global debate on the management of historic urban areas was completed with the adoption of the 2011 Historic Urban Landscape Recommendation. The litmus test of its success will be the capacity to adapt and apply the recommendations by bringing tools to that work. Regulatory systems, in all forms, can be applied to manage change effectively and reduce the heritage versus development struggle. Effectiveness in application will depend in part on the capability of directing development in full recognition of urban heritage and its importance for quality-of-life and the economic vitality of the future.

The Historic Urban Landscape process fosters wide efforts to *participate in developing and disseminating tools and best practices for the implementation of the Historic Urban Landscape approach*.[17] The measure of our success in applying the Historic Urban Landscape will be to the degree to which we can clarify objectives for urban heritage to activate the enduring values of urban conservation within the matrix of needs and desires for urban vitality.

---

[16] Convention for the Protection of Cultural Heritage in the Event of Armed Conflict, 1954 (The Hague Convention).
[17] UNESCO (2011) *Recommendation on the Historic Urban Landscape*: para. 22d.

Interview with Rahul Mehrotra
Constructing Cultural Significance

Professor of Urban Design and Planning, and Chair of the Department of Urban Planning, Harvard University, Cambridge, MA, USA

5 March 2013, Cambridge, MA, USA

Question: *There is a growing interest among architects and planners for the definition of new approaches to urban conservation, in consideration of the great diversity of situations and approaches that can be found in different parts of the world. As you are coming from India, how do you see this issue?*

Answer: I see urban conservation primarily as a planning instrument and I think that the detachment between urban planning and conservation has been detrimental for conservation and its evolution. In different regions of the world Conservation and Historic Preservation have evolved as a response to very different types of problems and urban conditions, such as the discussion prompted by Urban Renewal in America, or the quest for the reconstruction of memory in Europe after World War Two.

Conservation in rapidly developing urban contexts is essentially an instrument to modulate the rate of change. Correspondingly, societies respond in positive or negative ways to conservation according to the way they want to deal with (or not deal with) change. Conservation has developed first of all in Europe because societies wanted to slow down the pace of change and gave a great value to memory and to their historic traditions and image. This is why I say that conservation in these contexts is a planning instrument. It helped constructively reflect societies aspirations in terms of modulating the speed of change.

China, on the contrary, is an example where a country or society wants to accelerate change, and it is also a context where memory can be erased without major public reaction because of the political system. In places like India conservation has a completely different cultural meaning for society, mostly due to the religious traditions in India. Hindus believe in the cyclical nature of time, where everything is in constant change, and has a beginning and end. The idea of rebirth is a deeply set belief – everything is reborn. This enormously influences the attitude to conservation, as compared to societies where time is seen as a linear process, where the 'speed' of time modulates how fast things change.

John Brinckerhoff Jackson, one of the great landscape thinkers of the twentieth century, said very lucidly that ancient cultures, like Hinduism (and I extend this to the aboriginal people in Australia, where the elements are worshipped and there is an obsession with the original moment of creation often depicted in the stories of the dream time), give great importance to the original moment of creation, as proven by the many pictorial representations of the origin of the cosmos. On the contrary, more recent religions such as Judaism, Christianity and Islam tend to celebrate the moment of the covenant between man and the Almighty. The circumstances of the covenant are celebrated and codified in different forms (the Tables of the Law, the Crucifix, etc.). This difference influences the way conservation is conceived. For example in Mumbai, all the beautiful temples are demolished and rebuilt by the new generations, as the act of rebuilding or renewing of every generation, reaffirming its faith in the Almighty. In this case, the artifact is not as important as the abstract thinking or the symbol (the Phallus symbol in the form of the *Linga* for example) that codifies the meaning of creation.

I see this question as a fundamental one: conservation, at least in the form its institutions and canons have taken through time, has prompted a 'global' approach that does not factor in adequately the local context. There are some exceptions of course, like the Nara Document or the Burra Charter that try to deal with this issue. But we certainly need a 'bottom-up' approach that would allow us to root things in their context and not the other way around. The differences I referred to, linked to the religious approaches and the notion of time, are all part of the cultural nuances that need to be brought into the debate on conservation.

Question: *In the Historic Urban Landscape Recommendation, the issues you referred to have always been at the forefront. Do you think that the idea of the Historic Urban Landscape could stimulate the development of concepts in other cultures?*

Answer: Yes, I think so and I will refer to the case of Mumbai. I started working in Mumbai in urban conservation by accident, essentially because I had a great interest for the history of the city. I started working with conservation professionals that were for the most part trained in English school, and that had a strong sense of the physical fabric of the historic city and were obsessed by the idea of discerning and articulating the cultural significance of a place and then reinforcing it through preservation. But I was not convinced and my impulse was that cultural significance is not a static phenomenon as it changes and evolves. I felt we had to construct a new narrative on the historic city to make its contemporary relevance felt by the people, and connected to the planning process.

Through many experiences I realised that there was a fundamental difference between the situation in Europe or in America and the one in a country in a postcolonial condition: the custodians of the environment – in this case the Indians – and the creators of the environment – the Europeans – belong to two different cultures. So, the narrative that works in London, where the creators and custodians of the environment belong to the same culture cannot be used in a place where there is a disjuncture between the two cultures. The nostalgic narrative that is at the basis of European conservation is useless because it does not make sense to the local people.

We started working with a bottom-up approach, trying to talk to the people living in the historic districts to understand which were their experiences and their perceptions and more importantly their aspirations. In the Historic Fort area we created a series of associations within the larger district, to see if they could become the engine to drive a contemporary conservation process.

So we started constructing a new significance in the old areas, something that I think gives a great responsibility to us as architects, as we have to make sure that the appearance or the illusion of the historic architecture is kept intact while inventing new uses. In order to change the narrative we began creating new meanings in these areas, and we even changed their name, calling them 'associations' rather than 'historic districts'. In short, giving them personalities! So the famous Mumbai Gothic District was turned into an art district and this became the contemporary engine that drove the conservation process. We started organising art events that raised the funds to preserve the historic buildings: in short we began creating a new significance for the area. In Mumbai we obtained the first urban conservation legislation in the country in 1995. This did not mean that automatically things would be preserved. In fact, for the first few years nothing really happened on the ground. But I believe once we started constructing these narratives new imaginations were fuelled and we were lucky that we had enough concerned professionals in the city that could steer this delicate process on the ground. I believe if we had taken a top-down approach nothing would have happened!

Question: *An important aspect of conservation is the management process that you put in place. How do you see this process developing in the forms and at the scale required?*

Answer: The answer is not easy; we have to look at each case to find ways that are acceptable to the communities and the governments. Let's look for example at another case, the Taj Mahal. For five years we worked on a Master Plan for the Taj Mahal, we put a team together with many international experts and institutions. When we were preparing the Master Plan, the first thing we did was to understand what was the constituency we were responding to. And of course for the Taj Mahal this is a complex question! We realised very quickly that most policies concerning the Taj were aimed squarely at what we called the 'aesthetic' tourist, who is generally western. But this is a very small percentage of the people visiting this World Heritage site. It is interesting to see that the largest numbers of visitors to the Taj Mahal are pilgrims of many faiths on their way to or from Mathura (a famous Hindu site near Agra). The respect through gestures of reverence they offer at the grave of Shah Jahan or Mumtaz Mahal is akin to that to a saint or seer. The incredible range of visitors that pay homage to the monument is reflective of the aspirations of a secular and evolved nation with a deep respect for its heritage. This is in a way more a form of pilgrimage than tourism, and they are the absolute majority of the visitors.

We realised that this was a sacred space, much beyond the western understanding of significance through the narrow reading of the historicity of the site or even its Muslim appearance. This pushed us to think that what we had to represent in the Master Plan were all the multiple narratives that make the Taj a mind-blowing experience but also a complex object that means so much to different people.

Our response was to use these different lenses to discern the narratives to bring to the surface. We started researching and documenting extensively. For example, we looked at the signature of the masons. We thought this would be an incredible story to tell to visitors – the stories about the people who actually made the monument. Similarly we started defining the narratives about the original gardens. The site was originally an orchard – it was the representation of the garden of paradise. The lawns you see now were a creation of the British Governor Lord Curzon. But of course these were difficult issues to get past the Government's officials, because they began to question the very authenticity of representation of the site.

We thought we had the support of the greatest experts: Sir Bernard Feilden, the Getty Conservation Institute, World Monuments Fund, and other scholars, but the Government had no will to take on these complex questions – they were comfortable in the existing packaging of the Taj Mahal for the western tourist. So this is a case where our only constituency was the Government, which had no will to change anything, and so in short we had no patronage. But I am confident that the next generation will have the confidence to convince the Government to do what we have been unable to do. And hopefully some Government down the line will also have the confidence to take on these issues. I believe it will make the Taj Mahal a richer experience and its future will be more secure as it will be embedded in the imagination of the ordinary and local people.

Question: *India still has a vast and intact urban heritage, at a scale that bears no comparison with any other parts of the world, but it is very fragile, exposed and not protected. There is a serious risk that it may disappear under the push of modernity and globalisation, like it happened in China. What can be done?*

Answer: To make urban conservation possible, you need three conditions: 1) a supportive legislation; 2) an effective patronage – sometimes this is provided by the government and sometimes by civil society; and, 3) a robust group of individuals that can advocate for it, with the capacity to understand processes and build significance and the capacity to engage forces greater then themselves – international bodies, funding agencies, etc. Often in India as also in other parts of the world one or two of these components exist but all three don't align. The successful examples are the ones when this 'holy trinity' is aligned. In 1985 when I was still a student I went to Hyderabad to meet S.P. Shorey who had developed a programme for the conservation of the historic city with a Ford Foundation grant. This was a very classic plan that proposed to list all the buildings as a plan for urban conservation. I was very influenced by this encounter and it really inspired us in Mumbai.

When I became part of the movement in Mumbai I realised that what they were doing in Mumbai was myopic as they were only listing buildings. I tried to shift the debate to urban conservation as I felt that this is how we would embed the discussion in the urban planning debate. I learnt this from the experience of Shorey in Hyderbad. In 1995 we managed in Mumbai to get this legislation through and three or four years later the urban conservation and architectural projects proliferated all over the city. In the meanwhile, Hyderabad took many more years for this to happen even though the professionals there like Shorey had all the paper work done. But patronage did not take a lead there. In Mumbai the holy trinity aligned!

12 Devising Financial Tools for Urban Conservation

Donovan Rypkema

President of Heritage Strategies International, Washington DC, USA

In the coming years so much will be protected, but it will be increasingly difficult to manage it. Hence the decision of what remains protected and how, will partly become an economic decision. I just hope that conservation economists will then be ready.

Raymond Lemaire

Introduction

In the not-too-distant past 'heritage conservation' referred primarily to individual, majestic landmarks – cathedrals, palaces, large military fortresses, monuments to kings, memorial structures, and major libraries, museums and university buildings. These structures had more in common than their grandeur, aesthetics, or symbolic importance. First, they were usually owned by institutions or by some level of government. Second, many of them had a degree of public access.

But as the heritage conservation movement has grown and the definition of heritage broadened, added to the list of historic structures worthy of preservation have been more modest buildings, vernacular architecture, and often entire neighbourhoods. In the latter case the individual components of the neighbourhood might not make a 'landmarks' list, but as an ensemble they add to the distinctiveness of the

place and tell an important story about a city, its history and its evolution. The adoption by UNESCO of the *Recommendation on the Historic Urban Landscape* is a major step in the international recognition that not just the major landmarks, but their urban context, need to become a central feature of a comprehensive heritage conservation strategy.

These 'lesser landmarks' differ, however, from their major monument cousins in more than just size and grandeur. These buildings are often in private ownership and the access, at least to the interiors, is frequently limited to the individual owner and specifically invited guests or clients.

This broadening of the definition of heritage has also forced the recognition of an important fiscal reality: not even the most prosperous governments in the world have available (or the political will to provide) the financial resources necessary to restore and maintain all of the heritage buildings worthy of conservation. If the public coffers are not going to provide all of the funds necessary so that heritage buildings survive for future generations, where will that money come from? It will have to come from the private sector, from owners, and, to a lesser degree, from institutions. The corollary is that financial tools will be necessary to attract capital from those non-public sources. Consequently, this chapter will try to answer four questions:

1.   Why are financial tools required?
2.   What do financial tools do?
3.   What are the characteristics of the most effective financial tools?
4.   What are some examples of financial tools and how do they work?

Heritage Financial Tools are defined here as: mechanisms and programmes used to encourage or facilitate the investment of capital into heritage assets. Are all financial tools *incentives*? Are all incentives *financial tools*? The answer to both these questions is: most, but not all. For example, the use of a Public-Private Partnership is a financial tool, but may not have the characteristics of an incentive. Conversely, a direct grant to restore a heritage building is certainly an incentive but might not be considered a financial tool. But while the two concepts are not precisely synonymous, there is sufficient overlap in their meanings, and particularly their purposes, that a distinction will not generally be made in the pages that follow.

Why are Financial Tools Required?

To understand the need for financial tools, it is first necessary to understand the concepts of *cost* and *value* and further to recognise the multiple values that heritage buildings possess (see Box 12.1).

In real estate economics, *value* refers to the monetary amount that a property is likely to command if sold in the marketplace, or the capitalised stream of income that the property generates in net rents. The primary beneficiary of that value is the property's owner. It is also the *value* that a banker considers when the property is

provided as security for a real estate loan. Although frequently (and erroneously) used as a synonym, *cost* is a distinctly separate concept. *Cost* is the sum of the euros (or dollars or pounds or yen) that need to be spent to prepare a property to be sold, rented or occupied. *Cost* includes the price of acquisition, plus construction costs, fees, professional services, and other expenditures necessary to bring a property to the market.

To expect capital to invest on an on-going basis in a property when there is an economic gap is to ask them to act irrationally and against their own best financial interest. Ultimately capital will simply look for alternative investments where the amount of *cost* of the investment is at least equal to the *value* of the investment. Around the world when heritage buildings are left to deteriorate and sometimes abandoned, it is a clear statement that the current owners have reached the conclusion that the *value* restored and well maintained is less that the *cost* of restoring it.

But heritage conservation is driven by the fact that historic buildings have 'values' above and beyond their economic value – symbolic value, social value, environmental value, educational value, cultural value, aesthetic value, and others. However, the building owner is not the primary beneficiary of those values – the larger community is. Sometimes the 'community' means others in the same city; sometimes the benefitting community is the world's population. Think of the pyramids. The beneficiaries of the values of the pyramids are not the citizens of Giza or even the population of Egypt, but all of civilisation. In fact, it is even more than that. Generations yet unborn will ultimately benefit by the existence of the pyramids.

Considered in this fashion, the logic of creating financial tools for heritage becomes clearer:

1.   For multiple reasons the *cost* of restoring and maintaining a heritage building may exceed the *economic value* of doing so.
2.   However, there are additional *values* to the heritage of which the owner is only a negligible recipient.

Box 12.1

In market and quasi-market economies, when *value* exceeds *cost*, capital will usually act without intervention (that is incentives or other stimulus). But when *cost* exceeds *value*, capital (at least private capital) will either not act or act only with some incentive to do so. When *cost* exceeds *value*, the difference is referred to as the *gap*.

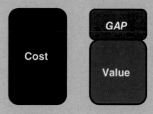

A primary purpose of financial tools, especially of the incentive variety, is to close the *gap* between *cost* and *value*.

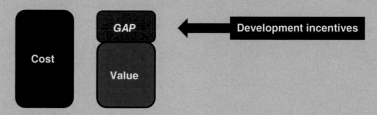

The point is this – there are values of heritage in addition to the economic value that accrues to the building owner. These multiple values might be represented as shown here:

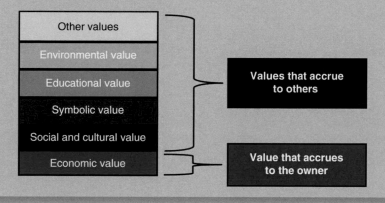

3. Therefore financial tools may be necessary to close the *gap* between cost and value in recognition that there will be *values* received by the larger community.

Of course for some properties in some markets there is no gap for the heritage building; the owner is receiving sufficient benefit from ownership that the cost of rehabilitation and mainte-

nance is fully retrieved through net income, sales price, and/or the value of occupancy. Is there ever a need for financial tools in that instance? Yes, in at least four circumstances:

• When a higher quality of restoration and greater sustainability of the heritage building can be obtained as part of the requirements for using the financial tool than would otherwise likely be the case.

- When the property is one where interior public access is important but would not be typically available. Again the specifics of the financial tool might require regular or periodic public access in exchange for the benefits the tool provides to the owners.
- When the zoning district within which the heritage building exists would allow a much larger building. It is not uncommon, particularly in city centres, for 20, or 30, or 40 storey buildings to be permitted. In those cases there is often great pressure to demolish a two or three storey heritage building. The use of a financial tool such as the Transferable Development Rights might be appropriate even though the current heritage building does not have a demonstrable financial *gap*.
- When the development of a particular building is seen as a catalyst for additional nearby development.

What Do Financial Tools Do?

Worldwide there are literally hundreds of financial tools developed to encourage investment in heritage. But there are actually only a finite number of impacts tools might have:

- **Reduce costs**: In some countries the Value Added Tax (VAT, elsewhere called sales tax) is waived for materials purchased for a heritage building restoration. This reduces the total cost of the project.
- **Reduce cash required**: A grant awarded for heritage building restoration would be an example of reducing the cash required. The labour and materials still cost the same, but the amount of cash the owner of the property has to provide through his/her own funds is reduced.
- **Increase income**: For commercial properties, the economic value is directly related to the amount of net income the property produces. An incentive that subsidises the rents for small business willing to locate in a herit-

age building could well increase income and thereby increase value. Increasing value or reducing cost will both have the effect of reducing the *gap*.

- **Reduce expenses**: Some countries levy lower property tax rates on heritage buildings. Since property taxes are an operating cost for a property, reducing that expense will have the effect of increasing net income and thereby increasing property value.
- **Improve financing**: In most of the world the feasibility of real estate investment (including in heritage buildings) is contingent upon the availability of borrowed money and the amount, interest rate, and loan term (payback period) of those funds. Low interest loans, favourable subordinate financing and loan guarantees are all examples of financial tools that may direct capital to heritage buildings.
- **Reduce risk**: In the investment world there is a relationship between risk and reward – the greater the risk, the higher return that investors will demand. It is the risk/reward relationship that results, for example, in bonds issued by the government of a country in financial difficulties commanding higher interest rates as compared to a more financially healthy nation. Some financial tools can be used to reduce risk on the investment in heritage buildings. A long-term lease by a creditworthy government agency for space in the heritage building could significantly reduce the risk. This reduced risk would mean the project would be looked on more favourably by both equity (ownership investment) and debt (borrowed funds) meaning that raising investment capital and securing a loan would both be easier.
- **Increase occupancy levels**: Having space ready and available for tenants (or buyers) but having inadequate demand in the marketplace is one of the fundamental risks in the ownership of real estate, including heritage buildings. Some universities have established special programmes of down payment assistance or forgivable loans if their faculty

will buy in a targeted historic area, usually near the campus. Over time this will increase demand for properties in that neighbourhood, reduce vacancy and, thereby, increase occupancy.

- **Clear title**: In many parts of the developing world there is not yet an adequate system of records and chain of title processes so that individual ownership of a particular parcel of land can be indisputably demonstrated. A process of 'clearing title' should be considered a financial tool, in that without clear title it may be difficult to secure either debt or equity to invest in the heritage building. Some governments have resolved this issue by directly acquiring the properties, clearing title, and re-conveying to the individual owners.

- **Improve investment environment**: Over the last two decades, the Inter-American Development Bank (IDB) has funded numerous heritage-based city centre revitalisation efforts in Latin America. Often one of the early uses of IDB funds is to improve both the infrastructure and the publically owned heritage buildings in the city centre. These public investments actually serve as a financial tool by improving the overall investment context within which an individual heritage building is located. There is the old cliché that the three most important things in real estate are location, location, location. When that location is improved through public investment, the feasibility of private investment is improved.

What are the Characteristics of the Most Effective Financial Tools?

There has now been enough experimentation and subsequent implementation of heritage financial tools that general characteristics can be identified as to when those tools are most effective:

- **Directed to a particular need**: Instead of thinking of financial tools as simply a set of programmes to encourage investment in heritage, tools should be developed that are directed to a particular need. On a modest scale, for example, if water damage through interior leaking is a major issue, then a micro loan for roof repair might be a useful tool.

- **Directly related to the gap**: When a financial tool takes the form of an incentive, its primary purpose is to close the gap between cost and value as discussed earlier. The incentive, then, should be directly related to the gap both in the amount of the incentive and the target of the incentive. If, for example, the gap is created by a requirement to add a sprinkler system for fire protection, then the incentive should be in the amount of the gap that particular expenditure generates and the funds specifically used for the fire protection investment.

- **Paired with appropriate regulations**: The user of a heritage financial tool is most commonly the owner of the building. But the resources to provide that tool (whether changes in public policies, incentives, specific programmes, etc.) most often come from the public sector. Therefore the public as well is entitled to a 'return' on their investment in the heritage building. That return might not be in financial payment, but rather in compliance with regulations or restrictions geared toward protecting the heritage building into the future. It is through this process that the quality and durability of the heritage conservation work can be assured.

- **Clearly communicated and actively marketed**: It is insufficient for a government simply to enact or authorise a given financial tool and then sit and wait for prospective users to take advantage of it. There needs to be an aggressive and effective programme to communicate and market the use of the tool. This can be done both directly to owners of specific heritage buildings, but also to groups who may be in a position to influence property owners: bankers, the real estate community, accountants, economic development professionals, city centre managers, heritage advocacy groups and others.

- **Simple in implementation**: This is one example of a principle where there may by necessity, be a distinction between a financial tool that is an *incentive* and one that is a transactional or ownership vehicle. An incentive like a property tax abatement, a low interest loan, or design assistance should be easy for a potential user to understand and relatively easy to utilise. For a transactional tool like a Public-Private Partnership, however, there will often be a degree of complexity that may preclude meeting the 'simple to implement' principle.
- **Within an overall public policy objective:** Financial tools should not be ad hoc or designed for a particular project. Rather the tools should be seen as part of an implementation strategy for an overall public policy objective. For example, a city council might establish as a policy objective, '*We want a heritage-based city centre revitalisation programme*' or '*We want to encourage the development of affordable housing in historic buildings*' and then specific financial tools are created, authorised, and funded to meet these public policy objectives.
- **Depoliticised**: Both the goals of heritage in general and the credibility of a specific financial tool in particular will be diminished if there is a sense that the use of the tool is limited to those with influence with government workers or elected officials. The criteria for the use of the tool should be established in a transparent and comprehensible fashion, then all those projects and/or persons who meet these criteria should be eligible.

Financial tools that are incentives, do not generally work well in countries where there is a low level of tax law compliance and a weak system of tax collections. If a property owner can reduce taxes by simply not paying them and face little risk of serious consequences, the choice might be made not to apply for incentives so as not to come under the scrutiny of public officials.

What are Some Examples of Financial Tools and How Do They Work?

There are hundreds of financial tools that have been developed in recent years to encourage investment in heritage buildings. Which tools can be used in which countries varies based on local enabling legislation, the political and economic systems, the priority of heritage as a public policy goal, and other variables. So, it is beyond the scope of this chapter to describe in detail dozens of financial tools, rather, six are discussed below which represent a range of tools that some countries have found effective in promoting heritage investment:

1. **Grants**: An award, usually in cash, to a project generally for a specific purpose and subject to certain requirements. Grants are usually made by a government entity, but sometimes by institutions or non-governmental organisations. A typical heritage grant might be for the restoration of a building's façade but with the stipulation that the work be done by approved contractors and in a manner consistent with good heritage conservation practices.

2. **Micro loans**: The difference between a grant and a loan is that the loan typically has to be repaid. *Micro loans* are usually made when the borrower might not qualify for a loan from a bank or other traditional lending institution because of no credit history, no equity capital available and/or because the size of the proposed project is simply too small to be of interest to a bank. Heritage micro loans might be made, for example, to the owners of small vernacular cottages to make roof repairs using materials and methods that are appropriate for that age and style of building.

3. **Property tax reductions**: In many parts of the world taxes are levied on real estate as one source of funding for the operation of government, usually local government. Generally (although not always) these are *ad valorem* taxes that are the amount of tax

increases based on the value of the property being taxed. This sometimes has the unintended consequence of discouraging investment in heritage buildings. An owner, for example, has a building worth €100,000 and is considering investing another €100,000 in its restoration. But the immediate result of that investment is a doubling of the property taxes. Heritage *property tax reduction* tools might address this by freezing the property taxes at the pre-restoration level for a finite time (seven years, for example) or by refunding the increased property taxes to the owner for further investment in maintenance and restoration.

4. **Tax credits:** This is a direct offset of taxes that would otherwise be payable. Tax credits usually are used in relation to income taxes, but could also be used for property taxes, VAT, or other forms of taxation. In the United States, for example, there is a twenty percent historic building tax credit against income taxes payable to the national government. If the owner spends $100,000 on the rehabilitation of a heritage building, and that work is certified at being consistent with good heritage practices, that owner would get a reduction of $20,000 to the amount that would otherwise be due as income tax.

5. **Transferable development rights (TDRs):** This is a regulatory tool that allows governments to authorise the transfer of allowable density from one building or site to another. When a three storey heritage building is located in an area that otherwise would permit 10 storey buildings, the owner is essentially left with seven floors of potential development, which cannot be used because of heritage restrictions. When a TDR is one of the available financial tools, that owner could sell that unused development right to another person who would then be permitted to add that amount of additional density to a building elsewhere, beyond what would otherwise be permitted. In most TDR legislation there will be identified a 'sending zone', often a heritage precinct, from which these *development rights* can be sold and a 'receiving zone', the area where the additional density will be transferred.

6. **Public-private partnerships (PPPs):** These are increasingly used around the world, particularly for major infrastructure projects like airports, highways, water systems, and power generation plants. But there is also a great opportunity for Heritage PPPs. This financial tool might be particularly useful for heritage buildings when the following are true:
   • There is a public need or a public benefit to be accrued through the arrangement.
   • There is a need for private investment capital.
   • There is a desire to leverage scarce public funds.
   • There is an interest on the part of both the public and private sectors to share risks.
   • The public sector lacks the development and/or management expertise to undertake the proposed project.
   • There is a desire on the part of the public sector to enhance the value of an asset owned by it.
   • The public sector wants to tap into the innovation of the private sector.
   • There is a need for ongoing public influence as to what happens to the facility.
   • There is a desire for a reversionary interest (an ultimate return of the facility to public control) at some time in the future.

Very commonly many or all of the above are true of heritage buildings, particularly those currently owned by some level of government. This does not mean that a PPP is always the answer or that using a PPP is either fast or easy. A government using this tool needs to have expertise in the use and operations of a PPP, however PPPs are currently an underutilised financial tool for heritage and more governments should consider their appropriateness.

Conclusion

Over the last three decades, both the successes and the failures of heritage conservation have taught us that a broader, more comprehensive approach must be taken. The Historic Urban Landscape approach is an important and positive shift in how cities need to think about and manage their heritage resources.

What are the likely outcomes as the Historic Urban Landscape approach is adopted at national levels and implemented into local implementation strategies? There are at least four:

1. It is not risk but uncertainty that is the deterrent to much real estate investment. As heritage conservation is integrated into a larger urban development context, the level of uncertainty as to permits, regulations, and the ability to proceed is reduced and investor confidence is increased.
2. This reduced uncertainty will have an even greater impact among those who currently own or are planning reinvestment in heritage buildings. The long term investment horizon that is often necessary for heritage buildings is enhanced with the confidence that the context, the view sheds, the streetscapes and other locational variables will not be rapidly changed in a manner that significantly and adversely affects the individual heritage building.
3. When uncertainty is reduced for equity investors there is a corresponding reduction in uncertainty for lenders. Banks and other financial institutions can be expected to respond positively about funding availability and the rates and terms those loans will command.
4. The speculative premium that is often generated in situations of rapid change and

uncertainty can be expected to diminish. It is this speculative premium – which attaches to land, not buildings – that is often the motivation for the demolition of heritage properties. While real estate speculators may not be happy with that consequence, long term owners, investors and proponents of stable economic growth will all be beneficiaries.

The adoption of the Historic Urban Landscape approach locally might be the ideal chance to simultaneously adopt transferable development rights legislation. This could well be a generational opportunity to shift currently unutilised development rights away from areas where the height and context are to be protected to areas where greater density might make more sense. Some imaginative cities will, no doubt, invent and enact additional carrot/stick legislation to achieve positive outcomes both in encouraging growth and new development in parts of the city while maintaining the quality and character of heritage areas. Some developers and their architects will find a niche market in new buildings designed to be compatible with, and sensitive to, a city's heritage and profit from the comprehensive Historic Urban Landscape approach.

To expect either the public sector to fund all of the heritage conservation that is required, or to expect private owners to restore and adequately maintain heritage buildings without useful financial tools is unrealistic. Central to the implementation of the Historic Urban Landscape approach will be the availability and effectiveness of the financial tools available. The bad news is that there are insufficient and inadequate tools currently available in many historic cities. The good news is that we have learned what works, how they work, and how they can be effective.

Case Study *A User's Guide for Heritage Economics*
Christian Ost

Economist and former Dean of the ICHEC Brussels Management School

The Historic Urban Landscape and the Macroeconomic Perspective for Conservation

Since the adoption of the UNESCO Convention for the Protection of Cultural and Natural Heritage in 1972, heritage conservation has been successful on a global scale. Yet, we realise today that many sites are still in danger, and that conflicts often arise between preserving the past and promoting future economic development.

Heritage conservation and economic growth have always followed similar paths, as the post-World War Two reconstruction era provided a strong incentive to preserve existing heritage from further deterioration. But it took some time for an international consensus to emerge, in what was mainly a western-centred initiative.

The analysis of economic long waves (initiated among others by Kondratiev and Schumpeter during the first part of the twentieth century) suggests alternating phases of strong and weak economic growth, and exemplifies how the 1970s were a turning point for economic historians. Indeed, the 1972 *World Heritage Convention* was adopted at the start of a declining growth in the Western world, concern for sustainable development, the rise of globalisation, and the emergence of free-market dominance over public economic initiatives.

Forty years later, the protection of cultural heritage has earned global recognition and legitimacy. But it is also a time to ask whether the problems stemming from the current economic crisis may challenge some of the protection principles originally adopted:

- Governments face the challenge of their financial commitment vis-à-vis the heritage. The debate between advocates of a free-market and those of government-backed intervention in cultural activities is stronger than ever.
- Globalisation has increased mobility and created a mass cultural tourism that sometimes threatens the monuments and sites.
- Historic cities face the challenge of rural exodus and urbanisation: the shift of consumption and production schemes throughout the world has brought the cultural assets into the agenda for poverty alleviation and sustainable development.
- Cultural capital is acknowledged as a powerful economic force in attracting investors and new business. The economic paradigm for development has been reversed: while economic factors used to bring welfare, today liveability and urban amenities generate economic benefits.

Economics as applied to heritage conservation developed during the 1970s, as a branch of environmental and cultural economics. The discipline has been enhanced by a growing need for tools to cope with the issues of managing and financing heritage in a context of crippling economic resources. Economics of conservation never aimed to challenge the principles of conservation, but just to help decision-makers in assessing and measuring impacts of projects.

Modern principles of heritage conservation have shifted from single monuments and sites with explicit universally recognised outstanding value, to integrated conservation, and to a more holistic perspective, best illustrated by the concept of the Historic Urban Landscape. Once again, conservation and economics have crossed paths. Methodological concern for economics when applied to conservation emphasises a comprehensive approach, or a macroeconomic perspective, which is perfectly consistent with the concept of Historic Urban Landscape. The new paradigm, based on the historic urban landscape, implies an intrinsic coupling of conservation and economics. Conservation economics is no more merely providing a toolkit to achieve cultural goals, but aims to decide with conservation specialists and urban planners which resources are to be allocated, and how.

Historic cities are considered today the most exemplary and challenging issues in heritage conservation. The forces of change that prevail in the framework of historic cities are mainly economically driven, and can be handled only with similar and consistent principles both from conservation and economics. The issue involving the historic urban landscape needs to be addressed in a macroeconomic perspective, which embeds adequately the systemic and collective dimensions of urban heritage.

The economics of conservation is about values. The distinction between use and non-use values, or between market values and values not captured by the market, enables one to identify and measure the beneficial or adverse impact of the forces of change on the various stakeholders in the historic city. An additional distinction is to be made between micro- and macroeconomic values, or between individual easy-to-localise values and public values widespread in the city and beyond, as a result of a chain of economic impacts. Economists of conservation are today well trained to measure individual use values, but the challenge of measuring non-use values and macroeconomic values remains problematic. This is similar to the challenge of conservation specialists who are coping with the historic urban landscape.

Assessing Economic Values with the Use of Indicators and Maps

Heritage economics addresses the concept of Historic Urban Landscape with the identification of the impact of urban heritage on stakeholders' values, in the city and outside. A practical way of assessing the impact on the stakeholders' values is by considering these key questions:

1. How much does the historic urban fabric contribute to the real estate market?
2. How much does cultural tourism contribute to local economic development?
3. How much is the cultural value of the city acknowledged by its residents and abroad?
4. How much do conservation works contribute to the local economy?
5. How much does heritage contribute to the budget (income/spending) of the city?
6. How much does heritage contribute to city branding, and attracting new investors?

By answering these questions, city authorities are able to identify the economic value of their urban heritage, not just in terms of individual benefits for private stakeholders, but from a collective and macroeconomic perspective that is consistent with the Historic Urban Landscape approach.

Heritage economics in an urban context aims to provide decision-makers with the widest range of information that includes all categories of stakeholders and all types of values, as far as they can be expressed in economic terms. It provides a comprehensive analysis of the close connections between heritage and economic flows of production, consumption, income, and jobs in an urban area. In doing so, heritage economics brings an understanding of the urban area additional to the Historic Urban Landscape approach, an understanding of cultural capital and its economic attributes.

Heritage indicators are key elements in assessing economic values. They are consistently used these days as an integrated approach in monitoring historic cities, and are considered a perfect tool to test city performance. Although they do not substitute the use of databases, they can simplify the interpretation of complex systems, in particular when direct documentation is costly or time intensive. They also can be displayed through maps.

Mapping of an indicator for cultural tourism (see question no. 2: How much does cultural tourism contribute to local economic development?)

Selected indicators will be measured by an adequate rating scale of impacts, either beneficial or adverse. Expert appraisals of the indicators will bring answers to the key questions such as: How much does the historic urban fabric contribute to the real estate market? Its contribution is assessed as negligible, minor, moderate, or major. The following three examples illustrate the mapping of an indicator, the appraisal of an indicator, and the selection of indicators for each key question, with the corresponding values and stakeholders.

The indicator is the *Number of visits over the carrying capacity of sites*, as measured by survey data. A value between 75% and 100% is an indication of a major contribution of cultural tourism. A value in excess of 100% indicates an unsustainable tourism development. Data can be broken down between places in the city, and mapped (Figure 12.1 is a fictional example). Combined with other indicators, it will give an appraisal of the total contribution of cultural tourism to the local economic development in the city (from negligible to major). Expert appraisal of the same indicator can substitute the lack of survey data.

Appraisal of indicators on city branding (see question no. 6: How much does heritage contribute to city branding, and attracting new investors?)

Table 12.1 selects five indicators and give a value using census data, survey, or expert appraisal. The measurement is based on a scale ranging from: 1 = Negligible impact, to 4 = Major impact. The indicators average value is 3.25, which means that the heritage has a *moderate* impact on city branding, and attracting new investors. The purpose of the assessment is not just to put emphasis on an absolute value, but to identify the strengths and weaknesses of the city in terms of city branding (the

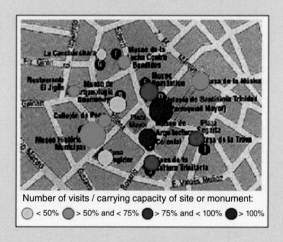

Number of visits / carrying capacity of site or monument:
○ < 50% ◓ > 50% and < 75% ◑ > 75% and < 100% ● > 100%

Figure 12.1 Mapping of visitors and carrying capacity of sites.

Table 12.1 Appraisal of indicators on city branding.

	1	2	3	4
Premium on rents and property prices for houses in the historic district (or any other relevant historic area)		x		
More or less foreclosures in the historic district during recession			x	
Increase in the number of residents in the historic district during the last five years				x
Increase in the number of businesses in the historic district during the last five years			x	
Heritage-related events organised in the city (open monument day, festival, etc.)	x			

1 = Negligible 2 = Minor 3 = Moderate 4 = Major

Table 12.2 Aggregate results for the assessment of the economic value.

Key questions	Indicators	Values	Stakeholders
How much does the historic urban fabric contribute to the real estate market?	Sales/Property prices/Rental values Available homes for sale/vacancy Housing affordability (etc.)	Use values Individual values	Residents, Property owners, Land owners, Real estate agencies
How much does cultural tourism contribute to local economic development?	Carrying capacity of sites /visitors admission fee/ ticket revenue Heritage-tourism related spending (hotels, shops, transportation, services)	Use values Individual values	Visitors, tourists, site managers, tourism agencies, local businesses
How much is the cultural value of the city acknowledged by its residents and abroad?	Existence value /willingness-to-pay/ propensity to visit/ option value/ bequest value/ willingness-to-pay for future generations	Non use values, collective values	Neighbours, passers-by, general public, international community, future generations
How much do conservation works contribute to the local economy?	Turnover of conservation companies Arts and crafts activities Jobs/ income related to built restoration	Use values Individual, collective values	Building industry, local jobs, conservation specialists, arts and crafts industry
How much does heritage contribute to the budget (income, spending) of the city?	Property tax revenues Jobs in local heritage administration Fiscal revenues from admissions/ visits Tax incentives/ subsidies	Use values Collective values	City authorities, regional and national authorities, Governmental agencies
How much does heritage contribute to the city branding, and attract new investors?	Property prices for heritage and heritage buildings Jobs from heritage-related events Sales for new investments and businesses	Use values Non use values, collective values	New business, investors, city at large

opportunity is to organise an annual event as a showcase for the heritage, as revealed with the last indicator).

As the aggregate results are compounded to give a comprehensive assessment for the economic value, the selected indicators are to be related to the categories of values and stakeholders, covering the six key questions.

Table 12.2 shows how key questions of the assessment are connected to indicators, values and stakeholders. This indicates how the economic value goes hand in hand with the Historic Urban Landscape approach, because of the comprehensive and macroeconomic perspective that has been taken. The final results aims not to give an individual value to the cultural building, but a global and macroeconomic value of the same built heritage embedded in a framework of diverse urban assets, and economic resources.

Addressing Urban Heritage Problems with the Help of Economic Tools

A method of economic indicators and maps contributes to reveal the changes in stakeholders' values, and to emphasise problems that may arise in the urban area, when the matching of conservation and

Table 12.3 Recommendations to city managers.

Key questions	Problems	Recommendations for decisions
How much does the historic urban fabric contribute to the real estate market?	Despite recognition for the cultural values of the heritage, there are not many identified economic values, because of the bad condition of the urban fabric, or buildings are not adapted to present needs.	Promote rehabilitation of heritage buildings to increase real estate (use) values. Maintain housing affordability. Develop social housing program. Improve public spaces, infrastructures, mobility.
How much does cultural tourism contribute to local economic development?	Few benefits from cultural tourism, in particular because of a lack of hospitality tourism, and adequate infrastructures and services. On the opposite, risk of not sustainable growth because of mass-tourism.	Attract more tourism by advertising the city (visitor centre, website). Improve the conditions of visits (guides, audio equipment). Promote investment in visit-related activities to increase indirect use values.
How much is the cultural value of the city acknowledged by its residents and abroad?	There is a lack of recognition for the heritage and its cultural values, hence, a deficit in terms of economic values, mostly of non use values. Could vary at local, national, and international level.	Make people aware of their heritage and of the values attached to it. Encourage open access for visit to privately-owned heritage buildings. Organise annual events about the city history.
How much do conservation works contribute to the local economy?	Heritage buildings are at risk because of a lack of conservation works, hence, the risk of losing present and future economic values. Lost opportunity of benefits from the works.	Promote conservation through fiscal incentives (income tax deductions, property tax exemption). Stimulate conservation works through subsidies. Redirect property tax to conservation.
How much does heritage contribute to the budget (income/spending) of the city?	Fiscal revenues from heritage buildings and tourism activities are negligible or ineffectively managed. Excessive spending for conservation. Conservation projects being less profitable than alternative projects.	Improve fiscal mechanisms, and fiscal governance. Raise taxes on visits, and tourism-related activities. Enforce development transfer rights. Enhance cost-benefit and multi-criteria analysis.
How much does heritage contribute to the city branding, and attract new investors?	The heritage is ineffective in attracting business opportunities, the urban historic landscape being assessed as a constraint rather than an asset for economic development. Heritage is not seen as cultural capital.	Coordinate city with heritage planning. Improve city infrastructure and public spaces. Provide communication networks and services for new business in historic area. Enhance city branding.

economic development seems to be ineffective. The information that has been gathered with the help of indicators addressing each of the key questions will provide the decision-makers with a list of problems to be identified, and decisions to be taken and implemented.

Table 12.3 gives for each of the key questions some of the identified problems and recommendations for decisions.

The historic urban landscape is the result of a layering of cultural and natural values and attributes. The macroeconomic perspective is embedded in the urban landscape, and reveals individual and collective connections inside the layering of values and between the attributes. The set of decisions suggested in response to the key questions should in any case be reinstated in a comprehensive perspective of the historic city.

There are several methods available to reveal the common ground for recommendations and decisions. In particular, a SWOT strategic matrix aims to define Strengths (= helpful attributes enabling to increase values, such as a restoration project), Weaknesses (= harmful attributes, such as a lack of financial means), Opportunities (= helpful external conditions, such as world heritage designation), and Threats (= potentially harmful external conditions, such as an excessive increase in mass tourism). As a complement to the suggested method, it will help to rank the different decisions described above, and to prioritise some important and strategic moves. A user's guide to heritage economics will provide the decision-makers in historic cities with relevant data and consistent tools to prevent inadequate development, and to foster sustainable projects that connect conserving the urban heritage with future initiatives.

Case Study *The World Bank's Tools for Urban Conservation*

MV Serra

Urban Planner with 40 years experience in Latin America, South and East Asia, and Africa in the fields of municipal development, housing finance and planning, municipal services and public utilities, urban upgrading, land policies and planning, and urban heritage.

Objectives and Instruments of the World Bank

The business of the World Bank is the promotion of development – urban conservation is increasingly angling to be part of its agenda. The Bank seeks to achieve its objectives chiefly through the provision of financial resources (foreign exchange) at special interest rates either to projects, in sectors that have been defined as national priorities, or to government agency budgets, focusing on public sector reforms. The Bank also provides, through the International Development Association, grants or credits, under especially favourable conditions – subsidised interest rates and longer grace and amortisation periods – to countries below certain levels of per capita income. Though increasingly less relevant as a provider of resources (particularly to countries such as the BRICS), the Bank remains prominent as a 'think tank' dedicated to advising countries on policy formulation and on the design of projects. In doing so, the Bank benefits from its access to best practices culled from across the globe.

Cultural Capital

Consonant with worldwide trends, the Bank has sought to adopt a more inclusive and effective definition of development. It has absorbed the growing concern with a more balanced distribution of development benefits and poverty alleviation. In fact, poverty alleviation, and consequently, a healthier, more educated, productive population, is seen, by the Bank, as crucial for the promotion of economic development. The Bank has also incorporated the idea that institutions, whether governments or laws and regulations, as well as civil society's organisations, norms and procedures (social capital) are crucially capable of hampering or promoting development. Stretching the definition of development farther still, given the expected worldwide finitude of physical resources (natural capital) and the limited capacity of the environment to absorb the growing discharges of human activities, the sustainability of the development process has come into question and measures to mitigate or compensate environmental damage have become one of the Bank's priorities.

Lastly, in what matters most to urban conservation projects, cultural capital (a concept akin to natural capital) gains increasing attention. Cultural capital refers either to unique heritage resources or to environments or objects, material or immaterial, capable of having either scientific content, displaying beauty, enhancing the sense of identity of a community, providing connectedness to the past, or contributing to social stability and cohesion. It is an understanding that cultural capital may lead to a more productive and developed society that allows the Bank to turn its attention to heritage initiatives and urban conservation projects.

Urban Projects

The Bank has a wealth of experience on urban studies, policies and projects, which will necessarily bear on its perspective on urban conservation projects. In the context of the limited resources characterising the countries and cities the Bank works with, it is typical that the supply of urban infrastructure and housing does not meet demand (or rather, needs). Consequently, the norm, in most of these cases is the presence of scarce public transportation, traffic congestion, extra-long commuting times, poorly serviced neighbourhoods, and, often, the massive presence of shanties. In this environment of scarcity, attention to the imbalance between need (or latent demand) and supply of urban infrastructure and housing has been a priority of the Bank, both in terms of policy advice and project financing. The Bank's support to urban projects has been mostly through single-sector financing, focusing largely on water and sanitation, transportation, and housing, since such projects give more scope for sector reform and, consequently, to ensuring strengthened, self-sustaining sectors. The Bank does finance, and increasingly so, the urban upgrading of slums and urban areas with poor infrastructure, through projects which are very close in conceptual and operational terms to urban conservation ones.

Urban Conservation Projects

With the Bank turning its attention to urban conservation projects, and from its perspective, such projects would mean the introduction of a set of spatially integrated economic, social and physical measures aiming at the overall improvement of an urban area endowed with cultural capital. Even though the main rationale for urban conservation projects may come from their contribution to the enhancement of existing cultural capital, their implementation would be expected to contribute also to the goals of poverty alleviation, the strengthening of social capital, and the attainment of improved levels of environmental sustainability.

Incorporating the Spatial Dimension in Project Design

Spatial projects make more sense than sector ones when the introduction of any service benefits from the simultaneous introduction of others and, together, the set of services tends to promote the improvement of the urban area as a whole. Spatial projects attempt the reproduction of the processes that constitute the very nature of cities. Cities are the result of a complex interaction of private markets – immersed in a web of market failures – and also, importantly, of state initiatives, both involving regulation and investments. This state-market interaction occurs through a technological menu of choices, affecting a city's adoption of specific urban services solutions. Of particular interest is the fact that cities are built with durable materials, which, as time passes, may become a constraint on the provision of required services, leading to deterioration, and, in some cases, eventually to abandonment, of the areas where they are located. This aging process leads to cities showing, at any point in time, the simultaneous presence of a vast array of urban and architectural patterns from different periods, some expressing certain uniquely historical periods. Society may, eventually, evaluate these urban and architectural patterns, from the point of view of their cultural capital, determining either their conservation or some sort of accommodation and compromise with future growth and change.

Building Blocks of Urban Conservation Projects

Consider, as a hypothetical example, the historical core of a city such as Fez, or Dubrovnik, Hanoi, or Rio de Janeiro. The Bank might see as eventually meriting support:

1. **Heritage**: These urban areas typically possess a number of structures deemed to be culturally, historically or aesthetically notable and in need of protection, safeguarding, improvement and, frequently, adaptation to new uses;
2. **Economic base**: These are often areas with a decaying economic base, requiring the definition of an enhanced role in the context of the larger city, and market-based incentives to develop this base, particularly for tourism or the protection of the traditional crafts/ trades, which may typify the area;
3. **Collective urban infrastructure network**: This refers to urban infrastructures power grids, water and sanitation, and roads – in most cases, these are in need of repair, upgrading and reconfiguration to both meet the needs of residents and to preserve or enhance cultural and cityscape values;
4. **Housing**: Often existing units fail to meet basic conditions for the use of families and economic activities, having limited sanitation facilities;

5. **Property titling**: In most cases, housing units have no proper definition of property, consequently limiting the growth of family wealth and physical mobility;
6. **Project finance**: Financing for the different project components, the design of cost recovery mechanisms, and subsidies will be needed;
7. **Governance**: These projects require definition of competencies and roles for all stakeholders involved, during and after project implementation, whether these be local governments or infrastructure agencies, the private sector, or community interests.

Economic Base and Poverty Alleviation

A number of potential unintended effects might frustrate the attainment of project objectives. Presumably, one of the expected results of an urban conservation project is a more pleasant and diverse urban area, endowed with better infrastructure, housing facilities and economic activities. There is an increasing belief (and some empirical evidence) that this leads to higher factor productivity and competitiveness of the project area, the young and gifted coming in and the poor being priced out. Activities related to tourism are a welcome source of income for locals but, beyond certain levels, may be damaging to the long-term sustainability of the nature and character of the project area, thus reducing the cultural capital it was set forth to enhance. Analysis of local population composition and evaluating benefits and costs that may accrue, especially to the poor, is crucial. Evaluation of benefits and costs must also reach non-residents, such as tourists, whether national or international, and, additionally, the residents of the larger urban context in which the project area is located.

Building Standards and Legal Provisions

Typically, physical components (infrastructure and housing) of urban conservation projects may require special design standards, both for buildings and neighbourhoods. This may arise, as well, from the need to protect the interests of the local population. For example, power grids may require an underground network, to avoid disrupting the site's historical character and the cityscape. Similarly, in the interest of preserving their heritage character, the adaptation and improvement of housing structures will demand regulation that requires the use of traditional materials and building technologies, no longer in use. But this may also require, in parallel, a quasi-tailor-made approach to housing unit upgrading, in the interest of meeting the needs of residents. The provision of property titles, in turn, may require complex legal reforms, plus complex administrative procedures and difficult interpersonal and inter-agency negotiations. All of this implies larger infrastructure and housing costs than are typical of 'green site' projects; increased complexity in the design of cost recovery mechanisms, and the likelihood that housing and infrastructure subsidies are required so as to make the project feasible.

Exploring New Governance and Financing Arrangements

Urban conservation projects are characterised by the presence of market failures, externalities and public goods, which require a mixture of incentives, regulation and investments. They affect a large number of stakeholders, and demand multiple source financing, complex cost recovery mechanisms and subsidies. Consequently, these are complex, labour-intensive projects, typically taking a long time to prepare, to implement, and to attain proposed objectives. They require a strong coordination effort and the presence of government as a catalyst. A lesson derived from the Bank's experience is that to achieve more expediency and timeliness in the implementation of projects with such characteristics, they need to experiment with new governance and financing instruments.

In this regard, Bank literature on the topic has been pointing to instruments such as the Special Purpose Vehicle (SPV), a consortium of financial institutions and private companies responsible for the implementation of activities within a Public-Private Partnership (PPP). SPVs are wholesalers, holding debt, benefiting from service fees, promoting construction, and contracting services. They may come tied to Land Value Finance (LVF), an instrument geared to recovering the capital cost of urban investments by capturing increments in land value accrued from the project. In the case of urban conservation, the need for regulation on heritage grounds, by limiting the intensity of land use beyond existing contours of buildings, may lower the potential for capturing such increments. Alternatively, the absence of such regulations may irrevocably damage the heritage nature of a site.

Conclusion

The Bank issues policy documents which are meant to serve as guides for the preparation of projects in diverse sectors such as housing, water, sanitation, education, energy and environment, to cite a few. Currently, it has no such document for cultural heritage or urban conservation. It has however increased the financing of projects with cultural capital characteristics (not necessarily urban conservation ones) in various regions of the world, thus expanding its own experience in this field and allowing for the gathering of empirical evidence of which approaches and solutions seem to work best. It has also requested preparation of various studies on topics related to culture, heritage and urban conservation – a step that can be seen as previous to the eventual issuing of a more comprehensive, focused policy piece.

13 Researching and Mapping the Historic Urban Landscape

Michael Turner[1] *and Rachel Singer*[2]

[1]*Architect and Professor at Bezalel Academy of Arts and Design in Jerusalem; and UNESCO Chair in Urban Design and Conservation Studies*

[2]*Graduate Student in Urban Design at the Bezalel Academy of Art and Design, Jerusalem*

A City is more than a place in space, it is a drama in time.

Patrick Geddes

Introduction

The Historic Urban Landscape Recommendation and its accompanying Resolution, adopted by UNESCO in 2011[1] have set the scene for the bringing together and integration of the urban complexities in a holistic manner. The approach embodying traditional knowledge includes the challenges, trials and errors that have accumulated over centuries, presenting examples of human beings taming nature, living with nature, creating places that demonstrate the way that human genius intertwines with traditions, the tangible and the intangible creating the spirit of place for liveability and well-being. In this way, the urban fabric while composed of architectural and archaeological monuments, sites and groups of buildings incorporates urban layering and

---

[1] General Conference 36 C/23 Item 8.1 – Proposals concerning the desirability of a standard-setting instrument on Historic Urban Landscapes:

1. Undertake comprehensive surveys and mapping of the city's natural, cultural and human resources;
2. Reach consensus using participatory planning and stakeholder consultations and to determine the attributes that carry these values;
3. Assess vulnerability of these attributes to socio-economic stresses;
4. Develop an integrated urban development and conservation strategy to integrate heritage values and their vulnerability status into a wider framework, including potential for change, non-changeable parts and areas of opportunity;
5. Prioritise actions for conservation and development;
6. Establish the appropriate partnerships and local management frameworks.

Reconnecting the City: The Historic Urban Landscape Approach and the Future of Urban Heritage, First Edition. Edited by Francesco Bandarin and Ron van Oers.

© 2015 John Wiley & Sons, Ltd. Published 2015 by John Wiley & Sons, Ltd.

opens doors to the culture and technologies of the future.[2] This approach lays out an array of steps for implementation that are at the forefront of urban conservation planning and heritage management, while the challenge for further developing and operationalising their application is the current goal.

In this debate, the authors wish to place emphasis on understanding cultural diversity and its relationship to historic urban landscapes.[3] Maintaining diversity and the ethnic and cultural balances in a dramatically changing environment is indeed a challenge that needs to be met between the consensus and the *casus belli*. As cities have a plethora of narratives and identities it is vital to offer an approach that allows their co-existence, and it is here that the layered approach can be applied as a methodology to incorporate cultural diversity, to be addressed in the field of integrative planning and urban management. Furthermore, a review of the EU Partnership for Peace PUSH (Promoting Dialogue and Cultural Understanding of Our Shared Heritage) project (2006–2009)[4] methodology for advancing concepts of shared, or even common heritage, and a discussion on cultural mapping techniques might be a key integrative tool referenced in the Historic Urban Landscape Recommendation.

The PUSH methodology proposal suggests that a shared understanding and appreciation of the significance of multicultural urban heritage by city residents themselves can support culture and development. The integration and management phase applied by the three participating institutions, each peer reviewed the theme developed by the other institutions (planning and design, cultural resource management and archaeology and history). In this phase these themes integrate to form a multilayered unit that can form the basis of the management mechanisms jointly developed by the institutions.

Its application through the Historic Urban Landscape approach attempts to resolve the gaps, and offers a method to absorb the knowledge and then tolerate the conflict through compromise and co-existence.

As the city is ever transforming with shared experiences of urban life, its work routines and public spaces offer a medium for interaction and urban vitality. Yet differing perspectives of the meaning of that historic urban landscape can intensify an inherited rivalry of identities. Transformations can translate into processes of evolution or revolution; the evolutionary course enabling urban resilience and better monitoring as opposed to potentially catastrophic results of revolutions on the urban fabric and its heritage, which can be particularly difficult to cushion. These metamorphoses might be the result of human interventions as in the Sacks of Rome and the contemporary damage to the Old City of Aleppo or natural disasters as in Bam, the Indian Ocean tsunami or the earthquake of L'Aquila, as well as earlier examples of Alexandria (365), the Great Fire of London (1666), Val di Noto (1693), and the San Francisco Fire (1906).

The concept of viewing cities as a series of themes that are divided into layers is a methodology developed by Ian McHarg in his seminal book *Design with Nature*. McHarg observes the intricate linkages between the city and its sur-

[2] Concept presented in a lecture by Michael Turner, 'A New International Instrument: The Proposed UNESCO Recommendation on the Historic Urban Landscape'. *The International Protection of Landscapes: A Global Assessment on the Occasion of the 40th Anniversary of the World Heritage Convention.* Florence, Italy, September 19–21, 2012.

[3] The Historic Urban Landscape approach considers cultural diversity and creativity as key assets for human, social and economic development, and provides tools to manage physical and social transformations and to ensure that contemporary interventions are harmoniously integrated with heritage in a historic setting and take into account regional contexts. Sourced from: UNESCO *Recommendation on the Historic Urban Landscape*, November 2011, http://portal.unesco.org/en/ev.php-URL_ID=48857.

[4] PUSH was funded by the European Union's Partnership for Peace Programme and the Norwegian Foreign Ministry, with additional support from the Beracha Foundation. The project was composed of teams of heritage experts, historians, designers and architects from Israel, Jordan and Palestine. For more information visit: www.pushproject.org.

roundings.[5] Layering is an instrument that can also be applied in the context of historic urban landscapes when introducing and merging additional discussions relating to integrative planning and diversity. It was the very basis for the development of Geographical Information Systems, the essence of information management.

Community participation, academic research and public awareness are all part of capacity building that is essential when working in culturally diverse settings. The importance of communities and their role was emphasised through their addition to the UNESCO Operational Guidelines as one of the Strategic Objectives.[6] Engaging different layers of players including academics, professionals, institutions and NGOs in dialogues and partnerships is very important in order to articulate the rules of the game, create a consensus on the content in addition to establishing a clear conceptual framework and a common language.

The Diverse City

One of the key concerns in urban conservation is managing conflicting situations both in time and place. This is highlighted in the 1994 Nara Document through the issue of cultural diversity, not only between people and places but also through time.[7] It is a two-way process of acceptance within the administrative and political structures providing the framework for the changing relationships between the players and stakeholders. Urban change has been very dramatic with large waves of immigration to cities, transforming the very essence of the city and its socio-political structure. This was not only intranational from the country to the city but involved global movements based on previous colonial structures and languages. This is very evident in cross-cultural migrations affecting our cities from North Africa to France; from Turkey to Germany; from the Commonwealth to the UK

---

[5] McHarg, I.L. (1969) *Design with Nature*. Garden City, NY: Published for the American Museum of Natural History by the Natural History Press.

[6] Community was added in 2007 at the session of the World Heritage Committee in Christchurch, New Zealand. The current Strategic Objectives (also referred to as 'the 5 Cs') are the following:

1. Strengthen the Credibility of the World Heritage List;
2. Ensure the effective Conservation of World Heritage Properties;
3. Promote the development of effective Capacity-building in States Parties;
4. Increase public awareness, involvement and support for World Heritage through Communication;
5. Enhance the role of Communities in the implementation of the World Heritage Convention. Sourced from: UNESCO and Intergovernmental Committee for the Protection of the World Cultural and Natural Heritage, 'Operational Guidelines for the Implementation of the World Heritage Convention,' 2012, http://whc.unesco.org/archive/opguide12-en.pdf.

[7] Sourced from: ICOMOS (1994) *The Nara Document on Authenticity*, Sections 5 & 6. http://www.icomos.org/charters/nara-e.pdf.

5. The diversity of cultures and heritage in our world is an irreplaceable source of spiritual and intellectual richness for all humankind. The protection and enhancement of cultural and heritage diversity in our world should be actively promoted as an essential aspect of human development.

6. Cultural heritage diversity exists in time and space, and demands respect for other cultures and all aspects of their belief systems. In cases where cultural values appear to be in conflict, respect for cultural diversity demands acknowledgment of the legitimacy of the cultural values of all parties.

7. All cultures and societies are rooted in the particular forms and means of tangible and intangible expression, which constitute their heritage, and these should be respected.

8. It is important to underline a fundamental principle of UNESCO, to the effect that the cultural heritage of each is the cultural heritage of all. Responsibility for cultural heritage and the management of it belongs, in the first place, to the cultural community that has generated it, and subsequently to that which cares for it. However, in addition to these responsibilities, adherence to the international charters and conventions developed for conservation of cultural heritage also obliges consideration of the principles and responsibilities flowing from them. Balancing their own requirements with those of other cultural communities is, for each community, highly desirable.

and a massive movement of rural to urban settlement across Asia.

This has dramatically changed our understanding of the meaning of local community and indigenous communities and raises questions whether immigrant communities can claim monuments in their new city to be 'their heritage' and how do the descendants of the original owners and creators of the site relate to such a claim. Heritage now belongs to place and not only people.[8]

There are both tangible and intangible aspects of culture, as underlined by the Japanese *Machinami* Charter, approved in 2000. The word *Machinami* itself refers to the physical and spiritual, the tangible and the intangible thereby highlighting the qualitative aspects of the built environment.[9]

The Nara Document is currently under review and the changes that have occurred in the past 20 years since its inception have resulted in the Himeji Declaration that recognises the need for a continued expansion on the debate relating to values, cultural diversity and heritage particularly in light of increasing globalisation, urbanisation and social change:

> The attribution of values to heritage is a social rather than a scientific or technical process involving multiple individuals and groups .... Since society is becoming more diverse, conflicts are emerging which lead to disputes within and between communities, governments and other stakeholders over heritage values and claims for authenticity. This reaffirms the need for greater emphasis to be placed on developing processes, tools and frameworks that can enable community participation in the negotiation of integrated heritage management strategies.[10]

In considering the future, the authors recognise that values seem to be a moving target –

> ... places may have a range of values for different individuals or groups and values are continually renegotiated[11] highlighting the paradox between the Universal and Particular. With changing social patterns and demographic movement, we have now to focus on the continuity of place rather than people. This capacity is strained with social change and population transformations. If the potential of the heritage is in its universal recognition then successful communities that survive these traumas are those that have managed to continue in some syncretistic or symbiotic form as the alternative parasitic situation will be doomed to failure. This might not only be between different cultures in the same time and place but the same cultures in different times allowing for new lives in old cities; this surely is evidence of the adage 'the survival of the adaptive'.

Building simultaneously on syncretism and symbiosis allows for a multi-cultural interpretation, providing situations of compromise (syncretism) and coexistence (symbiosis). While it is possible to view syncretistic forms as an illegitimate offspring of the original, something debased, perhaps they might be considered as a new evolving form in its own right? From a religious perspective any syncretistic form would be seen as a dilution of the source, so the acceptable limits of symbiosis might be a state of coexistence. The result may also be a mutation in as much as this is an act or process of being completely altered or changed.

The outcome would allow the transient community to own and take responsibility for the heritage of that place, and that the original owners would accept this transposition with the renegotiation of the new associated values.

Interpreting these attributes of the values at the urban scale provides the understanding and the guidelines needed for the conservation and development of the city. The contexts in which

---

[8] Concept presented in a lecture by Michael Turner 'UNESCO *Recommendation on the Historic Urban Landscape*.' *Quality Control and Conflict Management at World Heritage Sites*. Neue Residenz, Bamburg, Germany, July 14–15, 2011.

[9] See: Bandarin, F. and Van Oers, R. (2012) *The Historic Urban Landscape. Managing Heritage in an Urban Century*. Chichester: Wiley-Blackwell.

[10] Sourced from: ICOMOS, 'Himeji Recommendation (Draft).' *Heritage and Societies*, http://nara2014.wordpress.com/himeji-recommendation/.

[11] See ICOMOS (1999) 'The Burra Charter – The Australia ICOMOS Charter for Places of Cultural Significance.' http://australia.icomos.org/wp-content/uploads/burra_charter.pdf.

the monuments live are, in the words of Francis Bacon, the 'shipwrecks of time'.[12] These monuments remain while the people change. This links back to the question of '*how is the layering evident?*' New neighbourhoods are built over, built by, built instead of, absorbed in, occupied or even hijack the previous layers, periods and cultures. Based on the record of our present old cities these policies will provide the patterns for the new lives. Let us consider some examples of the city as project or process. The values of the city might be in the planning or building project as an 'object', exemplars of which are the city of Chandigarh or Bath or the plan of Brasilia.[13] However, most cities have layered values reflecting the processes that created them as in Old Delhi/Shahjahanabad, Mumbai, Bruges or Prague. One of the attributes might be in the building concept and form. At the next level, urban attributes may include: plot size, building lines, roof-scape, materials, ceremonies and events, the open spaces and vegetation. The value is in the process that created these places.

The potential of built heritage is in its identification within a comprehensive environment that will provide for spatial, in addition to temporal, sustainability. This needs to be attained through the consideration of process rather than project and the management of change through the cultural integration of socio-economic policies.

Methodologies and Tools

Many tools can be applied to implement the Historic Urban Landscape Recommendation, including:

- Visual analysis;
- Cultural mapping;

- Environment Impact Assessment (EIA) / Health Impact Assessment (HIA);
- Social,
- Political;
- Economic;
- Infrastructure;

and, the integrative approaches and layering in cultural mapping, which we will focus on.

Many of the current trends in planning and management are focused on integration and inclusion. While modernist planning was principally concerned with zoning as the solution of choice, resulting in the production of rather one-dimensional landscapes, by the 1960s the need for expanding the parameters had become apparent. Environmental, sociological and economic aspects were studied and included in planning practice and today these disciplines are considered integral aspects of the processes with multidisciplinary teams currently articulating many aspects of planning and managing the processes of change.

For effective urban heritage conservation and management of the Historic Urban Landscape approach, promoting the comprehensive mapping of all heritage assets within an urban environment and its larger setting is needed. This sets a new standard and work formulae based on multi-disciplinary actions, demanding a new knowledge with common denominators to allow for harmonisation. There is currently no consistent method for the way that data is collected to measure cultural impacts, which raises, questions regarding the necessary parameters. The way forward demands a regional political framework with additional tools for urban conservation with a more flexible language. This includes documentation of the city depicting information as extended urban layers and cultural mapping.

---

[12] See: Bacon, F. (1828) *Of the Proficience and Advancement of Learning, Divine and Human.* London: J.F. Dove, St. John's Square, 92.

[13] From the brief description of the World Heritage inscription: 'Urban planner Lucio Costa and architect Oscar Niemeyer intended that every element – from the layout of the residential and administrative districts (often compared to the shape of a bird in flight) to the symmetry of the buildings themselves – should be in harmony with the city's overall design.'

The shift from monuments to include the living city, the focus on the human scale, places and integrative methodologies is evident in much of the current urban planning discourse and is apparent also in the Historic Urban Landscape approach. The visual understandings developed by Lynch, Cullen, Appleyard,[14] among others, analysing the way people absorb and react to their surroundings have made room for emerging methodologies such as cultural mapping. Parallel to this, anthropological mapping was being developed by Amos Rapoport,[15] sociological trends as family kinship were being studied and mapped by Young and Wilmot[16] and terms as gentrification entered our vocabulary through the works of Glass.[17]

While the setting of the western city might be attributed to Vitruvius, the Chinese *Feng Shui* (= wind & water 風水) gives the conventional city meaning, and the *Shan Shui* (= mountain and water 山水) provides the city with spiritual and philosophical meaning following symbiotic thinking common in East Asia. These differences can be clearly seen in the manifestations of the grid, organic and diagrammatic cities as identified by Spiro Kostof.[18]

The first step in the implementation of the Historic Urban Landscape Recommendation advocates that Member States and local authorities *'undertake comprehensive surveys and mapping of the city's natural, cultural and human resources'*;[19] this is a critical indication for bringing culture into the realms of day-to-day living.

The term cultural mapping has gained popularity in recent years with multiple uses in popular contexts that includes a wide variety of meanings. As a methodology cultural mapping was examined by Peter Poole in a report for UNESCO.[20] In his report, Poole develops cultural mapping as a tool to strengthen cultural diversity in indigenous communities and trains locals in different types of mapping technologies so that they can create their own maps and protect their interests. The techniques include seven categories and cover low-tech methods such as sketch maps adding local information to existing topographical maps as well as high tech options that employs Global Positioning System (GPS), Geographic Information System (GIS),[21] and graphics software to produce geographically accurate maps. Poole asserts that the primary difference distinguishing cultural mapping is the intent and usage, though the information focus can be quite similar to other types of mapping.

Other experiments in cultural mapping differ greatly in terms of content and goals, Bianchini and Ghilardi,[22] use the term as an alternative method for place marketing and branding in Sweden. Their interpretation of cultural mapping uses both quantitative and qualitative techniques to detect and categorise sites and resources, reinforced by a series of underlying principles. A study in Ottawa by Julian Smith,[23] explores ways to map cultural landscapes, stressing that there are multiple identities and urban realities, and the maps illustrated in his article are individual

[14] See for instance: Lynch, K. (1960) *The Image of the City*. Cambridge: MIT Press. Cullen, G. (1961) *Townscape*. New York: Reinhold Publishing. Appleyard, D. Lynch, K. and Myer, J. R. (1964) *The View from the Road*. Joint Center for Urban Studies of the Massachusetts Institute of Technology and Harvard University. Cambridge, Massachusetts: MIT, Press.

[15] See: Rapoport, A. (1969) *House, Form and Culture*. Englewood Cliffs: Prentice Hall.

[16] See: Young, M. D. and Willmott, P. (1957) *Family and: Kinship in East London*. London: Routledge.

[17] See: Glass, R. (1964) *London; Aspects of Change*. London: MacGibbon & Kee.

[18] See: Kostof, S. (1991) *The City Shaped: Urban Patterns and Meanings through History*. London: Thames and Hudson.

[19] See: UNESCO *Recommendation on the Historic Urban Landscape*, November 2011, http://portal.unesco.org/en/ev.php-URL_ID=48857.

[20] See: Poole, P. (2003) Cultural Mapping and Indigenous People. Paris: UNESCO. http://www.ibcperu.org/doc/isis/11953.pdf

[21] Harvey, F. (1997) From Geographic Holism to Geographic Information System. *The Professional Geographer*, 49, no. 1: 77–8.

[22] See Bianchini, F. and Ghilardi, L. (2007) Thinking Culturally about Place. *Place Branding and Public Diplomacy*, 3, N. 4.

[23] See: Smith, J. (2010) Marrying the Old with the New in Historic Urban Landscapes. In Van Oers, R. and Haraguchi, S. (eds) *Managing Historic Cities*. World Heritage Papers N.27. Paris: UNESCO: 45–52.

and associative. UNESCO Bangkok has run cultural mapping training workshops that utilise GIS technologies, including a case study on *'Mapping and Revitalization of Endangered Languages in Thailand'*.[24] Janet Pillai offers a guide for conducting cultural mapping along with recommendations on how to assess and evaluate the findings while applying them to culturally sensitive planning processes.[25]

These examples of cultural mapping that have been applied across the world elucidate the developing nature of the concept that might be usefully employed in the Historic Urban Landscape approach, and the radically different uses that are currently attributed to a single term. As a recommended tool it is necessary to develop the urban focus of cultural mapping as an indicator of social networks and the guiding of potential changes accepting the challenges of urban transformation. The use of the layering methodology can aid in the comprehension of these intricate uses and politics of space.

A shift in conservation attitudes that emerged in the 1970s came on the heels of the nature conservation movement, a movement that has always been at the forefront of heritage protection. This can be seen in the evolution of the World Heritage Convention and the initial steps of environmental protection at the Stockholm Conference of 1972. Tools were developed based on scientific analyses and in this the Environmental Impact Assessment was created, which is now recognised worldwide. The scientific and quantitative nature is wider than the humanistic and qualitative culture, as it studies the universal, while culture focuses on the particular. In the cultural sphere a weaker edition emerged in the form of the Heritage Impact Assessment.[26] This could benefit from cultural mapping which is a critical tool to assess the two-way impact of heritage on development and vice versa. To achieve this we need to employ mapping techniques that track economic, social and cultural relationships that exist symbiotically in urban space. Cultural mapping must be further explored and expanded within the framework of conflict identification to pinpoint critical sensitivities and enable the facilitation of a sustainable historic urban landscape.

More recently, the concept of Cultural Commons can be seen to have its roots in environmental science. Cultural Commons refers to *'tangible and intangible forms of culture that can be understood as intellectual resources shared, produced and expressed by members of a community'*.[27] Cultures can be understood through the dimensions of space and community and the commons require maintenance and protection to avoid what is frequently known as the 'tragedy of the commons', the result of overuse of shared resources.[28] Cultural Commons engage in cultural mapping though it is not titled as such, the types of maps and matrices generated in this category includes the mapping of the Milanese Design between the 1950s and 1960s and an anatomy of French gastronomic districts.[29]

The PUSH Project pioneered a methodology of discussing and re-envisioning the cultural

---

[24] See: UNESCO Bangkok (2004) 'Putting Cultural Diversity into Practice: Some Innovative Tools'. *Cultural Mapping*, http://www.unescobkk.org/culture/cultural-diversity/cultural-mapping/cultural-mapping-at-unesco-bangkok/cultural-mapping-training/.

[25] Pillai, J. (2013) *Cultural Mapping: A Guide to Understanding Place, Community and Continuity*. Petaling Jaya: Strategic Information and Research Development Centre.

[26] For more on the Historic Urban Landscapes discussion in relation to Integrated Assessments and their integration of economic, social and environmental aspects and other planning tools refer to Bandarin, F. and Van Oers, R. (2012) *The Historic Urban Landscape: Managing Heritage in an Urban Century*. Chichester, Wiley-Blackwell: 159–166.

[27] See: Bertacchini, E. et al. (2012) *Cultural Commons: A New Perspective on the Production and Evolution of Cultures*, Cheltenham: Edward Elgar: 3.

[28] See: Hardin, G. (1968) The Tragedy of the Commons. *Science* 162, no. 3859 1243–1248.

[29] For examples see: Bertacchini, E. et al. (2012) *Cultural Commons: A New Perspective on the Production and Evolution of Cultures*, Cheltenham: Edward Elgar: 4,73, 101, 121, 137.

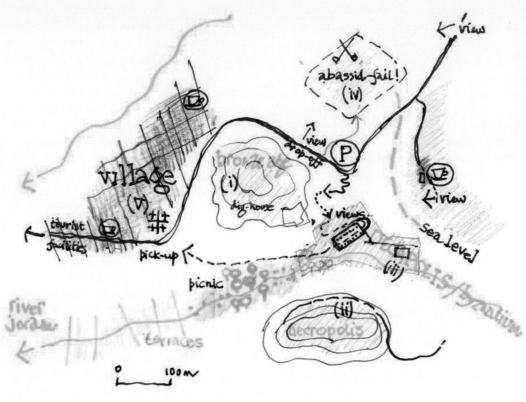

Figure 13.1 Cultural map as developed in the participatory workshop of the Roman Decapolis city of Pella, located in Jordan, a site included in the PUSH project.

heritage of the region as a common inheritance, rather than as competing national symbols or national tourist attractions for visitors from the outside (Figures 13.1 and 13.2).

The achievements of the PUSH Project suggest the positive potential of a similar approach to address the many challenges of urban heritage in the twenty-first century, multicultural world. In order to bridge conflicts stemming from issues relating to heritage it is important to first identify the different types of conflict associated with cultural diversity and heritage, including:

1. **Commemoration**: Conflict in which 'official' heritage commemoration and preservation laws becomes a source of hostility between those who are commemorated and those who are ignored. This is a particularly pressing policy issue in which immigrant communities are newcomers and old-timers, rich and poor, immigrants and natives over issues of ownership, zoning.

2. **Value**: Conflict in which particular heritage monuments and elements have great economic value for a portion of the population, but arouse hostility and opposition from other portions have conflicting values – such as the sacred sites claimed by indigenous peoples, or the revitalised historic centres of cities from which a pre-existing population (usually poor) has been displaced.

3. **Inheritance**: Conflict in which conflicting claims for the exclusive 'inheritance' of certain monuments or objects (places of worship, monuments, tombs, and historic sites) become the focus of direct conflict and sometimes violent struggles for possession

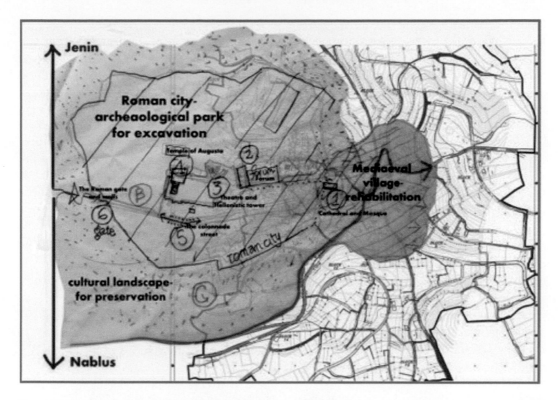

Figure 13.2 Cultural map as developed in the participatory workshop.

between contemporary ethnic and religious groups.[30]

The Role of University Research

University research has to be considered at the forefront of the dialogue and while these are usually individual or focused actions, they summarise the thinking at the grass roots level. Parallel to this are the enlightened agencies, institutions, governments and municipalities who have engaged in innovative actions. At the UNESCO Forum – Universities and Heritage,[31] where some 338 abstracts were submitted

opening up the debate on the Historic Urban Landscape with representation from 73 countries. The emphasis in that year was still on the issues of architectural interventions – as suggested by the Vienna Memorandum, while it was clear that the urban issues of transformations were becoming a more dominant theme.

The debate on the Historic Urban Landscape as to whether it was to be considered a new category was based very much on the concepts of Cultural Landscapes developed by Carl Sauer and the body of geographical knowledge brought to the debate through the works of Michael Conzen in analysing the town Alnwick and subsequently with Jeremy Whitehand under the

---

[30] This categorisation was created by Neil Silberman following the PUSH project.
[31] See: 2009 FUUH Seminar – University of Hanoi (Vietnam).
http://universityandheritage.net/SIFU/XII_Hanoi_2009/index.html.

headings of Urban Morphology.[32] It is interesting to note that this was the same year that the Cultural Landscapes were adopted into the Operational Guidelines of the World Heritage Convention, but the outcome was to see this as an approach clearly redefines the boundaries of research.

The Role of UNESCO Chairs

The global effect can be measured in the numbers and functions of UNESCO Chairs that have been instituted over the past decade. More especially, conservation is linked to a number of disciplines as urbanism, social sciences, sustainability in many forms, tourism and public policy. The many networks evolving are also evidence of the growing challenges for interdisciplinary action, which should be encouraged at all levels. There are some 174 chairs out of over 700 UNESCO Chairs that have references to the urban scene and that are potentially cross-sectorial. While it is not possible to categorise accurately the activities of each Chair, a summary in Table 13.1 clearly indicates the new emphases.

Table 13.1 Distribution of UNESCO chairs. Source: http://www.unesco.org/en/university-twinning-and-networking/access-by-domain/education/

Intercultural and interreligious dialogue	35	20.1
Cultural heritage	27	15.5
Sustainable development	27	15.5
Cultural diversity	17	9.8
ICT	15	8.6
Cities and habitat	13	7.5
Migration and multiculturalism	12	6.9
Cultural policy	7	4.1
Cultural development	6	3.4
Cultural tourism	6	3.4
Intangible heritage	5	2.9
Resolution of conflicts	3	1.7
Sustainable tourism	1	0.6
Total	**174**	**100.0**

Indeed, a new European Union (EU) proposal creating an academic research consortium for integrating research on the Historic Urban Landscape of some 12 institutions, initiated by the Bezalel Academy of Arts and Design, integrates subjects that include urban archaeology, smart cities, cultural heritage studies, rights to the city, urban design and conservation, cultural heritage studies, and the maintenance and management of urban areas with approaches to EIA/HIA and risk preparedness. Nevertheless, there are still many researches in urban methods and territorial analysis as, for instance developed by the Organisation for Economic Cooperation and Development (OECD), that are not integrated or harnessed towards the integrative process of the Historic Urban Landscape.

The Role of Category 2 Centres (C2C)

The UNESCO policy of developing the network of Category 2 Centres (C2C) was taken up by the World Heritage Committee and supported by the Culture sector at UNESCO and the World Heritage Centre. With some eight centres in approval, each centre has in addition to their regional role identified specific themes and subjects to be addressed at a global scale; these range from tourism (Norway), rock art (Spain) and to urbanism (Brazil). This emerging network might provide a critical mass and home for the various International Scientific Committees of the International Council on Monuments and Sites (ICOMOS), strengthening the professional knowledge, capacity building and academic research. While the centres in Brazil, Shanghai and Bahrain are clearly linked to urban issues and the Historic Urban Landscape recommendation, the wide range of other themes should give added value and integration for developing conservation policies.

---

[32] See Sauer, C. (1925) *The Morphology of Landscape.* Berkeley: University of California Press, Conzen, M. R. G. (1960) *Alnwick, Northumberland: A Study in Town-plan Analysis.* London: George, P. and Whitehand, J. W. R. (1992) *The Making of the Urban Landscape.* Oxford: Blackwell.

Conclusion

The Challenge for the Future. Linking Sustainability and Historic Urban Landscapes

The contemporary concept of sustainability was developed by the Brundtland commission, recognising that current development must take future generations into account.[33] Yet social justice needs a wider understanding of sustainability – who is sustaining what? A more equitable economic development is surely an important recipe for the resilience of the city. The rapid and intense city growth generates urban change has reached a point of metamorphosis and is threatening the very values that created the city. The challenge is that this new form of the city cannot be based on the same rules as past cities, which had remained somewhat static for centuries. The debate must continue at the political and administrative levels, so as to balance the national resources and create regional opportunities that prevent the head-on collision between conservation and renewal.

No longer can the conservation area be seen as an island of heritage in the city. Its sustainability is in the extension of its spheres of influence through the mapping processes. The urban pressures of speculation, socio-economic gaps and the resulting crass building are taking their toll on the fabric of the city, and an innovative approach is needed that will integrate the urban processes rather than the isolation, conceptually, politically and socially of conservation areas, this is the essence of sustainability.

It is the recognition of the shared values, between cultures and between times, that allows us to support the resilience in the city. As the human component of the city transposes, the 'minority spaces' and liminal places need to be understood in the context of the future. There is a continued need for the formulation of techniques for public engagement that can lay the foundation for stable co-existence of urban communities in the future as well as commemorate distinct cultural legacies.

These steps include mapping resources, and the identification of site-specific urban attributes, participatory planning and stakeholder consultations; vulnerability assessment in relation to socio-economic stresses, integrated urban development and conservation strategy, partnerships and local management frameworks (Figure 13.3).

Figure 13.3 describes an example of the integration methodology that can be applied to multi-cultural situations. In the first stage, data

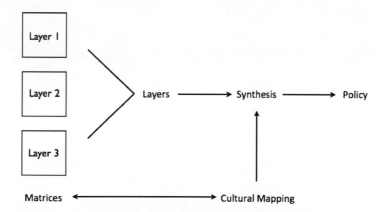

Figure 13.3 The integration methodology that can be applied to multi-cultural situations.[38]

---

[33] See: World Commission on Environment and Development (1987) *Our Common Future*. Oxford: Oxford University Press.

is collected and categorised into applicable layers. The matrices offer a method of organising the natural, cultural and human resource layers and understanding the parameters that should be included in the cultural mapping process. The cultural mapping has a two-way focus; it is taking into consideration in the initial choice of data sourced to create the layers and then combined with the layers to achieve a synthesis that can be translated into policy decisions. Many interdisciplinary partnerships and tools already exist and these should be linked and expanded. This methodology is an expansion of a method attributed to Geddes that describes a process that begins with a survey continues with an analysis and then graduates to the plan.[34]

The expanding global interest and research within diverse cultural environments can assist in developing dialogue through descriptive and analytical comparisons calling for an exchange of knowledge to harmonise metadata terminol-ogy and the setting of base lines for results based management. The lexicon of parameters between the various documents will need to be reviewed in the light of urban research, and include the values and attributes of the World Heritage Convention, the authenticity of Nara, the conflict of Burra, the landscapes of the European Convention and the sustainability of Brundtland together with the wider aspects of socio-economic franchise.

By linking the Historic Urban Landscape approach to social sustainability we provide a role for the non-government organisations (NGOs), thus expanding spheres of influence with the inclusion of previously marginalised communities. The opportunities provided by new communications, crowd-sourcing and technologies-for-all will be a factor for urban design as a key discipline to accompany and manage the processes of transformation as suggested by the Historic Urban Landscape approach.

---

[34] Hall, P. (1992) *Urban and Regional Planning*. London: Routledge: 43.

Interview with Rem Koolhaas
Heritage and the Metropolis

Founding Partner of OMA; Professor in Practice of Architecture and Urban Design at the Graduate School of Design, Harvard University, MA, USA

11 August 2013, Amsterdam, The Netherlands

Question: *The Historic Urban Landscape approach deals with urban conservation, but at the same time tries to establish a connection with the contemporary city. You dealt with this issue in your 'Cronocaos' exhibition. What is the lesson that came out of that experience?*

Answer: It was an important experience that allowed us to discover a number of fundamental issues that exist around preservation. For one thing, you find out that what is preserved is essentially the high quality buildings, the monuments, the palaces, the old and beautiful parts of a city, while there is no effort to preserve elements that are more banal or mediocre, sometimes even ugly. Historic preservation is highly selective, but it does not aim to preserve all the components that played a role in the city as a testimony of a given historical moment. During our research, we worked in Beijing, where a few exceptional monuments have been preserved, while much of the historic fabric has been wiped out. This obviously makes it difficult to connect the historic city and the contemporary city. In Beijing we tried to find a new way to look at the preservation of the city, not limited only to the most extraordinary places. We made a proposal to preserve – but entirely – some sections of the city, or even a regular grid of points, as examples of a given historical period to be maintained intact.

The real question you have to ask in dealing with preservation is what you want to achieve, and I think that the only thing that is legitimate is the idea of transmitting the past to future generations. This means that we cannot only transmit the monuments, because they do not speak about the conditions of life, about how things used to be and work. But this is a very difficult message to pass on. Many years ago I did a project on the Berlin Wall, when it was still there. I made a proposal to preserve it as an extraordinary piece of evidence of an historical period, the Cold War. And yet, when the Wall finally fell, the very moment it fell, everybody wanted to remove it as soon as possible. It was for me a real shock to see this fundamental testimony of the Cold War being wiped out from our memory. This shows that conservation is essentially a political choice. A similar case was the demolition of the People's Palace in the center of Berlin, the symbol of the DDR that was demolished to rebuild a fake version of the Palace of the Kaisers. All this made me aware of the fact that conservation has a strong political dimension.

Question: *This also opens the question who decides what is heritage and what is not. In the European tradition, normally it is the top, the Government, the expert, that establishes what is heritage. Very little comes from the community, from the people.*

Answer: Yes, that is the situation in most cases. However, we also noticed, in studying the Conservation Charters and the UNESCO texts that over time the language has become increasingly vague and allows for the inclusion of an ever-greater number of categories and types of heritage. This can be seen as an answer to a greater demand for heritage by the people. In a way, it is a form of populism.

Question: *At the same time, in many parts of the world, we are witnessing huge changes in the urban scene. Heritage is becoming relatively smaller and smaller, like islands of memory in an urban ocean, although this does not diminish the interest of the public, on the contrary. In an urbanised world, how can the demand for heritage be satisfied?*

Answer: Years ago I wrote an essay called 'The Generic City', where I look at this issue, namely how a small physical heritage can satisfy the demand of more and more people. The solutions are not easy, as everyone wants to touch the authentic heritage and this threatens its very survival, let alone the well-being of the communities living nearby. The only solution I envisage is the amplification of the heritage categories, the creation of more heritage. This is necessary to give a response to social needs. Look for instance at the case of Paris, which has a small historical core, but has become a large metropolis. The people living outside of the centre have a bad relationship with heritage because in their own environment nothing has been recognised as special or unique. Almost everyone agrees that the architecture of most of the suburbs of Paris is horrible, and does not deserve conservation. However, if we looked carefully, I am sure that we would be able to find elements that have a value and can become heritage in the future. For example, some public housing of the 1960s, done as a reaction to the 'bars' of the 1940s and 1950s, brought a lot of innovative ideas – that we now consider important and radical changes – that merit being considered from the heritage point of view. In a way, it is what happened in Le Havre when it was inscribed in the World Heritage List. Before being defined as 'heritage', nobody considered it important or beautiful. There are also many examples of contemporary interventions that add heritage value to a place that did not have it, like in the case of the Guggenheim in Bilbao or the High Line in New York. But in general in today's world, I see that preservation remains a very selective process, and that only a small part of what could be preserved is in fact saved. And I think that very often the destruction is not thought out carefully.

Question: *There are areas of the world where development is very fast, like China and other parts of Asia. What has been your experience in China with respect to the conservation of urban heritage?*

Answer: In China I have seen incredible changes in the past 10 years in the public awareness of heritage, and I think that the cliché that they destroy everything is not valid anymore, so, there are some reasons to be optimistic. I suggested quite strongly to look at the architecture and infrastructure of the Mao Zedong period, as well as to the work done by Soviet engineers and architects in the reconstruction phase. Some of it is very valuable, and could become heritage even if it is only 40–50 years old. Because of the numbers and the public demand, we need to expand the idea of heritage. That is the key to the story.

Question: *Which new categories of heritage should be considered in your view for the future?*

Answer: I think that infrastructure should be better evaluated, like parts of airports, parts of highways – I am sure they have a potential. For example, at St Petersburg's airport there is a small pavilion from the Soviet times in between two airline terminals. It is a fantastic building. There is even a map where you see all the air destinations existing at the time of the Soviet system, there were over 200 as compared to today's 70. The so-called 'Cosmic architecture' is another case of possible heritage. I think the Moscow metro could also be considered heritage. In conclusion I think that a new concept is emerging, a broader concept where the monumental aspect is important but is not the only one. In a way this concept is more people oriented, but at the same time it increases the problems of preservation. Traditional conservation focuses on the material substance of heritage, on the stone. But it's getting clearer that we don't have a way to preserve all the material substance, and we need to find ways to preserve the social and political values, the way people are attached to a place. In a way, this is what's happening in many emerging countries in the Arab region, in the Gulf for instance. In Doha, they have reconstructed the traditional souk artificially, and turned it into a centre for entertainment. Rebuilding an entire centre is an example of artificial construct that works because it has use value. Another case is the United Arab Emirates where big projects have been developed, with a mixture of Arab and western models. These are experiments we should look at, trying to go beyond our intellectual prejudices.

Question: *Do you see a form of Regionalism emerging in urban heritage? Will any region have a different model of urban conservation?*

Answer: No, I don't see this happening, I rather see in many countries and regions a state of 'cluelessness', like you would find in places like Indonesia or the Philippines. These countries have not yet defined their own vision on heritage, and this is where the action of UNESCO could be very productive. I think that Regionalism would introduce a fake national identity obsession; I would be more in favour of approaches that support forms of impermanence, which are very typical of Asia, where there is very little stone building and where other types of heritage could be valued, like for instance the system of canals of Bangkok or cultural landscapes linked to hydraulic management.

Question: *Looking at the future of conservation, do you see any effect that could come from the information society, something that could change the approach to heritage, as we know it today?*

Answer: I think that information technologies will have a great impact on heritage conservation. If you look at the current architectural scene, you see that fewer and fewer people see the actual architectural works; it's all played out in the media, on the Internet. A similar thing can happen for heritage, people will do more and more virtual visits and this will open new dimensions for conservation. In heritage, the personal experience of the visit is what matters. The question is: can this experience be replicated and up to which point?

Question: *You worked a lot on the role of megastructures in urban development. These are normally seen as a response to urban chaos, but do you see them also as a way to facilitate conservation?*

Answer: In fact megastructures are intended to perch over the city, so that nothing would be destroyed; they are completely compatible with the existing city. There is a very good case in Seoul in Korea, where a huge building has been built that is superimposed on the existing city, like a viaduct – it's a beautiful example of the concept. But also infrastructures will be very important in the future of conservation. Look for instance at the very successful case of Perugia in Italy, where a system of escalators has been built to transport tourists from the parking lots to the centre. It's a very positive example, that allows with a minimal compromise to guide the flow of people to the city.

Question: *The insertion of modern architecture in historic contexts is always problematic and raises continuous polemics. There are those who support the contrast with the existing environment, like the 'Friendly Alien' in Graz, and those that support integration, sometimes to the extreme of replica and imitation. How did you deal with this issue?*

Answer: A lot of our work dealt with historic contexts, in Europe and other regions. We were never interested in contrast, but also we did not want a simple compromise. In the case of the Dutch Embassy in Berlin, for instance, we did not accept the city planning guidelines that demanded a building to fill the block, and we started studying the particular position of the building that has many different perspectives. We then designed the building as a kind of 'machine' that allows people to perceive all the different environments. The building made itself sensitive to the place. We always try to do architecture with a real complexity, whereas contrast *per se* is almost never complex. Even the CCTV building in Beijing, with its slanted pillars and the ornamental frames, works very well with respect to the low-value context of the buildings around it.

Conclusion
The Way Forward: An Agenda for Reconnecting the City

Ron van Oers

Vice Director, World Heritage Institute of Training and Research for Asia and the Pacific (WHITRAP), Shanghai

> *Design is a funny word.*
> *Some people think design means how it looks.*
> *But of course, if you dig deeper, it's really how it works.*
>
> Steve Jobs

Managing the City as a Living Heritage

The Historic Urban Landscape is an updated heritage management approach based on the recognition and identification of a layering and interconnection of values – natural and cultural, tangible and intangible, international as well as local – that are present in any city. It is based also on the need to integrate the different disciplines for the analysis and planning of the urban conservation process, in order not to separate it from the planning and development of the contemporary city. In other words, the Historic Urban Landscape approach seeks to reconnect heritage precincts with the modern city; the city with its wider setting and hinterland; urban conservation with the process of city planning and regional development; new architecture with the historic context; and the different cultural traditions, including migrant and traditionally resident, with each other and with the socio-economic trends that are evolving in the contemporary city. All this is part of the day-to-day running of the city in order to respond timely and adequately to the dynamics of the twenty-first century urban condition, which is always in a state of flux and seemingly accelerating. With many competing demands for attention and investment of efforts and resources it can be hard to focus and determine priorities.

The Historic Urban Landscape approach focuses on the identity and values embedded in any city. In socio-cultural terms these are a key determinant of quality of life, while in economic terms they can be a strong component of competitiveness in the global marketplace.[1] Both

---

[1] PricewaterhouseCoopers (2005) *Cities of Opportunity.* New York City: 52.

Reconnecting the City: The Historic Urban Landscape Approach and the Future of Urban Heritage, First Edition. Edited by Francesco Bandarin and Ron van Oers.

quality of life and competitiveness are required to attract young talent and entrepreneurial citizens, as well as inward investment and businesses. In a process of mutual dependency, these actors will create a virtuous cycle characterised by a self-strengthening process that cares for the environment, ensures creativity and innovation, drives development, and further reinforces the identity of the city.

Similar reasoning regarding the nexus between culture and development is part of an international effort spearheaded by UNESCO in preparation of the upcoming review of the 2015 United Nations Millennium Development Goals (MDGs). The assessment of its achievements should lead to a formal recognition by the international community of the important role of culture in the development process, and thus for the inclusion of Culture in any post-2015 UN programme. On 5 December 2013 the United Nations General Assembly, at its sixty-eighth session, adopted Resolution A/C.2/68/L.69 on Culture and Sustainable Development, which recognises *that culture is an essential component of human development, that it represents a source of identity, innovation and creativity for the individual and the community and that it is an important factor in social inclusion and poverty eradication, providing for sustainable economic growth and ownership of development processes*.[2]

On the other hand, PricewaterhouseCoopers states in its fifth report on *Cities of Opportunity*, which analyses the trajectory of 27 global cities, that *a city's cultural influence becomes dominant only after it is reinforced by economic power, and cities such as Berlin, Istanbul, and Mumbai need economic strength to bolster their global cultural presence*.[3]

Thus, reciprocity is the key word, as each represents a conditioning factor for the other to occur and to flourish. After all, development without the conservation of key resources cannot be sustainable, while conservation cannot succeed without development to sustain its efforts. Indeed, as part and parcel of sustainability, a city's identity based on heritage values and local culture generates opportunities for urban development that, once in motion and when properly managed, will support and further strengthen this identity.

Identity and Sense of Place

A city's identity resides in its inherited past, as well as in its present condition. It comprises the multitude of meanings expressed through its built fabric, historic and contemporary, as well as its traditions and attitudes that have been deposited over time by successions of resident communities. Local communities in the present are inspired by it and continue that tradition in their collective way of living and individual performances. In the case of Paris, for instance, and in full recognition of Baron Haussmann's grand design in the second half of the nineteenth century, it is the local communities however who continue to sustain that remarkably beautiful city replete with world-class monuments, cultural institutions, culinary delights and fashion trends, as the example *par excellence* of the art of living. In writing about Rome, Francine Prose exclaims: *'One of the joys of Rome is its ability to make you feel as if there is a place that has spent centuries accumulating layers of beauty and history, patiently waiting just for you to arrive'.*[4] Daniel Bell and Avner de-Shalit in *The Spirit of Cities: Why the Identity of a City Matters in a Global Age* put similar arguments forward in support of other world cities, which *'invest thought, time, and money in protecting their unique ethos and preserving it through policies of*

---

[2] United Nations (2013) *Resolution A/C.2/68/L.69*, New York: 2.

[3] PricewaterhouseCoopers (2012) *Cities of Opportunity*. New York City: 73.

[4] 'Exploring the secret corners of Rome', *International Herald Tribune*, 7–8 September 2013: 23.

design and architecture and through the way people use the cities and interact with them'.[5]

The complex layering of natural, cultural, built, intangible and local heritage, superimposed on and interacting with each other, and connecting the physical with the socio-economic and cultural environment, is what can be labelled as the historic urban landscape of a city – it is the critical underpinning of its sense of place, or *genius loci*. A thorough identification, inventory and assessment of natural, cultural and community resources of the city, therefore, through a process of cultural mapping, stakeholder consultations and participatory planning, among others, should provide the basis for a better understanding of the city's character and identity, and its broader sense of place. This should inform management of the city, including policy-making and enforcement towards the conservation of key resources (among which its urban heritage), as well as towards the planning and design of new extensions and infill. These policy domains should touch each other, so that heritage and local culture can be put to work for the improvement of living conditions for local communities without damaging or destroying it, which would deprive future generations of the enjoyment and inspiration, as well as the means to their sustainable development.

Culture shapes people and people make places. Through an assessment of local cultural traditions in their urban setting, an additional layer of intangible values over and above the merely architectural and urban (historic monuments and larger architectural ensembles) can be recognised. This field remains largely unexplored up to this day.[6] However, increasingly the inter-dependency between tangible and intangible heritage is being recognised, as put down in 2004 in the Yamato Declaration on Integrated Approaches for Safeguarding Tangible and Intangible Cultural Heritage.[7] This document advocates for the recognition that the values associated with monuments and sites should also be considered as intangible cultural heritage (as defined under UNESCO's 2003 Intangible Heritage Convention) when they belong to the living heritage of present-day communities. Hence the identification of intangible cultural heritage in the urban context needs involvement of local communities. Its preservation and promotion, however, requires a connection between different fields of public policy linking education and skills training in the arts and crafts to local tourism development, for example, or support to local small and medium-sized businesses, or to festivals and trade fairs as part of promotional events for the city, to name some opportunities.

Local Heritage and Corporate Image

Without such an integration of different policies and their approaches, synergies in safeguarding and management of tangible and intangible cultural heritage are hard to achieve. This would hamper a full realisation of the potential of both heritage categories to contribute to the enjoyment and further cultural development of the city. When considering the cultural development of the city, which should build upon and strengthen its identity, there is a need to distinguish between recently established 'traditions', which serve primarily a commercial purpose,

---

[5] Bell, D and de-Shalit, A. (2012) *The Spirit of Cities: Why the Identity of a City Matters in a Global Age*. New Jersey: Princeton University Press: 5. The role of cultural heritage in creating livelihoods, in reviving communities and in establishing identities is further highlighted in: Bigio, A. G. and Licciardi, G. (2010) *The Urban Rehabilitation of Medinas – The World Bank Experience in the Middle East and North Africa*. Urban Development Series N.9, Washington DC: The World Bank.

[6] See for example: Akins, J. and Binson, B. (2011) Transmission of Traditional Lanna Music in Chiang Mai: Continuity and Change in a Contemporary Urban Environment. *City, Culture and Society* 2: 243–254.

[7] Nara: Japanese Agency for Cultural Affairs; UNESCO (2004). Available at: http://portal.unesco.org/culture/en/files/23863/10988742599Yamato_Declaration.pdf/Yamato_Declaration.pdf; see also: Bandarin, F. and Van Oers, R. (2012) *The Historic Urban Landscape. Managing Heritage in an Urban Century*. Chichester: Wiley-Blackwell: 51.

and those where city culture has grown out of authentic local traditions into a regional or international phenomenon that attracts scores of visitors from around the globe. Examples such as St. Patrick's Day in Chicago, Carnival in Rio de Janeiro, the arrival of the new sake in Shirakawa-Go, the Oktoberfesten in Munich, or the Kumb Mela in Varanasi, are all genuine celebrations of often age-old traditions in the setting of the contemporary city. They constitute more than just another pretext to party: they strengthen social cohesion and transmit an attitude towards relationships in the city as well as the values of a society. These are powerful messages to attract positive attention and visitors, citizens and businesses into the city.

Next to a growing importance of the intangible heritage of local communities, seemingly contradictory, another phenomenon can be observed: urban heritage has come to be bound more to place than to people. The hyper-mobility of goods, ideas and people that characterises globalisation in the twenty first century puts a premium on fixed assets to place, and urban heritage is paramount among these. Cities, more than rural areas, are places of transition, where people and their ideas move in, out and through. This creates the contradictory dynamics, which according to Rem Koolhaas are characteristic of the city, since 'the essence of metropolitan culture is change – a state of perpetual animation – and the essence of the concept of 'city' is a legible sequence of various permanencies'.[8]

While this does not diminish the need for local community consultation and participation in heritage identification and protection, however, it does carry some consequences. According to Aspa Gospodini, the more traditional perception of local culture and built heritage doesn't appeal to new immigrants as it hardly

represents a multi-cultural identity.[9] This means an active process of information, education and engagement is needed to involve newcomers, whether immigrants or temporary visitors, in the conservation and celebration of the city, both historic and contemporary.

A prime example of such a strategy concerns the management of the Tower of London in the United Kingdom, which was inscribed on the World Heritage List in 1988. During the 1990s and in the development of a World Heritage Site Management Plan for this key monument, Historic Royal Palaces incorporated a policy of engagement with immigrant population groups (primarily from South Asia) in the boroughs directly adjoining the Tower of London. These neighbourhoods are among the poorest in London, where the overwhelming majority belongs to an immigrant community. These communities had no cultural relationship with, nor socio-economic interest in, the Tower's exceptional status as World Heritage. For these boroughs a physical upgrading as well as social engagement programme was developed as part of the monument's management scheme.[10] After all, the key to passing on a tradition to the community is by making it relevant to them.

In this vein, between 2003 and 2006 the office of the Mayor of London (then Ken Livingstone) set up a major initiative called the Mayor's Commission on African and Asian Heritage (MCAAH). Its aim was to include the cultural heritage of the city's immigrant communities from Asia and Africa into the management of London's heritage spaces, in a manner that sought to support the African and Asian community-based heritage sector, to integrate it into the national story, and to utilise and develop their contributions to heritage tourism and to marketing London as a global city.

[8] Koolhaas, R. (1994) *Delirious New York*. New York: The Monacelli Press: 296.

[9] Gospodini, A. (2004) Urban Space Morphology and Place Identity in European Cities – Built Heritage and Innovative Design. *Journal of Urban Design* 9(2): 225–248.

[10] Historic Royal Palaces (2007) *Tower of London World Heritage Site Management Plan*. Historic Royal Palaces: Hampton Court Palace Surrey.

An important finding of the Commission reported that:

> ... *mainstream definitions of heritage often present the notion that heritage must be tangible to provide documented acknowledgement of historical existence and cultural traditions. African and Asian communities often engage with their heritage as part of their daily activities and rituals. This notion of living heritage presents challenges to the mainstream interpretation: many approach heritage through cultural activities and customary practices.*[11]

Thus in relation to the preservation of intangible cultural heritage, as viewed by traditional communities in particular, it is not only the knowledge that's being passed on, but it's also about providing the context for someone's own experiences to happen and to be nurtured.

The City as Repository of Urban Experiences

Next to heritage becoming place rather than population bound, hyper-mobility introduces another challenge, that is of a new architecture in the historic context. Our fast and consumerist society, where the average shelf-life of products has diminished significantly in tune with the innovations in Information and Communication Technologies (ICTs), has stopped building cities. It presumably requires a time horizon too far removed from instant gratification. Or, as Peter Mörtenböck and Helge Mooshammer posit, *'from the perspective of the global market economy,*

the city is increasingly becoming a construct that can be dismantled and shifted'.[12] Instead, now we only build buildings.

But in as much as the virtual world of ICTs can have a profound impact on innovations in developing new urban models – away from destructive patterns of urban development, for instance, which consume too much resources including urban heritage – in the real world architectural practice should still be concerned with the design of buildings that reflect a society's social and cultural values. Architectural form, after all, involves the question of the production of relationships. Next to searching for new ways to respond to current needs and wishes, including architecture's role in responding to the sustainability agenda, this should involve also a respect for the inherited wisdom of previous generations of builders, as expressed in time-honoured construction techniques, locally sourced materials, and climate-responsive spatial patterns and architecture.

A selection of recent articles related to the incompatibility of new architecture in historic context reveals that this issue is far from being resolved and, moreover, that apparently even the finest in the trade experience difficulties in getting the balance right.[13] Surely, these are critiques written by journalists and what do they know, one perhaps tends to think. However, maybe the most outspoken critic of contemporary architectural practice comes from the inner circle of architects. Duo Dickinson in *ArchitectureBoston* laments that:

---

[11] Mayor's Commission on African and Asian Heritage, London: Greater London Authority, July 2005:40; for the full report, visit:
http://www.academia.edu/421539/Delivering_Shared_Heritage_Report_Mayors_Commission_on_African_and_Asian_Heritage; for the case study, see: Arokiasamy, C. (2012) Embedding Shared Heritage: the Cultural Rights of London's African and Asian Diaspora Communities. *International Journal of Heritage Studies*, Vol. 18, No. 3, May: 339–345.
[12] Mörtenböck, P. and Mooshammer, H. (2008) *Contested Spaces. Networked Cultures*, Rotterdam: NAi Publishers: 81.
[13] See for instance: 'Stuck in neutral', *Financial Times*, October 2012:16; 'The Netherlands Tries Hard to Make a Splash', *International Herald Tribune*, December 2012:11; 'Zaha Hadid Piques Preservationist Anger in Beijing and Belgrade', October 2013, at:
http://www.blouinartinfo.com/news/story/936646/zaha-hadid-piques-preservationist-anger-in-beijing-and?utm_source=BLOUIN+ARTINFO+Newsletters&utm_campaign=23a1b2b5a0-Daily+Digest+8.7.13&utm_medium=email&utm_term=0_df23dbd3c6-23a1b2b5a0-82964634; as well as 'The Icon (and Iconoclast)', *The Wall Street Journal*, March 2013: W9.

'architects are now taught to design buildings for the approval of other architects, our version of preaching to the choir'. The issue that he poignantly raised concerns architectural education and training, which deal with 'abstracted design methodologies grounded in academic theory' that hardly go beyond the aesthetic. But, he continues:

Buildings must also deal with the world of social context, as almost all buildings are in proximity to others and exist in a cultural reality. Whether taught or not, anthropological, sociological, ergonomic, and historical realities of architecture are in the mix of design criteria by which non-architects judge buildings. Sadly, these non-aesthetic criteria have been back-burnered, lip-serviced, or outright ignored in architectural education. Abstract artistic sensibilities have become the pre-eminent point of departure in architectural education and in those buildings venerated in academia and the professional press.[14]

If architectural education and training does not include instruction and reflection on the socio-cultural context or historical achievements of previous generations of builders, including of vernacular dwellings,[15] how then can we expect any sensitivity, even from the most gifted among the architects, towards the values and potential of urban heritage? It came all the more as a surprise (or, did it?), that the 2012 Pritzker Prize, the highest international recognition in architecture, was awarded to Wang Shu, the first Chinese architect ever to receive this. He is critical of China's rush to urbanisation and wholesale importation of Western-style architecture, which distorts age-old physical and social structures, as he reveals in the interview in this book (conducted just before he received his prestigious award). Wang Shu is recognised foremost for his modest, 'anti-starchitecture' approach, which

includes renovation of existing buildings and reuse of building materials.[16] All this signals the need for a renewed encouragement and confidence in the reconnection and integration of different disciplines and professional practices that converge in the historic city.

Integrating Disciplines and Professional Practices

The conservation and rehabilitation of the historic city needs to go beyond the expert opinion of the conservation specialists only. There is an urgent need to involve a broader constituency comprised of other groups than the usual suspects, such as residents, youth, entrepreneurs, urban planners and managers, artists and the media, among others, in order to forge collaborative alliances, reduce conflicts and optimise creative use of the historic city. Moreover, this should be extended toward providing direction for the planning and design of the city's contemporary parts.

Following a long and solid tradition built up by the Aga Khan Trust for Culture's Historic Cities Program,[17] planning for intervention in historic cities or inner city areas – labelled by His Highness the Aga Khan as *'multi-input development areas'*[18] – should be based on established priority objectives as regards the physical, legal, economic and social context, with community support and a programme to achieve them. This comprehensive approach requires broader engagement, both as regards professional inputs from a variety of disciplines, as well as from a greater diversity of stakeholders, who need to be

---

[14] Posted online in Volume 13, No. 4 on 10 November 2010, at http://architectureboston.wordpress.com/2010/11/10/why-modern-architectural-education-is-archaic/.

[15] See for this argument also: Van Oers, R. (2006) Preventing the Goose with the Golden Eggs from Catching Bird Flu – UNESCO's Efforts in Safeguarding the Historic Urban Landscape. *Cities between Integration and Disintegration: Opportunities and Challenges*, ISoCaRP Review 02, Sitges: 21.

[16] See also: 'The Local Architect/Wang Shu', at http://www.archdaily.com/212424/the-local-architect-wang-shu/, February 2012; and 'Modern Architect with a Traditional Vision', International Herald Tribune, August 2012: 18.

[17] See http://www.akdn.org/hcp/.

[18] See the interview in this book.

consulted on the particulars as regards the relationship between people and place. Local people need to describe and make practical their emotional attachment to a place through the identification of physical and associative attributes. More than gathering essential input for decision-making as regards interventions, this process serves to build ownership and trust. These are critical components when aiming for a significant degree of consensus that can move initiatives and projects forward.

This book presents a selection of the most relevant disciplines and professional practices concerned with urban conservation and management, but they cover by no means the complete spectrum. While all of the disciplines have an important part to play in the process of urban conservation and development, currently there is little integration between them. This leaves the field of urban heritage management seriously compartmentalised, much to the detriment of the historic city and its potential to contribute to the development of the contemporary city. With a view to de-compartmentalise and reconnect the city's human, social and cultural capital, some of the more successful cities of today, including Turin, Portland, Bilbao, Seattle and Barcelona, to name a few, have created a one-stop-shop for the management of their urban environment.

Since the adoption of Agenda 21 in 1992, when the United Nations framed sustainable development and put forward the need to position local communities centre-stage in the decision-making process, a gradual shift in development thinking has occurred away from a focus on *managing* cities to one on *governing* cities. The latter places greater importance on the quality and sensitivity of local government and signals a shift towards 'participative culture'.[19]

Tim Campbell (2012) explains that one of the central elements related to good governance involves fostering a process toward collective learning (as opposed to acquiring specific skills that are the focus of twinning arrangements and international municipal cooperation). The key idea behind this, he outlines:

> ... is that cities or parts of them learn as a unit. They begin to reach agreement faster, and they show a widespread common understanding of themselves. They achieve a collective city identity. All of this is based on common shared values, and this in turn on exposure to and working through the differences among key stakeholders in the city.[20]

He demonstrates how networks already operating in most cities are used to foster and strengthen connections, to achieve breakthroughs, and to catalyse and convert information and knowledge into high-value 'clouds of trust'. These are the raw material of innovation. When realising the potential, as well as the stakes, one wonders why not all cities are learning from each other?

As a remarkable example of good governance, for instance, Lagos in Nigeria has increased its tax receipts and is currently using the extra cash flow to overhaul its infrastructure to prepare for a population growth from 11 million now to 19 million by 2025.[21] This is in sharp contrast to Jakarta in Indonesia, whose greater metropolitan area has a population of 28 million but remains one of the world's few major cities without a rapid-transit system. This results in a clogging of its streets by 9.9 million motorised vehicles for hours on end each working day.[22] Or the plans for an overhaul of *La Defense* in Paris, which in the eyes of planning professionals has remained a failure in terms of urbanism,[23] while Canary Wharf in London (often compared to *La Defense* because both were conceived as decentralised

---

[19] Cumberlidge, C. and Musgrave, L. (2007) *Design and Landscape for People – New Approaches to Renewal*, London: Thames & Hudson: 16.

[20] Campbell, T. (2012) *Beyond Smart Cities – How Cities Network, Learn, and Innovate*, New York: Earthscan/Routledge: 41.

[21] See: 'Lessons from Lagos'. *Financial Times*, 29 June 2012: 11.

[22] See: 'For Jakarta Commuters, Relief is Distant Prospect'. *International Herald Tribune*, 1 August 2013: 4.

[23] See: 'Seeking a Human Touch for Paris's Financial Hub'. *International Herald Tribune*, 1 August 2013: 15.

hubs away from the city centre), is enjoying tremendous growth as part of a city axis that extends to King's Cross in London's centre.[24] The King's Cross district is booming, where average house and apartment prices have risen 41% over the past five years, in part helped by the restoration and rehabilitation of the gothic train station of St. Pancras, which spurred on a regeneration of the entire district.[25]

Singapore, the tropical island nation and home to thousands of native edible plants, is reinventing itself as an urban farming hub to produce and consume its own organic food with, among others, the recently opened Kranji Heritage Trail, which includes 34 independent farms and agriculture-related businesses.[26] Such re-focus and reinvention might contain important insights for the city of Detroit, which filed for bankruptcy on 18 July 2013. It has currently 78,000 abandoned buildings and is seeking how to reinvent itself economically, geographically, demographically and culturally.[27] A similar problematic situation, but of a completely different nature, exists in Cairo after the revolution, where urban planning has become a do-it-yourself affair.[28] Moscow, meanwhile, has opted for the establishment of a world-class educational facility, the Strelka Institute for Media, Architecture and Design, which uses the city as *a canvas for world-famous architects and urban planners to host discussions on a liveable design*.[29]

However, there are several reasons for cities to be hampered in collective learning, among others small city size, adverse policy, and phase changes, to name a few. As Tim Campbell further explains:

Going beyond smart cities means understanding how cities construct, convert and manipulate the matrix of relationships that grow naturally in urban environments. For instance, getting the most out of networks is a matter of managing tight and loose bonds and fostering intersections, for instance the public and private, the marginal and civic. But management is not just a matter of public policy. It is also a matter of private will, of businesses and corporations as well as marginal communities together with local policy makers recognising mutual self-interest and joining a wider circle of stakeholders to engage in trusting relationships that are the gold standard of learning cities.[30]

Future Challenges of Urban Conservation

Urban conservation has become a policy and practice in most European countries since the Second World War. Initially linked to the needs of post-war reconstruction, it has subsequently evolved into a fully fledged component of urban planning and management. Several countries have adopted specific legislation in the 1960s and 1970s that are to this date the main support to conservation practices. Overall, these policies have been successful in preserving the physical fabric of the historic cities of Europe, while at the same time their economic and social fabric has been altered, due to gradual social and economic transformations, and the emergence of tourism as the prime user of historic areas.

Following the European models, other regions of the world (e.g. Latin America and the Arab region) have developed conservation plans and policies that have been instrumental, at varying degrees, for the conservation of important heritage sites. Different is the situation of Africa and Asia, where conservation policies have been occasional and where urban heritage has been

[24] See: 'Trouble in Metroland'. *The Economist*, 24 August 2013:26.

[25] See: 'Station Helps put London District on the Right Track', *International Herald Tribune*, 23 December 2011: 13.

[26] See: 'Improbably, Singapore Goes Locavore'. *International Herald Tribune*, 13–14 July 2013.

[27] See: 'Fish Farms, Start-ups and the Search for a Way to Fix Detroit'. *International Herald Tribune*, 4 September 2013: 18.

[28] See: 'Re-doing Cairo, from Bottom up'. *International Herald Tribune*, 29 April 2013: 2.

[29] See: 'Redesigning Moscow'. *The Moscow News*, No. 50, 3–5 July 2012: 16.

[30] Campbell, T. (2012) *Beyond Smart Cities – How Cities Network, Learn, and Innovate*. New York: Earthscan/Routledge: 13.

largely affected by rapid and abrupt social and economic transformations.

UNESCO's World Heritage Convention of 1972 has been the main tool to promote urban conservation practices and values in the global sphere, and in fact today urban heritage represents the largest category of the World Heritage List (about 300 sites out of just over 1000). In the past 20 years urban heritage has become the focus of several development and conservation strategies, supported by the United Nations system, among which UN-Habitat, UNESCO and the World Bank, and a number of private sector actors, among which the Aga Khan Trust for Culture has been the most active and successful. This has allowed the development of important experiences and has fostered the advancement of local protective legislation, regulation and practices.

Fifty years of conservation experiences have created a vast inventory of practices and policies. In spite of the uniformity of the doctrinal principles descending from the professional Charters and UNESCO Regulations, these practices vary from place to place, and have evolved along the years, reflecting a shift of perceptions of heritage values and an evolution of the needs, linked to the demands of social and economic actors. However, the different practices developed in Europe and in other regions are not fully documented. There is still a lack of national and international research in the field of urban conservation.

In many regions of the world, urban heritage is subject to intense pressures due to neglect, decay and population shifts, development of central areas, and building substitutions and alteration, among others. In certain countries marked by high rates of development (i.e. China and South East Asia) a large part of the traditional urban structures and heritage has been irreparably lost. Perhaps the Indian subcontinent is the largest remaining 'reservoir' of authentic urban heritage, although this is extremely vulnerable and not adequately protected by local or national regulations. In other regions and countries (i.e. Latin America and Africa) similar processes are underway. In many countries of the Arab region the present unstable situation has exposed urban heritage to destructions due to conflict and rapid alterations related to the loss of public control over the built environment – the situation of historic Cairo being perhaps the most striking.

Globally, with the notable exception of the sites protected under UNESCO's World Heritage Convention (and even so) many historic centres throughout the world are threatened and may disappear in the coming one to two decades, should the development rates experienced in the past 30 years continue unabated. Among the most important forces of change at work, tourism is certainly one of the strongest. Many historic urban landscapes have become the focus of tourist development, due to their attractiveness and quality, and to the relatively small investments required for use changes and adaptation. While this is not *per se* a negative process, there is a risk of social exclusion and population substitution (gentrification), as has happened in many cases, both in the developed and emerging world.

Today, many countries and cities are asking for assistance to protect and value their urban heritage and are willing to develop policies and actions to address the issue. International actors have already demonstrated their interest and commitment. However, several challenges need to be tackled in the development of an effective global action:

- The lack of a good database on the history and practices of urban conservation in the different regional contexts, in order also to understand and value cultural diversity in the approach to conservation.
- The lack of an updated 'map' of the challenges and critical cases, in order to orient policies at international and national levels.
- An insufficient knowledge of the existing legislation in different countries and regions, in order to orient governments toward successful practices and partnership models.

These global challenges can and should be dealt with in the coming years, in order to respond to the needs expressed by many countries to develop adequate legislation and professional practices for the protection of the quality of their urban heritage and of the identity of their historic cities. The programmes and projects associated with the application of the Historic Urban Landscape approach are an integral part of this response.

The Critical Path: Historic Urban Landscape Action Plan

In addition to collective learning, as introduced earlier, a convergence of different policies and actions is necessary in order to respond to international needs, to create synergies between different urban actors, to optimise efficiency and effectiveness of conservation efforts, and so stay ahead of the game. And with the imminent impacts of a changing climate this game is becoming less playful, perhaps even deadly serious, with little margin for error.

In order to facilitate the implementation of the new UNESCO *Recommendation on the Historic Urban Landscape* by local governments and city councils, and to seek a reconnection between disciplines and professional practices to create synergies, a six-step Historic Urban Landscape Action Plan was elaborated. The action plan was included in the General Conference Resolution on the Historic Urban Landscape and suggests the following set of actions:[31]

> While stressing the need to take account of the singularity of the context of each historic city and urban settlement, which will result in a different approach to its management, nevertheless six critical steps can be identified for Member States to consider when implementing the Historic Urban Landscape approach. They would include the following:
> 1. *Undertake comprehensive surveys and mapping of the city's natural, cultural and human resources (such as water catchment areas, green*

> spaces, monuments and sites, view sheds, local communities with their living cultural traditions).
> 2. *Reach consensus using participatory planning and stakeholder consultations on what values to protect and to transmit to future generations and to determine the attributes that carry these values.*
> 3. *Assess vulnerability of these attributes to socio-economic stresses, as well as impacts of climate change.*
> 4. *With these in hand, and only then, develop a city development strategy (CDS) or a city conservation strategy (CCS) to integrate urban heritage values and their vulnerability status into a wider framework of city development, the overlay of which will indicate (a) strictly no-go areas; (b) sensitive areas that require careful attention to planning, design and implementation; and, (c) opportunities for development (among which high-rise constructions).*
> 5. *Prioritise policies and actions for conservation and development.*
> 6. *Establish the appropriate partnerships and local management frameworks for each of the identified projects for conservation and development in the CDS/CCS, as well as to develop mechanisms for the coordination of the various activities between different actors, both public and private.*

As becomes clear from the above, the six-step Historic Urban Landscape Action Plan stands out for its simplicity: it was drafted in such a manner that it could appeal to local governments and city councils in different parts of the world, including those with only limited resources and institutional experience.[32]

Historic Urban Landscape: A Stepped Approach

The Historic Urban Landscape Action Plan acknowledges the need to map and assess the local economic context, and subsequently to

---

[31] UNESCO (2011) General Conference Resolution [2011] 36 GC/41 Historic Urban Landscape, Paris: 1.

[32] For this reason also the World Heritage Centre developed and published, with financial support from the Government of Flanders, a user-friendly brochure entitled *New life for historic cities – The Historic Urban Landscape approach explained,* which is available in all six UNESCO languages at: http://whc.unesco.org/en/news/1026/.

elaborate a CDS, which is *'critical to both good city management and economic performance'*, as promoted by the Cities Alliance.[33] The benefits of engaging in a CDS process go far beyond the drafting of a document alone, and include the participation of a wide range of public, private and civil society leaders, who normally do not have many occasions to meet and work together, and the cross-sectoral communication that ensues in order to align the city's social, economic and environmental interests.[34] The Historic Urban Landscape Action Plan, therefore, aims to align itself with this process. However, it suggests a critical adaptation. Instead of dealing with heritage and local culture only as stand-alone assets of strength (such as in a SWOT-analysis), it proposes to use these as pivots of the process – indeed, since they are part and parcel of the city's identity and paramount to staging a successful developmental process, as explained at the start of this chapter.

For example, the City Development Strategy for Sana'a, the 2000 year old capital of Yemen with around 6000 buildings of heritage significance in its inner city and declared a UNESCO World Heritage Site in 1984, makes reference to this rich and important heritage, as well as to its long tradition of handicrafts and artisan manufacturing.[35] However, the report also clearly states that:

> Old Sana'a still lacks a comprehensive development plan that utilises its unique potentials for economic revitalisation and integrates it with the overall urban context. The new proposed master plan should come up with a clear strategy and action plans on how to integrate the Old Sana'a with the rest of the municipality, protecting its heritage and image from modern development. Such a comprehensive plan must also address conservation and revitalisation of a number of other historic and heritage areas in and around Sana'a.

At the end of the report, in the chapter on Implementation – Agendas for Action, the only further reference made says: *'This strategy will complement the Sana'a Master Plan updating exercise, which is to be carried out by international consultants'*. Hence, we are back at square one: despite all good intentions, there is still no integrated approach.

Therefore, the first three steps of the Historic Urban Landscape Action Plan deal with an extensive identification, mapping, assessment and discussion of the city's values and sense of place; where this actually crystallises in structures, places and rituals, through tangible and intangible heritage and their respective physical and associative attributes, or carriers; what the community as a whole considers essential not to lose when further developing the city, by a reasonable degree of consensus (which is hard to quantify, but should be aimed at through a process of consultation with participation and approval of a majority of stakeholders); and what the vulnerabilities are in terms of socio-economic as well as climatic impacts.

The latter introduces the need to perform Environmental Impact Assessments (EIA)[36] and Heritage Impact Assessments (HIA) when preparing broader development as well as detailed intervention plans.[37] When a clear picture of the city's unique selling points has emerged, where it can be found and in what form and function, and its vulnerability to change has been determined, *then* all this information can

---

[33] The Cities Alliance (2007) *Understanding Your Local Economy – A Resource Guide for Cities*, Washington DC: The World Bank.

[34] *Aden: Commercial Capital of Yemen – Local Economic Development Strategy*, Aden Governorate Local Council and The Cities Alliance, n.d.

[35] *Sana'a – A City Development Strategy*, Capital Secretariat of Sana'a and The Cities Alliance: 7, 22, 31.

[36] See: Bandarin, F. and Van Oers, R. (2012) *The Historic Urban Landscape. Managing Heritage in an Urban Century.* Chichester: Wiley-Blackwell: 159–163.

[37] See: Bakker, K. (2013) Using Heritage Impact Assessment as Tool in the Historic Urban Landscape Approach: The Case of the Mambo Msiige in Zanzibar Stone Town. In Van Oers, R. and Haraguchi, S. (eds) *Swahili Historic Urban Landscapes*. Paris: UNESCO: 50–62.

be integrated into a vision for the future of the city and subsequent elaboration of a CDS. The next steps relate to broadening constituencies and reconnecting the various disciplines and professional practices needed for the day-to-day running of the city, of both its historic and contemporary parts in a synergetic manner, through establishing policies and actions for conservation *and* development.

It is critical here to consider that, next to defining new operational guidance and activities, it will be important as well to review the overall policy landscape to detect overlaps, including potential contradictions, and identify gaps. As often has been shown, while a ministry of culture is defining important avenues for preservation and regeneration of local cultural heritage assets, the ministry of public works, or development, or tourism, in a completely independent manner, has just decided to move forward with a major intervention scheme that significantly impacts on those same heritage assets. An example of this concerns the recent case of the Mambo Msiige in Zanzibar, where the local government is deeply involved in applying the Historic Urban Landscape approach to the overall planning and development of Stone Town and its environs, while in an independent fashion an individual hotel construction project is being executed in one of the prime locations in Stone Town, triggering the alarm of the World Heritage Committee. Such incoherence with potential damaging effects needs to be detected and avoided through cross-sectoral communication and information sharing between different ministries. At the same time such closer interministerial cooperation can also result in uncommon synergies, as was the case for instance in the restoration and rehabilitation of the historic town of Genadendal in South Africa, where a hitherto non-used rural housing subsidy could be utilised to financially complement a restoration and rehabilitation scheme for cultural-historic significant parts of this settlement in the Western Cape.[38]

Finally, an essential element as discussed in this paragraph concerns the establishment of the necessary partnerships and management frameworks to guide and integrate each and every significant action affecting the historic city. The London Tower Environs Scheme mentioned before remains a fine example of an integrated management framework.

Having discussed the different steps of the Historic Urban Landscape Action Plan, subsequently the main challenge is to integrate this action plan into the city's (or nation's) planning framework. This may involve significant consultations, with adaptations, this time at higher government levels. As this framework may be favouring certain dominant cultural groups in society and their preconceptions, such an integration will likely constitute a major overhaul of the framework itself, its relevant legislation and all related components, such as the structure and conduct of Public-Private Partnerships (PPPs), etc. While challenging indeed, it will also constitute an important occasion to progressively move forward to the next stage in making the planning framework a true instrument of twenty first century forward-looking, pro-active, inclusive and participative decision-making.

Interdisciplinary Context and Operational Coordination

The integration of professional disciplines and practices will have consequences for the role of the urban heritage specialist. With an increase in complexity, an expanding circle of stakeholders, and a wider divergence of interests – which are a natural and common element in urban management today and thus nothing to be particularly alarmed about – the role of the urban

---

[38] See: Du Preez, H., Van Oers, R., Roos, J. and Verhoef, L. (eds) (2009) *The Challenge of Genadendal, South Africa*. Amsterdam: IOS Press.

heritage specialist is broadening. Next to typical expertise in interventions, with specialised knowledge of urban conservation history, restoration methodologies and technologies, and materials science, the professional will have to include an attitude and practices of a general facilitator, skilled in guiding people and interest groups, orchestrating processes, understanding different viewpoints, making them heard and practical to broader audiences. Such broadening requires new sets of skills previously not necessarily associated with urban conservation. After all, *'civic space cannot be static, but must be able to accommodate conflict, frictions, debate, difference and multiplicity'*[39] – and the corollary of cultural diversity is dissent, hence the need for negotiation and conflict resolution.

All this means, that heritage studies and research should be broadened, which is what several universities currently are exploring. Such is the idea, for instance, behind the establishment of a new inter and trans-disciplinary research and PhD programme by the UNESCO Chair in Heritage Studies at Brandenburg Technical University at Cottbus in collaboration with the Institute of Heritage Studies at the International Academy for Innovative Pedagogy, Psychology and Economy (INA) of the Free University of Berlin, in Germany. It aims to investigate the different facets of heritage, what it stands for and means to different groups of stakeholders and actors. Precisely by broadening inter- and trans-disciplinary cooperation in heritage research new insights could be gained, and how these could be used in updated programmes and activities for heritage preservation.[40]

Also at Durham University in the United Kingdom a new Master of Arts in International Cultural Heritage Management has started that aims to incorporate the complex challenges of development in new ways and means of cultural heritage custodianship.[41] Next to these, however, an increase in specialised research and knowledge in a variety of urban conservation-related disciplines remains also needed.

In considering these overall dynamics, what has become paramount is a need for the provision of *'the intellectual space where theory and practice can find a shared platform'*[42] to develop a dialogue and link good practice in urban management to a profound understanding of the city's sense of place. Moreover, understanding gained from the experience of practice can also contribute to the advancement of theory.[43]

A 20-Point Research Agenda for Planners and Designers

As stated earlier, the range of disciplines and professional practices involved in urban conservation could be supplemented by various other fields of thought. Nevertheless, the scope of this paragraph relates to the disciplines covered in this book, with several suggestions related to fundamental or applied research, which build upon the arguments and insights put forward by the contributing authors.

Overall, and with a particular view on sustainability, a critical research area would involve the transformation from an industrial, carbon-rich to an ecological, low-to-no carbon civilisation. In order to make this transition, next to surviving the current environmental crisis, an important first step would involve a significant growth in knowledge and technology to enable us to

[39] Cumberlidge, C. and Musgrave, L. (2007) *Design and Landscape for People – New Approaches to Renewal*, London: Thames & Hudson: 17.

[40] See: Albert, M-T., Bernecker, R. and Rudolff, B. (eds) (2013) *Understanding Heritage – Perspectives in Heritage Studies.* Heritage Studies Volume 1. Berlin: De Gruyter GmbH.

[41] See https://www.dur.ac.uk/archaeology/postgraduate/taughtprogrammes/maichm/.

[42] Nasser, N. (2013) The Contemporary City: Speaking the Same Language in Design and Theory. *Urban Morphology*, Volume 17, Number 1: 52.

[43] This is promoted by Willowbank's School of Restoration Arts, in Ontario, Canada, for example – see http://www.willowbank.ca/content/beta/home/school-of-restoration-arts/index/.

manage our built environment as an ecological system. Our progress as a society will hinge in large measure on changing attitudes and behaviour as regards resources consumption, as well as the establishment of a new balance with nature. This should address:

1. Our technological capability to mimic nature in its recycling and reuse of resources.
2. How cities can become stable ecosystems in themselves, in particular how to achieve zero-impact on the surrounding environment.[44]
3. What intellectual, ethical and societal leaps forward need to be made to reach the understanding that learning to live in harmony with nature will constitute a necessity as well as an enrichment of our future urban society.

The Historic Urban Landscape approach aims at re-establishing the connection between management of the historic environment, contemporary urban development and the geological context, in order to ensure a higher degree of sustainability and risk control, as well as harmony and continuity in urban forms, building structures and materials. This conceptual and operational connection demands a stronger integration between sectors and disciplines, and the development of *ad hoc* knowledge and management tools should be guided by questions on:

4. How geology determines urban forms and guides urban development processes.
5. How to integrate geological approaches and urban management practices.

The *Recommendation on the Historic Urban Landscape* aims to give relevance, among other things, to the 'time dimension' in the management of historic cities, and recognises in the archaeological methods a principal source of knowledge and management practice. A landscape approach aims to integrate this dimension in the urban planning, conservation and development process. However, several issues need to be discussed and resolved before this can become an operational reality, including:

6. The role of archaeology in understanding modern urban development processes, including its compatible uses and integration with the contemporary urban landscape.
7. The harmonisation of management processes.
8. The presentation and aesthetic treatment of archaeological resources.

As a methodology aiming at enabling the understanding of the process of formation and transformation of the historic context, urban morphological analysis proves to be an essential tool for the management of the historic urban landscape. However, the operational capability of the method seems limited, primarily because it is a time-consuming exercise. In order to improve on this, several issues need to be explored, including:

9. Obtaining a clearer picture of the advantages and limits of the operational use of urban morphology in the planning and management of historic cities.
10. How new ICTs can assist in the development of urban morphological analysis as a tool for the understanding of urban transformations.
11. How to integrate urban morphological analysis within the comprehensive approach to urban conservation as suggested by the *Recommendation on the Historic Urban Landscape*.

---

[44] For instance, the concept of the 'zero-waste city', which includes a 100% recycling rate all recovery of all resources from waste materials, is explored through case of Stockholm, Sweden, and Adelaide, Australia. In Uz Zaman, A. and Lehmann, S. (2011) Urban Growth and Waste Management Optimization Towards 'Zero Waste City'. *City, Culture and Society*, 2: 177–187.

Although the Historic Urban Landscape approach, by aiming to re-introduce local cultural traditions into territorial planning and urban design interventions, would reduce environmental vulnerabilities and increase resilience, it would need to be embedded in a larger framework of policies and strategies, both at the national and local level, in order to be fully efficient. In order to better assist local authorities with the task in hand, among others, the following issues need to be addressed:

12. The establishment of an inventory of risks for the city, which can then serve to develop an integrated municipal protection plan.
13. The question how climate change, planning law and practice need to work together in order to achieve climate-resilience for the city.
14. How governments have been responding to date to climate change adaptation, in particular as regards facilitating the development of mitigation strategies and tools for their cities.

The Historic Urban Landscape approach aims at giving a proper place to intangible heritage values in the process of interpretation, planning and conservation of historic cities. This is to ensure that a full understanding of the role that intangible heritage values can play is associated to planning methodologies and decisions related to the management of the urban environment. Issues to resolve include:

15. Which are the intangible heritage values to be found in historic urban landscapes?
16. Which values are usually represented and conserved, and which are not?
17. How intangible heritage values affect, or are affected by, urban management practices.

In addition to giving a proper place to intangible heritage values, the Historic Urban Landscape approach also stresses the need to pay greater attention to the role that local communities play, both in the process of conservation and in maintaining the sense of place, the importance of their values and knowledge, as well as their creative capacity. In order to be able to fully utilise their knowledge, skills and strengths to capture the new economic opportunities, among others the following issues need to be explored:

18. The role of demographic and social change in urban conservation and the need to constantly redefine 'communities' and 'local population groups'.
19. The benefits to be had, economically, socially and culturally, when poor and marginalised segments of society are put back into business through the restoration of their dignity and a rebuilding of their pride.
20. How new forms of socio-economic development based on creativity can be promoted through urban conservation.

Cities have always been at the core of economic development and innovation. As more than half of the population of the world now lives in cities, their global role has expanded and is due to grow even more in the future. Urban heritage today can play a very significant role in positioning culture and cultural activities at the core of the development process. As the United Nations are currently discussing the new International Development Agenda for Post-2015, culture and cultural heritage, and particularly urban heritage, have been pushed to the forefront by several international Development Agencies and the World Bank[45] as enablers and

---

[45] See for instance: UNESCO (2013) *MDG-F Culture and Development Knowledge Management Project*. Paris: UNESCO. For the point of view of the World Bank see: Licciardi, G. and Amirtahmasebi, R. (2012) *The Economics of Uniqueness. Investing in Historic City Cores and Cultural Heritage Assets for Sustainable Development*. Washington DC: The World Bank.

drivers of social and economic development,[46] including in a recent UNESCO-UNDP Report on creativity.[47]

The fundamental and applied research questions are formulated in the overall context of the Historic Urban Landscape approach, meaning that they would increase knowledge and understanding of the general principles related to the concept of the Historic Urban Landscape and the implementation of the 2011 UNESCO *Recommendation on the Historic Urban Landscape*. Additionally, there is a need to develop per geo-cultural region, or even per country, specific frameworks for the application of the Historic Urban Landscape approach, each defined by particular questions and issues that relate to the local context.[48]

After all, if there is one overriding message that can be distilled from the pages of this book, it would be that the Historic Urban Landscape has different meanings in different places that are shaped by different environmental, economic, social and cultural conditions – these should be recognised and respected in order to fulfil its potential to reconnect the historic city to its urban context and to the dynamics of the urban century.

[46] *The Hangzhou Declaration Placing Culture at the Heart of Sustainable Development Policies Adopted in Hangzhou, People's Republic of China, on 17 May 2013*. The Declaration issues a nine points Agenda dealing with the role of culture in the development process. Concerning cities, the Declaration states: '*Harness culture as a resource for achieving sustainable urban development and management*'.

   A vibrant cultural life and the quality of urban historic environments are key for achieving sustainable cities. Local governments should preserve and enhance these environments in harmony with their natural settings. Culture-aware policies in cities should promote respect for diversity, the transmission and continuity of values, and inclusiveness by enhancing the representation and participation of individuals and communities in public life and improving the conditions of the most disadvantaged groups. Cultural infrastructure, such as museums and other cultural facilities, should be used as civic spaces for dialogue and social inclusion, helping to reduce violence and foster cohesion. Culture-led redevelopment of urban areas, and public spaces in particular, should be promoted to preserve the social fabric, improve economic returns and increase competitiveness, by giving impetus to a diversity of intangible cultural heritage practices as well as contemporary creative expressions. The cultural and creative industries should be promoted, as well as heritage-based urban revitalization and sustainable tourism, as powerful economic sub-sectors that generate green employment, stimulate local development, and foster creativity'.

[47] '*The tangible, built heritage has long been recognised for its value as a driver of development. In fact, it was the first cultural domain to be considered 'bankable' in the contemporary sense. Already in the 1970s, both the United Nations Development Programme (UNDP) and the World Bank were beginning to justify investments in 'historic preservation on purely economic grounds, and by the late 1980s it had become commonplace to speak of a 'heritage industry' that had emerged, notably in symbiosis with the growing tourism industry. Today, many countries and/or cities twin the culture and tourism sectors under a single ministry or department; monuments and museums alike are increasingly recognised as important sources of income and builders of a city's image. The adaptive re-use of historic monuments as public buildings is often cost-effective and helps rejuvenate the economic base of older parts of the city, generating both income and employment, as well as tapping into increasingly important tourism flows. These approaches also often support top- down commercial operations that disrupt neighborhoods and their monumental heritage, and drive out the poor. This is clearly the dark, destructive side of heritage in development initiatives. They often promote elitist commercial operations while dislodging the delicate relationships between prevailing economic levels, neighborhood life, the traditional urban fabric and the monumental fabric that has existed nestled within it, albeit precariously*'. UNESCO-UNDP (2013) *Creative Economy Report, 2013 Special Edition. Widening Local Development Pathways*. New York and Paris: UNDP; UNESCO: 44–45.

[48] A Road Map for the implementation of the new UNESCO Recommendation and application of the Historic Urban Landscape approach in China has been developed at Tongji University's Advanced Research Institute for Architecture and Urban-Rural Planning, with institutional assistance of the World Heritage Institute of Training and Research for the Asia-Pacific region (WHITRAP) in Shanghai – see: Van Oers, R. and Pereira-Roders, A. (2013) Road Map for Application of the HUL Approach in China. *Journal of Cultural Heritage Management and Sustainable Development*, Volume 3, Number 1: 4–13.

Index

Note: page numbers in *italics* refer to illustrations; those in **bold type** refer to Boxes, Case Studies, and Interviews; roman numbers refer to text in the Preface; footnotes are indicated by suffix 'n' (e.g. 12n)

access for disabled   277
access and participation for communities   **247**
adaptation measures for climate change   120–1, **128**
adobe bricks   49
   characteristics   *50*
adobe buildings   53, *54–5*
aerial surveys   251
Aga Khan, His Highness [interview]   xvii, **240–4**
   [quoted]   322
Aga Khan Development Network   **240**
Aga Khan Trust for Culture (AKTC)   325
   Historic Cities Programme   98, 322
Agenda 21   323
agriculture, beginnings of   47–8
Ahmedabad [India]   217
Aleppo [Syria]   *95*, 278, 302
Alexander, Christopher   88
altruism   **127**
Amiens [France]   206
Amsterdam Declaration   xiii, 211
Amsterdam [Netherlands]
   City Restoration Company   **174**
   continuity of conservation approach   **173–5**
   Municipal Department for Conservation and Restoration of Historic Buildings   **173–4**

Anderson, Stanford   225
Annapolis [Maryland, USA]   274
anthropological mapping   306
anticipatory planning   **178**
Antiquities Act [USA, 1906]   270
Antônio Prado [Brazil]   216
Appadurai, Arjun   15
APPEAR project   **40**
Appleyard, D.   306
archaeological knowledge, impact on urban communities   37, 42–5
Archaeological Landscape Management Strategy (ALMS)   40
archaeological remains
   balance between *in situ* preservation and excavation   20
   in situ preservation and mitigation strategies   30–4
   incorporation into new buildings   *24, 28*, 29–30, 36–7, *37–9*
   integration into urban environment   35–7
   in open sites   *26, 27, 37*
   under open-air shelters   *29, 37, 39*
archaeological sites, presentation within modern urban landscapes   25–30
archaeology   19–45
   and restoration   **243**
   in World Heritage Site management plans   21

architectural
   micro-analysis   90–1
archives   43–4
Ashihara, Yoshinobu   **261**
Ashworth, G.J.   31, 190, 191
Asia, urban population growth   182
Askew, M.   186–7
Asmara [Eritrea]   208
Assisi [Italy]   **164**, 214, **268**
   master plans for   **164**, 214, 215
associative cultural landscapes   **146**, 189, **197**, 224
Astengo, Giovanni   **164**, 213–14, 215
Athens Charter   4, 208
Athens Conference [1931]   2
Athens [Greece], archaeological remains   35
Australian heritage practice   189
authenticity
   cultural heritage   134–5
   historic buildings/ monuments   **176**
Aymonino, Carlo   88

Bakhtapur [Kathmandu Valley, Nepal]   57, *58*
Bam [Iran]   53, *54*, 302
Bamiyan Valley [Afghanistan]   49, 51–3
Bandarin, Francesco   1–16
Bangalore [India], Hindu temple   *252*

Reconnecting the City: The Historic Urban Landscape Approach and the Future of Urban Heritage, First Edition. Edited by Francesco Bandarin and Ron van Oers.
© 2015 John Wiley & Sons, Ltd. Published 2015 by John Wiley & Sons, Ltd.

Bangkok [Thailand], Chao Phraya
    River   186–7
Barcelona [Spain], archaeological
    remains   37, *38*
Bath [UK]   215–16
Baudelaire, Charles
    [quoted]   205
Bauhaus School   207
Beijing [China]   216–17, **313**,
    315
Beirut, preservation versus
    development   **22–4**
Bell, Daniel   318–19
below-ground remains/above-
    ground fabrics
    distinction   24
Belur [India]   *254, 255*
Belvedere Memorandum
    [Netherlands]   15
Benevolo, Leonardo   5
Benjamin, Walter, [quoted]   19,
    249
Berger, Alan   9
Berlage, Hendrik Petrus   **174**,
    207
Berlin [Germany]   *45*, **313**,
    315
Berlin Wall   **313**
Bern [Switzerland]   *90–2*
Berque, Augustin   **148**
Bianca, Stefano   75–111
Bianchini, F.   306
Bigio, Anthony Gad   113–28
Bologna [Italy]   5, 13
    case study   **107–11**
Bowles, S.   **127**
Brandenburg Technical
    University   329
Brazilian Charter [1987]
    212
bricks   49–59
    fired clay bricks   *49, 50,* 53,
        55–9
    production of   49, 55
    sun-dried bricks   49–53
British market towns   209–10
Brundtland Commission   311
buffer zones [for World Heritage
    properties]   12n, 217, 259
building methods, changes over
    time   168

building standards, and urban
    conservation projects   **299**
build-up of archaeological
    deposits   25–6
Bukhara [Uzbekistan]   *29*
Bungamati [Kathmandu Valley,
    Nepal]   *137, 139*
Burra Charter [1979,
    1999]   162n, 189, **279**

Cairo [Egypt]   *101–2, 223,* **241**
    Darb al-Ahmar programmes
        101–2, **242**
calligraphy   **106**
Calvino, Italo [quoted]   1
Campbell, Bruce   **128**
Campbell, Tim   323, 324
Canada, *see also* Montreal;
    Ottawa; Toronto
Canadian First Nations   **146**, 224
Canberra [Australia]   *185,* 187,
    195–7
    Academy of Science
        building   *196–7*
    and Historic Urban Landscape
        approach   **195**
Caniggia, Gianfranco   210–11,
    217, **261**
Cape Town [South Africa]   16
Cappadocia [Turkey], cave
    dwellings   *62*
car-free zones   122
case studies
    Bologna   **107–11**
    Cultural Diversity
        Lens   **245–8**
    heritage economics - user's
        guide   **291–6**
    Tokyo   **261–8**
    traditional Chinese view of
        nature   **148–59**
    World Bank's tools for urban
        conservation   **297–300**
'catalogue heritage'   191
Champaner-Pavagadh [Gujarat,
    India]   44
chaos, as urban condition   7
Chartres [France]   254
Chemetoff, Alexandre   9
Chester Plan   **165–6**
Chin, L.H.   202

China
    Historic Urban Landscape
        approach in   **158**
    limits of historic
        preservation   **158**, 313
    urbanisation   **157**, 322
    see also Beijing; Hangzhou;
        Jinan; Nanjing; Yangzhou
Chinese approach to urban
    conservation   **103–6**
Chinese historic cities   **149–57**
    development planning   **158**
    identity loss   **158**
    landscape conservation
        in   **157–8**
Chinese view of
    landscape   **148–9**
Churchill, Winston [quoted]   xiii
CIAM (*Congrès internationaux
    d'architecture moderne*)   4,
    208
    Athens Charter   4, 208
cities
    as cultural
        landscapes   179–202
    as 'organic' entities   89, 194,
        224
city branding, contribution of
    heritage   **293–4**, 295
city conservation strategy
    (CCS)   326
city development strategy
    (CDS)   326, 327
    example of use [Sana'a]   327
Civic Amenities Act [1967]   **165**
civic engagement   4, 31–3
civic engagement tools   xvii,
    221–48
Civita di Bagnoregio [Italy]   *60*
clay-based human
    settlements   49–59
Clean Water Act [USA,
    1972]   276
climate action plans   120–1
    Edinburgh   123–5
climate change
    adaptation measures   120–1,
        128
    effect on natural hazards   114
    impacts on cities   114
    mitigation   121–2

resilience   120–1
Scottish Government's agenda
    125
climate change plans, and historic
    cities [Edinburgh
    example]   122–5
cognitive mapping   189, 202,
    225–6, **239**
    examples   **236–8**
'collage cities'   6–7, 102
combined heat-and-power (CHP)
    schemes   125
community-based design and
    development   231, 234–5
community participation   **247**,
    303, 320
community vision planning   274
complex urban environments,
    geology   79–84
Confucianism   **149**
Connecticut River Valley [USA],
    design manual   274
conservation area, Chester
    [England]   **165**
conservation and development
    agency   169–70
    coordinating committee   170
    organisational
        framework   170–1
    technical planning
        office   170–1
conservation easements   271
conservation plan, questions to be
    answered   162–3
conservation planning, basic
    premise   162
conservation practice
    workshops   200
'continuing and associative
    cultural landscapes'   **146**
continuity of conservation
    approach [in
    Amsterdam]   **173–5**
continuity [of cultural knowledge/
    practice/skills]   139, 141–2,
    146, 243
Conzen, M.R.G.   209–10, 211,
    217, 309
core zones, World Heritage
    properties   259
Corner, James   9

'Cosmic architecture'   **314**
cost, meaning of term   284
Council of Europe
    definition of 'cultural
        landscape'   212
    European Charter of the
        Architectural Heritage
        [1975]   211
    Faro Convention [2005]   219
Courajoud, Michel   9
craftsman-designers   226, 234
Crawford J.H.   122
creativity
    and cultural diversity
        lens   **247–8**
    UNESCO–UNDP Report
        [2013]   332
Cronocaos exhibition   13, **313**
Cullen, [Thomas] Gordon   180,
    209, 306
cultural analysis   254
Cultural Asset Maps   252
cultural capital   **291**, 297
Cultural Commons   307
cultural continuity   **243**
cultural diversity   212–13
    valuing   **245–8**
Cultural Diversity Lens [case
    study]   **245–8**
    constructing   **246**
    evaluation tool   **245**
    thematic structure   **246–8**
cultural evolution   **126**, **127**
cultural geography   180, 209
cultural heritage
    and cultural diversity lens
        247
    protection of   **291**
cultural heritage impact
    assessment   258–9
cultural landscape model
    183–6
cultural landscapes
    characteristics   187–9
    cities as   179–202
    continuing and
        associative   **146**
    definition(s)   **145, 148**, 212
    as expressions of local
        identity   **145–7**
    origin of term   183, 309

relationship to historic urban
    landscapes   **146–7**
    studies in Germany   183,
        209
    UNESCO categories   179, 184,
        186, 189, 224
    as World Heritage
        category   **145–6**, 310
    see also associative cultural
        landscapes
cultural mapping   202, 224–6,
    229, 254–6, 306–7
    examples   *230–2, 255*, 307,
        308, 309
cultural tourism   34, 223
    contribution to local economic
        development   **292–3, 295**
culture–nature interaction   184,
    186, 260
    in Canberra [Australia]   187,
        195

Dakar [Senegal]   **241**
Daly, P.   182
Daoism   **149**
Dark Sky compliance   277
Dastur, Arish   229
Davidson, Justin   190
De Carlo, Giancarlo   4
de-Shalit, Avner   318–19
Delhi [India], Nizamuddin
    Dargah   *134*
demolition and renewal   85, **103**,
    105, 107, 161n, 175, **279**
    see also Modern Movement
designation, historical areas/
    sites   33
designer–builders   226, 234
Detroit [USA]   324
Dias, Marina Tavares [quoted]
    129
Dickinson, Duo   321–2
digital elevation models
    251
disabled access codes   277
disaster preparedness   277–8
disaster resilience, in Kathmandu
    Valley   142
DiStefano, Lynne   227
diverse perspectives, and cultural
    diversity lens   **247**

documentation
  of experience   223
  of intangible heritage
    values   138
  of physical form   222–3
'dogmatic' approach to urban
    planning   **240–1**
drawings   251
  see also kolams
Dresden Elbe Valley cultural
    landscape   194
Dubrovnik [Croatia]   79, *81*
Durham University, Master's
    programme   329

Easter Island [Chile]   72, 74,
    77–8
ecological urban living   227,
    228–9
Ecological Urbanism   3, 10, **178**
  in urban management   14
economic 'long waves'   **291**
economic value(s)
  assessing with use of indicators
    and maps   **292–4**
  'values' in excess of   284, *285*
economics of culture, and cultural
    diversity lens   **247–8**
Edinburgh [Scotland]   122–5
  Biodiversity Action Plan
    125
  emissions reductions
    actions   125
  fire risk   123–4
  flood risk   123, 124–5
  Old and New Towns   122
  skyline   274
  tourism revenue   122–3
  Water of Leith management
    plan   124–5
  World Heritage site action
    plan   125
  World Heritage status   122
education
  architects   322
  planners   202
  urban heritage specialist   329
Eliot, T.S.   224
Engelhardt, Richard A.   227,
    245–8
engineering practices   43

English Heritage
  Historic Landscape
    Characterisation (HLC)
    program   15
  on reburial   32n
entelechy   86
environmental impact
    assessments (EIAs)   259,
    307, 327
environmental planning   6
equilibrium, and urban
    ecology   234
Erve, Marc van der   **146**
European Architectural Heritage
    Year [1975]   211
European Charter of the
    Architectural Heritage
    [1975]   211
European Landscape Convention
    (ELC)   15
European Union (EU),
    integration of research on
    Historic Urban
    Landscape   310
excavation   26, 43n
  and *in situ* preservation   20
experts, role in community-based
    design and
    development   231, 234

Falini, Paola   **268**
Farjadi, Homa   **177**
Faro Convention [2005]   219
Feng Han   **148**
Feng Ji Zhong   **103**
feng shui   **150**, 306
Ficulle [Italy]   *72*, 74
fieldwork, importance of   97
financial tools for urban
    conservation   xviii, 283–300
  characteristics   287–8
  effects   286–7
  examples   288–9
  reasons for use   284–6
financing of conservation
    programmes   168
favourable rates/
    conditions   286
Finland, standardisation of new
    developments   208–9
fire risk, Edinburgh   123–4

fired clay bricks   *49*, 53, 55–9
  characteristics   *50*
First Nations communities   **146**,
    224, 229
flood risk   *119, 120*
  Edinburgh   124–5
form-based planning
    codes   274–5
Fort Wayne [Indiana, USA]
    277
fossil fuels, increasing
    scarcity   **126**
France *see* Amiens; Chartres; Le
    Havre; Paris

Gabellini, Patrizia   **107–11**
Garden City Movement   207
Gazzola, Piero   212
Geddes, Patrick   2, 6
  [quoted]   301
Genadendal [South Africa]   328
generational changes   **242**
genius loci   5, 129, 194, **195**, 213,
    319
gentrification   12, 190, 306, 325
geographical information systems
    (GIS)   35, 224, 257, 303,
    306
geology   xiv–xv, 47–84
  clay-based human
    settlements   49–59, 84
  complex urban
    environments   *49*, 79–84
  hard rock based human
    settlements   *49*, 67–8, 71–9,
    84
  soft rock based human
    settlements   *49*, 59–67,
    67–70, 84
geospatial mapping   251
geospatial referencing   257
Ghilardi, L.   306
Ghurid dynasty   51n
Giovannoni, Gustavo   2, **107**,
    164, 207
Glass, R.   306
global positioning systems
    (GPS)   306
Global Risk Data Platform   116,
    119
global warming   113–14

globalisation, effects   2, 10, **291**, 320
Goethe, Johann Wolfgang von   86
Gospodini, Aspa   320
governance   168–9
  and urban conservation projects   **299**
Graham, B.   190, 191
Grand Paris planning exercise [2007]   11
grants, heritage building restoration   286, 288
Gratz, Roberta Brandes [quoted]   161
'green' regulatory systems   275
greenhouse gas emissions   113
  factors affecting amount per capita   121
  reduction of   113, 121
Greffe, X.   184
Griffin, Walter Burley   187
Gropius, Walter   207
Gubbio Charter [1960]   214

Hague Convention [1954]   278
Hangzhou [China]   151, *154–5*
  Historic Urban Landscape approach   **197**
  loss of historic urban landscape structure   **158**
  Qianjiang new town   *159*
  West Lake   104n, **148**, **150**, **151**, *152*, 194, **197–9**
  Zhongshan Road/ Street   **104–5**
Hangzhou Declaration   332n
Hanoi [Vietnam]   230–1, *233*
hard rock based human settlements   *49*, 67–8, 71–9
  rounded natural blocks   *72*, 74, 79, *80*
  square blocks   68, *71*, 72, *73–8*, 74
Hargreaves, George   9
Harper, Anita   224
Hasegawa, Takashi   **261**
Haussmann, Georges Eugène   318
Hayden, Dolores   227

heritage
  Aboriginal conception of   **146**
  increase in categories   **314**
  who decides what is   **313**
heritage conservation, meaning of term   283
heritage conservation planning   191–2
heritage economics, user's guide [case study]   **291–6**
heritage financial tools   283–90
  characteristics   287–8
  effects   286–7
  examples   288–9
  reasons for use   284–6
heritage impact assessment (HIA)   138, 258–9, 307, 327
heritage indicators   **292–4**
heritage management process, ten-steps guide   189
heritage and the metropolis, Koolhaas on   **313–15**
heritage tourism   34
'heritagisation'   191
Himeji Declaration [2012]   304
Hino City [Japan]   **266–7**
historic 'architectural assets', protection of   208
historic buildings
  re-conditioning of   **165**
  'values' in excess of economic value   284, *285*
'historic centre'   13, **107–8**, 209
historic cities
  and climate change mitigation   121–2
  and climate change plans [Edinburgh example]   122–5
'historic city' concept   12, 13, **107–8**
Historic Landscape Characterisation (HLC) program   15
historic macro-analysis   90
historic preservation
  legislation   270–5
  selectivity   **313**
  see also National Historic Preservation Act

Historic Preservation movement   270
historic urban areas
  early appreciation   205–6
  international recognition of   211–13
  protection of   208
  public management of   168–9
Historic Urban Landscape
  Action Plan   xviii, 218, 326–8
  definitions   181, 222, 224, 227
  first defined   212
  key aspects of analysing and planning   163–8
  UNESCO Recommendation [2011] on   3, 4, 12n, 13, 14, 20, 21, 121, 135, **148**, 205, 213, 218–19, 227, 234, 235, **245**, 270, 273–4, 283, 301, 326
Historic Urban Landscape approach
  basis   4–7, 129, **177**, 213, 215, 250, 317
  and Canberra [Australia]   **195**
  in China   158, **197**
  conceptual framework   48
  dialogue with city planners and urban designers   194–202
  and economics   **291**, 292
  geo-diversity classification   *49*
  implementation of   16, 278, 326
  and intangible heritage values   129–30
  and local cultural significance   **280**
  methodologies and tools used   305–9
  [Lisa] Prosper on   **147**
  and regulatory systems   270, 273–4, 278
  relevance within Urban Morphology concept   98–102
  research agenda for planners and designers   329–31
  as response to global challenges   11
  Road Map   332n

Historic Urban Landscape
    approach (*cont'd*)
    [Filipe Duarte] Santos
        on   **127–8**
    tools used   250
    traditional and customary
        management systems   260
historical analysis and
    mapping   252–3
history/ecology fused survey
    method   **266–8**
Hoffmann kiln   59
Holocene climatic optimum   47
Hosagrahar, Jyoti   249–68
housing, and urban conservation
    projects   *102*, **174**, 207, 216,
    298, 299
Hugo, Victor   206
human–nature relationship
    deterioration of   **157**
    traditional Chinese
        view   **148–9, 150, 158**
hybridisation concept   **177**
Hyderabad [India]   **281**

ICOMOS (International Council
    on Monuments and
    Sites)   212
    Brazilian Charter [1987]
        212
    Burra Charter [1979,
        1999]   162n, 189
    Charter for the Conservation
        of Historic Towns and Urban
        Areas [1987]   212
    on limits of change   24
    Xi'an Declaration [2005]   183
iconic city identity   *223*, 253–4,
    274
identity [of city]   318–19
Ildan, Mehmet Murat
    [quoted]   47
immigrant communities,
    effects   303, 304
impact assessment, intangible
    heritage values   138
in situ preservation and
    mitigation strategies   30–4
    technological advances   42
incentives [for heritage building
    restoration]   284, 288

India
    approach to urban
        conservation   **279–81**
    public–private partnership
        agreements in   **243**
    public spaces   131, *134*
    urban conservation
        legislation   **280**
    urban heritage   325
    see also Ahmedabad;
        Bangalore; Belur; Delhi;
        Hyderabad; Mumbai;
        Srirangapatna; Varanasi
Industrial Revolution   85, 206
information and communication
    technologies (ICTs)   42, 321
    impact on heritage
        conservation   **315**, 321
infrastructure
    heritage categories   314
    and urban conservation
        projects   167, **298, 299**
infrastructure analysis   257
infrastructure developments,
    archaeological
    implications   34
infrastructure
    urbanisation   **177–8**
Institute of Historic Building
    Conservation   34
institutional continuity   97
intangible heritage   14, 129, 130
    compared with tangible
        heritage   130, **146**
    dynamic nature   134–5, 144,
        146
    of local communities   320
    meaning of term   139–42,
        319
    public spaces and
        boundaries   131–4, 135
    role in building disaster
        resilience   142
    UNESCO [2003]
        Convention   319
intangible heritage values
    defining in historic urban
        landscapes   130–5
    documentation of   138
    and Historic Urban Landscape
        approach   129–30

impact assessment   138
impact of urbanisation   135
    integrating into urban
        planning   142–3
    recognition in urban
        management systems   136,
        138
integrated planning   161–3
integration of disciplines and
    professional
    practices   322–4
Intelligent Systemic Design   9
Inter-American Development
    Bank (IADB)   287
interdisciplinary research and
    studies   329
Intergovernmental Panel on
    Climate Change
    (IPCC)   **128**
International Development
    Association (IDA)   **297**
interviews
    [His Highness The] Aga
        Khan   xvii, **240–4**
    [Rem] Koolhaas   xviii,
        313–15
    [Rahul] Mehrotra   xvii–xviii,
        279–81
    [Mohsen] Mostafavi   xvi,
        176–8
    [Lisa] Prosper   xv, **145–7**
    [Filipe Duarte] Santos   xv,
        126–8
    Wang Shu   xv, **103–6**
inventory of cultural
    resources   251–2
Italy
    study of urban history   4–5,
        88, 210, **261**
    see also Assisi; Bologna; Civita
        di Bagnoregio; Ficulle;
        Orvieto; Ravenna; Rome;
        Sardinia; Sassi di Matera;
        Sorano; Syracuse

Jackson, John Brinckerhoff   5,
    279
    [quoted]   179, 189
Jacobs, Jane   4, 227
    [quoted]   221
Jakarta [Indonesia]   323

Jang Yung He **104**
Japan 216
　architectural research **261–8**
　see also Hino; Kyoto; Tokyo
Jigyasu, Rohit 129–59
Jinan [China] **151**, *153*, *156–7*,
　156–7
Jobs, Steve [quoted] 317
Jokilehto, Jukka 205–19
Jorge, F. 202

Kashi Mandala 131
Kasthamandapa [wooden temple,
　Kathmandu] 55–6
Kathmandu Valley [Nepal]
　architecture 55–9, *136–7*
　clays 56, 57, 59
　disaster resilience 142
　Guthi system 142
　uncontrolled urbanisation
　　in *141*
　World Heritage site *56*, *57*,
　　136, 217
　see also Bakhtapr; Bungamati;
　　Patan
Kawazoe, Noboru **261**
knowledge and planning
　tools xvii, 249–68
Koetter, F. 6–7, 102n
kolams [drawings] 225
Kondratiev, Nikolai **291**
Koolhaas, Rem xviii, 7, 13,
　313–15, 320
Kostof, Spiro 16, 306
Krier, Leon 88
　[quoted] 269
Krier, Rob 88
Kuhn, Thomas **128**
Kyoto city [Japan] 16, 216
Kyoto Vision [2012] 219

Lagos [Nigeria] 323
Lalibela [Ethiopia] 65–7
　churches 65–7, *67–70*
land ownership/use
　and historic urban landscapes
　　planning 166
　legal regulations directing
　　affecting 270–5
Land Value Finance (LVF)
　300

'landscape'
　as cultural construct 184, *185*
　meaning of term **148**, 183–4,
　　188
　traditional Chinese
　　view **148–9**
landscape paintings,
　Chinese *150–6*, **150**
landscape planning 6, 11
Landscape Urbanism 3, 8–10
　in urban management 14
landslide risk *120*
Latour, Bruno 260
layering in urban
　environment xiv–xvi, 227,
　　231, *233*, 255–6, 302–3, 305,
　　319
　and archaeology xiv, 19
　and geology 79–84
　Historic Urban Landscape
　　approach 181, 222, 225,
　　227, 317
　morphological analysis xv,
　　4–5, 90, **261**, **263–6**
Shanghai 192
Le Corbusier 4, 161n, 208
Le Havre [France] 215, **314**
Lefebvre, Henri 209, 227
legal regulations
　indirect impact on urban
　　heritage 275–8
　land-based 270–5
Léger, Fernand 202
Lemaire, Raymond [quoted]
　283
Lia Jia Kung **104**
Liang Sicheng **103**
Lidar surveys 251
light pollution 277
Lincoln [England] 34
linkages, decision-makers and
　urban communities 44
Lipe, William 44–5
listening, importance of 202,
　240
living heritage, city as 317–18
Livingstone, Ken 320
local communities, intangible
　heritage 320
local government staff levels,
　effects on archaeology 34

London [England]
　Canary Wharf 323–4
　King's Cross District 324
　Mayor's Commission on
　　African and Asian Heritage
　　320–1
　medieval palace remains 27,
　　28
　MoLA Bloomberg site 40, *41*
　no. 1 Poultry site 40
　Roman amphitheatre 27, *28*
　Roman floor levels 26
　Rose Theatre 33
　Tower Environs Scheme 320,
　　328
　Tower of London 320
Lynch, Kevin 5–6, 13, 44, 180,
　188, 209, 224, 226, 306

McGill, G. 21
McHarg, Ian 6, 8, 15, 302
Machinami Charter
　[2000] 216n, 304
Machu Picchu [Peru] 72, *73–6*
Maki, Fumihiko **261**
mandala, representation of urban
　landscape in 131, 224–5
mapping 250, 306
　see also anthropological
　　mapping; cognitive mapping;
　　cultural mapping
Maretto, Paolo **261**
Margottini, Claudio 47–84
Martini, Viviana 218
masonry blocks *50*, 67–8
megastructures in urban
　development **315**
Mehrotra, Rahul
　[interview] xvii, **279–81**
Meinig, Donald 183–4
'mental mapping' 44
'meta-planning' 16
metamorphosis 86
Meyer, Hannes 207
micro loans 288
Mies van der Rohe, Ludwig
　207
mines infrastructure **177**
Mitchell, W.J.T. 184
mitigation strategies 31
　technological advances 42

mixed land use, zoning to
  encourage   122, 260
modern architecture in historic
  contexts   **315**, 321
Modern Movement   2, 4n, 85,
  161n, 207, 215
Moggridge, Hal   194, 253n, 274
Montreal [Canada], triplex
  apartments   228
'monument-centric' approach to
  heritage conservation   21,
  241
monuments
  as focus points   **241**
  as museums   222
Mooshammer, Helge   321
morphological analysis   209,
  257
morphology
  origins and implications of
  term   86–7
  see also Urban Morphology
Mörtenböck, Peter   321
Moscow [Russia]   **314**, 324
Moses, Robert   227
Mostafavi, Mohsen   xvi, 10,
  176–8
Multi-Input Area Development
  (MIAD) approach   **242**,
  244, 322
multidisciplinary approach to
  planning   305
multi-ethnic/multi-religious
  cities   96
multi-hazard risk   *117–18, 119*
Mumbai [India]   **279, 280, 281**
Mumford, Lewis [quoted]   85
Muratori, Saverio   4, 88, 210,
  217, **261**
museum-like approach to urban
  conservation   161, 190
museums
  archaeological knowledge
  and   40, 43
  monuments as   222

Nakano-ku [Tokyo suburb]   **265**,
  266
Nakazawa, Shinichi   **264**
Nanjing [China]   *150*
  Xuanwu Lake   150

Nara Document on Authenticity
  [1994]   **279**, 303, 304
National Historic Landmark
  documentation   273
National Historic Preservation
  Act [NHPA, 1966,
  USA]   270–1
natural processes, integration
  with urban design/
  conservation   6–7, 14
nature–culture interaction   184,
  186, 187, **195**, 260
Neolithic Revolution   47
Nepal *see* Kathmandu Valley
'Nepali' architecture   55–9
New Urbanism   **177**, 274
New, W.H.   224
Nietzsche, Friedrich   95n
Nolli, Giambattista, map of
  Rome   225, *226*
non-governmental organisations
  (NGOs), role of   200, 260,
  312
Norberg-Schulz, Christian   5
normative framework, evolution
  of   205–19
normative frameworks, response
  to challenge caused by urban
  development   213–16

occupancy levels [in heritage
  buildings]   286–7
O'Donnell, Patricia   269–81
Okuno, Takeo   **261**
Olmsted, Frederick Law   6, 273
'oneness with nature'   **149**
Orongo village [Easter Island,
  Chile]   72, 74, *77–8*
Orvieto [Italy]   63–5, 65n
Ost, Christian   **291–6**
Ottawa [Canada], Byward Market
  area   *230–3*
outreach strategies   43

Pachauri, Rajendra K.
  [quoted]   113
painting, Chinese   **106**
Panjabi, S.   188
panoramic photographs   251
Paris [France]   11, **178**, 194, 206,
  314, 318, 323

parks, protection of   273, 278
partial excavation [due to modern
  development
  footprint]   26–8
participation by communities,
  and cultural diversity
  lens   **247**
participatory planning and
  implementation
  strategies   171–2
  in Amsterdam   **173–5**
passive temperature-regulating
  systems   122
Patan [Kathmandu Valley,
  Nepal]   *59*, 132–3, 135,
  136–7, 138, *140*
Pella [Jordan]   *308*
Penang Heritage Trust
  (PHT)   200
Penang [Malaysia], George
  Town   194, 200
Perugia [Italy]   **315**
philosophy, Chinese   149
photogrammetry   251, 254
physical mapping   **239**
physical records   222
physical survey   162
Pillai, Janet   307
place, sense of   5, 44, 190,
  319
planning controls
  and archaeological
  remains   **32**, 33
  effect on urban integrity   85
planning institutions
  management of historic urban
  areas by   168–70
  in partnership
  arrangements   171–2
  see also conservation and
  development agency
Planning (Listed Buildings and
  Conservation Areas) Act
  [1990]   **166**
Planning Policy Guide 16
  (PPG16)   **32**
planning tools   260
plot-by-plot cadastral
  analysis   90–1, 93–4
Poole, Peter   306
'The Porous City'   11

Portugal, adaptation measures for climate change **128**
poverty alleviation, and urban conservation projects **299**
poverty survey, in Cairo **242**
preservation, conflict with development 21–5
PricewaterhouseCoopers, *Cities of Opportunity* [2012] report 318
Pritzker Prize [2012] **322**
processional/ritual/sacred paths/routes 131, *132*, 135, *139*, 255
professionals, and intangible heritage 144
property tax, reduction for heritage building restoration 286, 288–9
property titling 287, **299**
Prose, Francine 328
Prosper, Lisa [interview] xv, **145–7**
protective zoning 259–60
public access, in exchange for financial benefit 286
public institutions
management of historic urban areas by 168–70
see also conservation and development agency
public management of historic areas 168–9
public meetings 171
public participation, in planning and implementation strategies 171–2, **173–5**
public–private partnerships (PPPs) **243**, 284, 288, 289, **299–300**, 328
public services
and historic urban landscapes planning 167
partnership arrangements 172
public spaces
in India 131, *134*
in Kathmandu valley *56*, 135, *136*, 142
protection of 273, 278
public utilities, and historic urban landscapes planning 167

Pugin, A.W.N. 207
Punekar, A. 192
purchase of development rights (PDR) laws 271
PUSH (Promoting Dialogue and Cultural Understanding of Our Shared Heritage) methodology 302, 307–8

quality control, in archaeological work 33
quality of life, factors affecting 142, **240**, **242**, **243–4**, 317
quality of restoration, effect of financial tools 285–6

radiocarbon dating 47n
Rapoport, Amos 306
Ravenna [Italy], archaeological remains 37
'reading the city' method **261**
application in Japan **262**
'reading the territory' method **268**
REAP (Rapid Ethnographic Assessment Procedure) methods 202
reciprocity 318
'reconnecting the city'
meaning of term 3
post-war attempts 3–7
recording systems 43
recycling of architecture 88
Regionalism **314–15**
regulatory systems xvii–xviii, 269–78
definitions 269–70
rehabilitation of historic buildings 172
reinforced concrete 48, 84
remote sensing techniques 251
repository of urban experiences, city as 321–2
Ricoeur, Paul 249
Rio de Janeiro [Brazil] 216
risk reduction, for heritage building restoration 286
Riverside [Illinois, USA] 273
road network density, effect on climate change 121

rock masses
standardised description/classification 59–60, 67
see also hard rock; soft rock
Rome [Italy]
archaeological remains 35, *36*, *82–4*
Augustus' Forum 79, *82*
basilicas *82*, 83
building materials 79, 83–4
Colosseum 79
cultural landscape mapping 225, *226*
Forum 79
identity of city 318
master plans 215
San Clemente Basilica *82*
Theatre of Marcellus 79, 83, *83–4*
Trajan's Market and forum *36*
Rossi, Aldo 5, 88, **176**
Rössler, Mechtild 186
Rowe, C. 6–7, 102n
Ruskin, John 206
Rypkema, Donovan 283–300

Sachs, I. 131
'sacred space' 131
'sacred territories' 131–2
safety codes 277
St Petersburg [Russia], airport building **314**
sales tax, waived for heritage building restoration 286
Saltaire [England] 207
Samarkand [Uzbekistan] *100*
San Francisco, revitalisation of urban core **178**
San Juan [Puerto Rico], historic district 273
Sana'a [Yemen] 327
Santos, Filipe Duarte [interview] xv, **126–8**
Saragossa/Zaragoza [Spain] *39*
Sardinia [Italy] **263**
Sassi di Matera [Italy] *63*
Sauer, Carl 309
Sayer, Jeffrey **128**
scenario planning 274
Schlüter, O. 209
Schumpeter, Joseph **291**

Scottish Climate Change
        Programme   125
seeing, ways of   222–4
sense of place   5, 44, 190, 319
Seoul Declaration [2007]   182
Serra, M.V.   **297–300**
Shahr-i-Zohak fortress and town
        [Afghanistan]   49, 51–3
shan shui landscapes   **150, 151,**
        306
Shanghai [China]
    Bund waterfront   192
    canal towns   192
    Suzhou River   191–2
Shorey, S.P.   **281**
Singapore   200, 324
Singer, Rachel   301–15
Singh, Manmohan   **243**
Singh, Rama P.B.   192
Sintra [Portugal]   **128**
Siravo, Francesco   161–78
site management systems,
        traditional and customary
        systems   260
Sitte, Camillo   2, **174**, 207
skylines   *223*, 253–4, 274
slums, resource uses   228–9
Smith, Julian   189, 221–48, 306
social capital, in traditional
        communities   131
social exclusion   325
socialism and philanthropy   207
socio-cultural content   91–3
socio-ecological systems   **128**
socio-economic analysis   256–7
socio-economic information, and
        historic urban landscapes
        planning   166
sociology   306
soft rock based human
        settlements   *49*, 59–67
    directly built into soft
        rock   *61–3*, 65–7
    soft-rock building
        construction   60, *60, 61,*
        63–5
Sorano [Italy]   *61*
spatial anthropology   **262–3**
Special Purpose Vehicles (SPVs)
        [financing arrangements]
        299
speculative premium   290

Spirn, A.W.   8
Spizzichino, Daniele   47–84
squatter settlements   **241**
Srirangapatna Fort
        [India]   255–6, *256*
steel frame buildings   48, 84
Stein, Clarence   **103**
Steinberg, Saul   224
Steinitz, Carl   15n
Stone Age   47
stormwater management
        techniques   276–7
strategic planning   258
    and archaeology   33–4, 35, 44
street elevations   254
Strengell, Gustaf   207
Sub-Saharan region, adobe
        buildings   53, *55*
Suginami-ku [Tokyo
        suburb]   **265**
sun-dried bricks   49–53
    characteristics   *50*
surveying tools   250–1
surveys   251, 306
sustainability   311
    and Historic Urban
        Landscapes   311–12
sustainable communities   213
sustainable community-based
        design and
        development   234
sustainable development, and
        archaeology   34
sustainable diversity   229–31
sustainable livelihoods, intangible
        heritage and   143
sustainable planning, and historic
        areas   **178**
sustainable tourism, ten
        steps   189
sustainable transport   122
sustainable zoning and
        development codes   275
SWOT analysis   258, **296**
symbiosis   304, 306
syncretism   304
Syracuse [Sicily, Italy]   *25, 26, 27*

Taj Mahal [India]   **280–1**
tangible heritage, compared with
        intangible heritage   130,
        146

Taut, Bruno   207
tax credits, for heritage building
        restoration   286, 288–9
Taylor, Ken   179–202
Temple Trusts   260
territorial approach to
        planning   143
three-dimensional (3D)
        models   251, *252*
Timurid dynasty   51n
title [to property/land], clearing
        of   287, **299**
Tokyo [Japan]
    case study   **261–8**
    structure compared with
        Edo   **262**
    suburbs   **263–6**
toolkits [for urban heritage
        management]   xvi–xviii,
        203–300, 216–19
    civic engagement tools   xvii,
        221–48
    financial tools   xviii, 283–300
    knowledge and planning
        tools   xvii, 249–68
    regulatory systems   xvii–xviii,
        269–78
topographic analysis   89–90
Toronto [Canada], Kensington
        Market   **235–9**
Total Station Surveys   251
tourism
    cultural tourism   143–4, 223
    heritage tourism   34
    and local communities   190,
        229, 325
traditional builders/
        craftsmen   168
traditional Chinese view of nature
        [case study]   **148–59**
traditional and customary
        management systems   260
transfer of development rights
        (TDR) legislation   271, 290
transferable development rights
        (TDRs)   286, 289
transportation policies, and
        historic urban landscape
        planning   167–8, 277
travel demand management
        122
Troll, Carl   8

Tunbridge, J. 31
Tung, A.M. **173, 174**
Turner, Michael 301–15
'typology' concept 210–11

uncertainty reduction, for
    heritage building
    restoration 290
UNESCO
    Category 2 Centres
        (C2C) 310
    Chairs 310
    Convention [2005]
        on…Diversity of Cultural
        Expressions 212
    Convention [1972] for the
        Protection of Cultural and
        Natural Heritage **245, 291**
    cultural landscape
        categories 179, 184, 186,
        189, 224
    Forum - Universities and
        Heritage [2009] 309
    Intangible Heritage Convention
        [2003] 319
    Recommendation [1976] on
        Historic Areas 211–12, 214
    Recommendation [2011] on
        Historic Urban
        Landscape 3, 4, 12n, 13,
        14, 20, 21, 121, 135, **148**,
        205, 213, 218–19, 227, 234,
        235, **245**, 270, 273–4, 283,
        301, 326
    Universal Declaration on
        Cultural Diversity
        [2001] **245**
    on urban population
        growth 181
    World Heritage
        Convention 12, **145, 148**,
        219, **291**, 307, 325
    see also World Heritage sites
UNESCO–UNDP Report on
    creativity [2013] 332
United Nations (UN)
    Conference on the Human
        Environment [Stockholm,
        1972] 307
    Environment Programme
        (UNEP), Global Risk Data
        Platform 116, 119

Millennium Development
    Goals 318
    Resolution on Culture and
        Sustainable Development
        [2013] 318
Universal Declaration on Cultural
    Diversity [2001] **245**
university research, role
    of 309–10
urban agriculture 229
urban archaeologies, future
    developments 42–4
urban communities, role of
    202
urban conservation
    Chinese approach to **103–6**
    conflicts in 303–5
    critique of present
        approach 13
    financial tools for xviii,
        283–300
    future challenges 324–6
    Indian viewpoint **279–81**
    instruments developed
        for 209–11
    limitations 2, 3
    repositioning 11–16
    World Bank's tools for [case
        study] **297–300**
urban conservation projects
    economic considerations **298**,
        299
    financing
        arrangements **299–300**
    governance **299**
    heritage areas 298
    housing considerations **298**,
        299
    infrastructure
        considerations **298, 299**
    spatial projects **298**
    World Bank financing of **298**,
        300
urban cultural landscape 184
urban development
    19th century 206–7
    synthesis via archaeology 43
urban ecology 228, 234
urban form analysis 4–5
urban heritage management,
    compartmentalisation
    of xiii, 323

urban heritage specialist, role
    of 328–9
urban landscape
    mapping/measuring/
        visualising 250–1, 306
    protecting/enhancing/
        improving 257–60
    reading and
        interpreting 251–7
urban management, factors
    affecting 3
urban management systems,
    recognition of intangible
    heritage values in 136, 138
Urban Morphology
    advantages and
        problems 94–8
    architectural
        micro-analysis 90–1
    of city in landscape
        setting 194
    difference from modern
        planning methodologies 89
    historic macro-analysis 90
    and linguistic syntax 89
    methodology and
        procedures 88–94
    relevance within Historic
        Urban Landscape
        approach 98–102, 310
    scope xv, 87–8
    socio-cultural content 91–3
    topographic analysis 89–90
urban morphology management,
    in Paris [France] 194
urban planning
    development and impact
        of 206–9
    'dogmatic' approach **240–1**
    factors/forces affecting **240**
    and historic city **177**
    intangible heritage values
        integrated in 142–3
    limitations 1, 3, 136
    traditional principles replaced
        by modern system 138
urban regeneration projects 8
urban resilience 114
    measuresto increase 115
urban risk 114
urban risk assessment 114
urban values 5–6

urbanisation
    effect on cultural heritage **291**
    intangible heritage values
        affected by 135
Uzbekistan
    projects *99–100*
    see also Bukhara; Samarkand

Valletta Convention [1992] 30
value, meaning of term 284
value added tax (VAT), waived
        for heritage building
        restoration 286
value of city centres 175
'values' in excess of economic
        value 284, *285*
van der Erve, Marc **146**
van Eyck, Aldo 4
van Oers, Ron 181, 317–22
Varanasi [India] 131–2, *133*,
        192
    Panchakroshi pilgrimage
        route 131–2, *132*
    see also Kashi
Varzia [Georgia], cave
        dwellings *61*
Vaux, Calvert 273
Venice Charter [1964] **145**
Vermont Act 250 [1970] 272–3
vernacularism 190, 202
Victorian house(s), flexibility of
        uses **178**
Vienna [Austria] *93–4*
Vienna Memorandum 180, 212,
        309
    discussions at Montreal Round
        Table [2006] 181
Vigan [Philippines] 200, **201**
visioning 258
visual analysis 253
visual impact assessment 258–9
Vitruvius 306

Wagner, Martin 207
Waldheim, Charles 8, 10
Wang Guoping **105**
Wang Shu xv, **103–6**, 322
'water city', in Japan **262–3**,
        266–7

water management plan,
        Edinburgh 124–5
water pollution, approaches to
        prevention 276–7
Watson, G.B. 181
Weber, Max **126**
Whitehand, J.W.R. 211, 309
Williams, Jim 42
Williams, Tim 19–45
Willmott, P. 306
Willowbank Centre for Cultural
        Landscape **145**, **146**
Winichakul, Thongchai 224
Winter, T. 182, 188
World Bank
    Aga Khan's discussions
        with **243–4**
    financial tools for urban
        conservation [case
        study] **297–300**
    objectives **297**
    review of risk exposure of
        World Heritage
        cities 116–19
    urban conservation
        projects **298–300**
    urban projects **297–8**
World Heritage cities 115
    car-free areas 122
    exposure to multiple
        hazards 115–19
    flood risk *119*, *120*
    landslide risk *120*
    multi-hazard risk *117–18*,
        119
    Urban Morphology
        approach 98–102
    vulnerability to physical
        hazards 116
World Heritage Committee
    definition of 'cultural
        landscape' 212
    Expert Meetings 12n, 217–18
    on revision of definition of
        historic cities 12n
World Heritage Convention 12,
        145, **148**, 219, **291**, 307, 325
    Operational Guidelines 259n,
        303

World Heritage cultural
        landscapes **145–6**, **148**
    categories 179, 184, 186, 189,
        224
World Heritage sites *25–7*, 49,
        51–3, 53, *54*, *55–9*, *63*, *65–7*,
        67–70, 72, *73–6*, 79, *81*,
        93–5, *97–8*, *100–2*, **104**n,
        128, 132–3, 135, *136–7*, 138,
        140, **164**, **173**, 194, **197–9**,
        214, 215–16, 278, **314**, 320,
        327
    archaeology in management
        plans 21
    'areas of influence' 12n, 217,
        259
    buffer zones 259
    core zones 259
world population growth
        126
World War 2, impacts of 208,
        209
Wright, Frank Lloyd,
        [quoted] 249
Wu Liang-Yong **103**

Xi'an Declaration [2005] 183

Yamato Declaration [2004]
        319
Yanase, Yuji **263**
Yangzhou [China], Slender West
        Lake 150, *152*
Young, M. 306
Yuen, B. 200

Zabid [Yemen] 217
Zanzibar [Tanzania], Stone
        Town *97–8*, 328
Zarrudo, Eric 200
'zero-waste city' 330n
Zetter, R. 181
Zhuijiajiao [China] 192, *193*,
        202
zoning 271–2
    disadvantages 138, 229, 272,
        305
    historic sub-districts 259–60,
        274

WILEY Blackwell

From the Same Authors

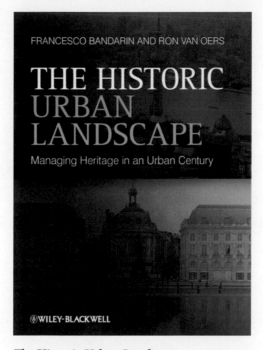

FRANCESCO BANDARIN AND RON VAN OERS

THE HISTORIC URBAN LANDSCAPE

Managing Heritage in an Urban Century

WILEY-BLACKWELL

The Historic Urban Landscape:
Managing Heritage in an Urban Century
Francesco Bandarin and Ron van Oers
9781118932728

…the crux of the book's message: 'transition from the classical paradigm of conservation to the one of managing change' and addressing acceptable levels of change…This is an important book, destined one hopes to be essential reading for those involved in urban conservation globally: scholars, practitioners, managers, students.

Ken Taylor
Journal of Landscape Research, 2012

Heritage has become a recognized economic driver, and the authors are proponents of the integration of urban conservation into urban development. The Historic Urban Landscape methodology, therefore, becomes a tool for the management of change.…[This book] is well written, straightforward and easy to understand. Its concepts are an evolution of current practices in a rapidly changing world impacted by urban migration and globalization. The Historic Urban Landscape will change in a major way accepted notions of historic cities' preservation.

Pamela Jerome
Association of Preservation Technology Bulletin, 2012

…with this volume, Francesco Bandarin and Ron van Oers manage to make a most valuable contribution to this body of literature with a thoughtful and practical approach…I would recommend the book to scholars of urban planning, but also as a must read for any policy-maker and officers of local authority working in the field of heritage conservation and heritage site management.

Laura Pierantoni
SPANDREL, 7 (2013)

Keep up with critical fields

Would you like to receive up-to-date information on our books, journals and databases in the areas that interest you, direct to your mailbox?

Join the **Wiley e-mail service** - a convenient way to receive updates and exclusive discount offers on products from us.

Simply visit www.wiley.com/email and register online

We won't bombard you with emails and we'll only email you with information that's relevant to you. We will ALWAYS respect your e-mail privacy and NEVER sell, rent, or exchange your e-mail address to any outside company. Full details on our privacy policy can be found online.

www.wiley.com/email

WILEY

13-56637

Printed and bound by CPI Group (UK) Ltd, Croydon, CR0 4YY

28/10/2024

14581378-0001